777.

Rock Slope Engineering

ISBN 0 900488 36 0

First edition, 1974.
Revised second edition, 1977.

Printed in England by Stephen Austin and Sons Limited, Hertford

Rock Slope Engineering

Revised second edition

Evert Hoek

D.Sc(Eng), Ph.D., M.Sc., B.Sc., F.I.M.M., C.Eng., P.Eng.

Principal, Golder Associates Limited,
224 W 8th Avenue, Vancouver, B.C., Canada.

Formerly

London University Professor of Rock Mechanics,
Imperial College of Science and Technology, London, England.

John Bray

Ph.D(Eng)., B.Sc.

Senior Lecturer in Rock Mechanics, Imperial College
of Science and Technology, London, England.

The Institution of Mining and Metallurgy, London

1977

Preface to first edition

In designing the very large excavated slopes which are becoming increasingly common in both mining and civil engineering projects, the engineer is faced with two conflicting requirements. On the one hand, vast sums of money can be saved by steepening the slopes, thereby reducing the amount of material to be excavated. On the other hand, loss of life and serious damage to property can result from failures induced by excessive steepening of a particular slope. How does the engineer achieve an optimum design - a compromise between a slope which is steep enough to be economically acceptable and one which is flat enough to be safe?

Because the rock mass behind each slope is unique, there are no standard recipes or routine solutions which are guaranteed to produce the right answer each time they are applied. A practical solution is built up from the basic geological data, rock strength information, groundwater observations and a good measure of engineering common sense. These ingredients are blended in different proportions for each case and the only assistance available is a collection of tools and techniques which will help the engineer to collect the information quickly and efficiently and to process it in an orderly manner.

This book sets out to describe these tools and techniques and to illustrate their application to practical problems by means of a number of worked examples. As far as possible, the text has been kept free of mathematics and a number of simple design charts and graphical methods have been included to enable the non-specialist engineer rapidly to obtain approximate answers to his problem. These approximate answers are frequently adequate but there will be situations in which the engineer will wish to call upon a geotechnical specialist for assistance. Having attempted to solve the problem for himself, the engineer will be in a strong position to communicate his needs to the geotechnical specialist and to work out, with the specialist, the most practical engineering solution.

The authors make no apology for the fact that the book has been printed by offset lithography from typescript and that some of the drawings and photographs are not perfect. The intention has been to produce an engineering handbook at minimum cost rather than a fine example of the printer's art. Wide margins have been provided for the reader's notes and the authors hope that each copy will wear out from hard use rather than decay in decorative inactivity.

London,   Evert Hoek
November, 1973.   John Bray

Preface to revised second edition

Since the first edition of this book was published three years ago, the authors have had the opportunity of assessing how well the contents have measured up to the demands of practical engineering and of teaching. While these demands seem to have been met reasonably well, it became evident that some corrections, alterations and additions were required and it was decided to produce this revised second edition.

In addition to improvements in typographic quality, several chapters have been extensively revised and two new chapters and two new appendices have been included. One of these new chapters deals with toppling failure while the other discusses the problems of blasting. The new appendices deal with a method for computing the stability of rock wedges and with the factor of safety calculation for a reinforced rock slope.

Many of the improvements in this edition have come about as a result of constructive criticism and comment from a large number of individuals who have taken the trouble to write to the authors about the book. While the authors have no intention of revising this book again, at least not for several years, they would still like to hear from anyone who has any comments or suggestions for further improvements.

London
January, 1977

Evert Hoek
John Bray

3

Acknowledgements

This book is the outcome of a four year research project on slope stability in open pit mines carried out at the Royal School of Mines between 1968 and 1972. The project was sponsored by the following companies:

Anglo-American International (UK) Ltd. on behalf of six member companies.
Bougainville Copper (Pty) Ltd.
Consolidated Gold Fields Ltd.
English China Clays Ltd.
National Coal Board Opencast Executive
Palabora Mining Company Ltd.
Rio Tinto Espanola S.A.
Rio Tinto Zinc Corporation Ltd.
Roan Selection Trust Ltd. on behalf of two member companies in Zambia.

The following member companies of the Australian Mineral Industries Research Association.
Broken Hill Proprietary Company Ltd.
Conzinc Riotinto of Australia Ltd.
Electrolytic Zinc Company of Australasia Ltd.
Mount Isa Mines Ltd.
New Broken Hill Consolidated Ltd.
North Broken Hill Ltd.
Western Mining Corporation Ltd.

Mr. M.J. Cahalan, of the Rio Tinto Zinc Corporation's Research Secretariat, acted as co-ordinator of the project.

The authors wish to acknowledge the generosity of these companies and also their willingness to provide information and practical assistance whenever it was requested.

The research was carried out by a team of staff members and research students at the Royal School of Mines and the important contributions made by all of these persons is gratefully acknowledged.

This second edition has been prepared by E.Hoek while working for Golder Associates Limited and the company's generous contribution of his time is acknowledged. Many practical contributions have also been made by the staff members of Golder, Hoek and Associates in England and Golder, Brawner and Associates in Canada.

A number of individuals have made important contributions in the form of critical comments, detailed discussions or the provision of specific data. It would be impractical to list all of these individuals but the most significant contributions were made by:

Dr Ted Brown of Imperial College, London,
M Pierre Londe of Coyne and Bellier, Paris,
Prof. Dick Goodman of the University of California, Berkely,
Prof. Branko Ladanyi of Ecole Polytechnique, Montreal,
Mr John Ashby of Golder, Brawner and Associates, Vancouver,
Mr Ken Matthews of the University of British Columbia,
Dr Nick Barton of the Norwegian Geotechnical Institute, Oslo.

The bulk of this manuscript was typed by Mrs Theo Hoek who also assisted in the preparation of many of the drawings and photographs. Her help and encouragement over the years is warmly acknowledged.

Contents

CONTENTS

CONTENTS

Chapter 1: Economic and planning considerations

Introduction

This book is concerned with the stability of rock slopes, with methods for assessing this stability and with techniques for improving the stability of slopes which are potentially dangerous. Rock slope failures, or the remedial measures necessary to prevent them, cost money and it is appropriate that, before becoming involved in a detailed examination of slope behaviour, some of the economic implications of this behaviour should be considered.

A number of authors[1-8]* have discussed the influence of slope angle upon the design and economics of open pit mining and the interested reader is referred to these publications which deal with the subject more fully than is possible in this introduction. One of the most obvious facts to emerge from these discussions is that, in order to reduce to a minimum the amount of waste rock which has to be excavated in recovering an ore body, the ultimate slopes of the mine are generally cut to the steepest possible angle. Since the economic benefits gained in this way can be negated by a major slope failure, evaluating the stability of the ultimate slopes is an important part of open pit mine planning.

Stewart and Kennedy[1] show that it is not only the steepness of the ultimate slopes in an open pit mine which has an influence upon the overall profitability of the operation. On the basis of cash flow calculations, they show that there is frequently considerable economic advantage to be gained from using steep slopes during the initial stripping programme. These authors also emphasise the fact that there are several factors, in addition to stability, which determine the steepness of the slopes in an open pit mine. Large mining equipment cannot be operated on narrow benches, haul road grades have to be kept within limits imposed by the optimum operating conditions of trucks or trains and this generally means flatter slopes and in some cases, local mining regulations define maximum bench heights and widths.

While the overall slopes are clearly important in terms of the economics of the entire mining operation, the stability of individual benches is usually a matter of more immediate concern to the engineers responsible for the day-to-day mining operations. Slope failure in a bench, which carries a main haul road or which is adjacent to a property boundary or an important installation, can cause severe disruption to the mining programme. It is also in these relatively small failures, which can occur with very little warning, that lives can be lost and equipment damaged.

The stability of an individual bench is controlled by local geological conditions, the shape of the overall slope in that area, local groundwater conditions and also by the excavation technique used in creating a slope. These controlling factors will obviously vary so widely for different mining situations that it is impossible to give general rules on how high or how steep a bench should be

---

* Numbers in parenthesis refer to the list of references given at the end of each chapter.

to ensure that it will be stable. When the stability of a bench, which is important in a particular mining operation, is suspect, its stability must be assessed on the basis of the geological structures, groundwater conditions and other controlling factors which occur in that specific slope. This book is devoted to providing the engineer or geologist with techniques for carrying out such an evaluation.

Economic consequences of instability

Possibly the best introduction to the subject can be given by an example which includes a consideration of the most important factors which control rock slope behaviour as well as the economic consequences of instability.

In the slope illustrated in Figure 1, two major discontinuities have been exposed during the early stages of excavation. Measurement of the orientation and inclination of these discontinuities and projection of these measurements into the rock mass shows that the line of intersection of the discontinuities will daylight in the slope face when the height of the slope reaches 100 feet. It is required to investigate the stability of this slope and to estimate the costs of the alternative methods of dealing with the problem which arises if the slope is found to be unstable.

The factor of safety* of the slope, for a range of slope angles, is plotted in Figure 2 for the two extreme conditions of a dry slope and a slope excavated in a rock mass in which the groundwater level is very high. It will become clear, in the detailed discussions given later in this book, that the presence of groundwater in a slope can have a very important influence upon its stability and that drainage of this groundwater is one of the most effective means of improving the stability of the slope.

A slope will fail if the factor of safety falls below unity and, from Figure 2, it will be seen that the saturated slope will fail if it is excavated at an angle steeper than 64°. The dry slope is theoretically stable at any angle but the factor of safety of approximately 1.2 is not considered sufficiently high to ensure that the slope will remain stable. In most mining situations, where a slope is only required to remain stable for a relatively short period, a factor of safety of 1.3 is normally regarded as the minimum acceptable value. For more permanent slopes such as those which carry the haul road, a factor of safety of 1.5 is more appropriate.

In this example, a factor of safety of 1.3 is considered adequate and this means that, if no other steps are taken to stabilise the slope, it would have to be excavated at an angle of 46° for the saturated condition or 55° if it is dry in order to give this value.

An estimate of costs can only be obtained if the tonnages to be excavated or cleared-up, if failure occurs, are

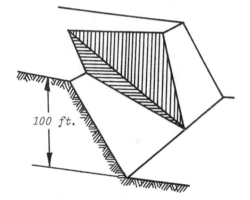

Figure 1: Geometry of wedge failure in example of bench stability analysis.

Details of wedge geometry and material properties used in this analysis :
Discontinuity surfaces upon which wedge slides both dip at 45° to the slope face, giving a symmetrical wedge. The surfaces both have friction angles of 30° and a cohesive strength of 1000 lb./ft². The unit weight of the rock is 160 lb/ft³.

100 ft.

---

*The definition of this and of other terms used in the stability analysis is given later in the book. A detailed knowledge of the method of analysis is not necessary in order to follow this example.

Example of a wedge failure in an
open pit mine bench.

Figure 2 : Variation of Factor of
Safety with slope angle.

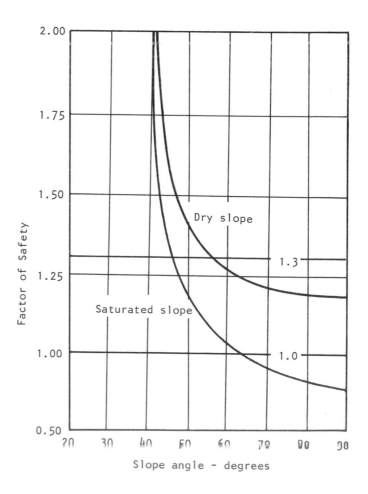

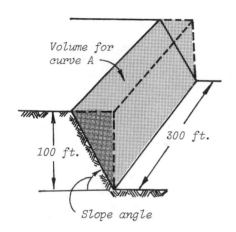

Volume for curve A

100 ft.

300 ft.

Slope angle

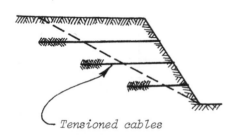

Tensioned cables

calculated. This has been done for a range of slope angles and the results are plotted in Figure 3. In calculating the tonnage involved in flattening the slope, it has been assumed that 300 feet of slope face has to be excavated. In many cases, flattening the slope would also influence benches above that under consideration and much larger tonnages than those given in Figure 3 would be involved.

Also included in this figure are two curves giving the external load, applied by means of cables installed in horizontal holes drilled at right angles to the slope face and anchored in the rock behind the discontinuity planes, required to give a factor of safety of 1.3 for both dry and saturated slopes.

The cost of the various options which are now available to the engineer will depend upon the geographical location of the mine, availability of specialist services for installation of drainage or of tensioned cables and upon local labour costs. In deriving the costs presented in Figure 4, the following assumptions were made:

a. The basic cost unit is taken as the cost per ton mined from the face. Hence, line A in Figure 4 is obtained directly from line A in Figure 3.

b. The cost of clearing up a slope failure is assumed to be $2\frac{1}{2}$ times the basic mining cost. This gives line B which starts from a slope angle of 64°, theoretically the flattest slope at which failure could occur.

c. The design and installation of a drainage system involves a fixed cost of 75,000 units, irrespective of slope angle (line E).

d. The cost of tensioned cables, installed by a specialist contractor, is assumed to be 10 units per ton of load. This gives lines C and D.

On the basis of a set of data such as that presented in Figure 4, the engineer is now in a position to consider the relative costs of the options available to him. Some of these options are listed below.

a. Flatten slope to 46° to give a factor of
    safety of 1.3 under saturated conditions.
    (Line A)   Total cost: 116,000 units

b. Flatten slope to 55° and install drainage
    system to give a factor of safety of 1.3
    for a dry slope. (Lines A and E)
                                Total cost: 159,000 units

c. Cut slope to 64° to induce failure and
    clear up failed material. (Lines A and B)
                                Total cost: 166,000 units

d. Cut slope to 80° and install cables to
    give factor of safety for saturated slope
    of 1.3. (Lines A and C)
                                Total cost: 137,000 units

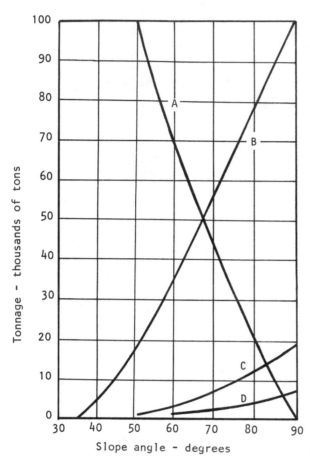

Figure 3 : Excavation tonnages and cable loads.

Line A – *Tonnages excavated in flattening slope 100 ft. high × 300 ft. long.*
Line B – *Tonnage to be cleared up if wedge failure occurs.*
Line C – *Cable load required for a factor of safety of 1.3 for a saturated slope.*
Line D – *Cable load required for a factor of safety of 1.3 for a dry slope.*

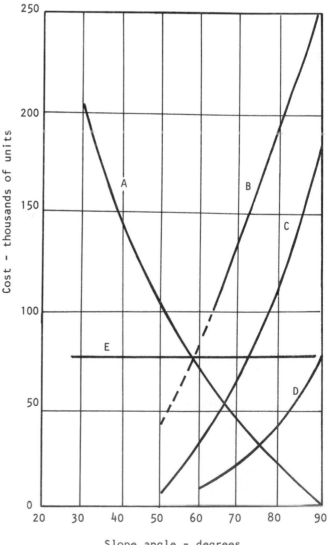

Figure 4 : Comparative cost options.

Line A – *Cost per ton mined from face – from line A in Figure 3.*
Line B – *Cost of clearing up a slope failure.*
Line C – *Cost of installing cables in a saturated slope.*
Line D – *Cost of installing cables in a dry slope.*
Line E – *Cost of draining slope.*

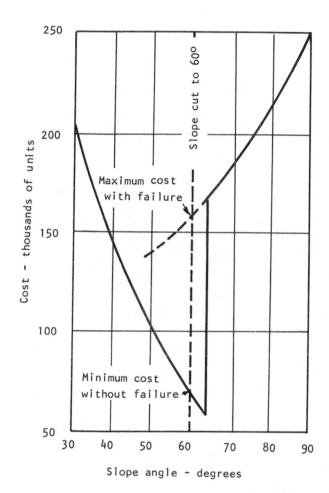

Figure 5 : Cost associated with mining the
slope to an angle of 60° and
accepting the risk of a failure.
The cost before failure is given
by line A in Figure 4. The cost
of excavating the slope and clearing
up a failure is given by the sum of
lines A and B.

e.   Leave slope vertical, install drainage
system and cables to give factor of safety
of 1.3 for dry slope. (Lines A, D and E)
                         Total cost: 155,000 units

f.   Cut slope to 60° on the assumption that it
may not fail and make provision for clearing
up failure if it occurs. (Figure 5)
                    Maximum total cost: 159,000 units
                    Minimum total cost: 70,000 units

It must be emphasised that these estimates are hypothetical
and apply to this particular slope only. The costs of these
and other options will vary from slope to slope and no
attempt should be made to derive general rules from the
figures given.

On the basis of the estimates listed above, most open pit
engineers would probably decide to flatten the slope to 46°
and thereby would eliminate the problem. The cost of this
option is lower than the others considered with the exception of the minimum cost of f. Flattening the slope to 46°
has one important advantage over the other option in that it
does not carry with it the possibility that, having spent a
great deal of money on remedial measures, the slope could
still fail as a result of some unforseen combination of
circumstances. The total cost, if this were to occur, would
be very high.

In some situations, flattening the slope to the extent
required to give an adequate factor of safety under all
conditions may not be possible. Under these circumstances,
one of the other options would have to be considered.

In recognition of a decision which is frequently made in
opencast mining, although not often with the background of
knowledge available in this example, the option listed as
item f and illustrated in Figure 5 has been included.
Figure 2 shows that failure of the slope is likely if it is
cut steeper that 64° and if it is saturated. Assuming that
this condition only arises during exceptionally heavy rains
which may only occur once in the next 10 - 20 years, i.e.
hopefully not during the anticipated life of the slope, the
open pit engineer may decide to cut the slope to 60° and to
accept the risk of failure as part of his mine planning. If
he is lucky and failure does not occur, the total cost will
have been kept to 70,000 units. On the other hand, if
adequate provision has been made to accommodate and to deal
with failure, the total cost of 159,000 units incurred if
failure occurs is still within the range of costs of the
other options.

The publications of Kennedy et al[9],[10], dealing with the
prediction and successful accommodation of a major slide at
the Chuquicamata mine in Chile, have demonstrated that
accepting failure as part of a planned mining operation is
feasible provided that the risk of loss of life and damage
of equipment can be minimised. Knowledge of the likely
behaviour of the slope, derived from a stability analysis
such as that given above, is essential if any measure of
control over the consequences of failure is to be achieved
in this situation.

Planning stability investigations

A typical open pit mine may only suffer two or three slope
failures during its operating life. How can the isolated
slopes which are potentially dangerous be detected in the
many miles of slopes created in a large mine?

The answer lies in the fact that certain combinations of
geological discontinuities[*], slope geometry and groundwater
conditions result in slopes in which the risk of failure is
high. If these combinations can be recognised during the
preliminary geological and pit layout studies, steps can be
taken to deal with the slope problems which are likely to
arise in these areas. Slopes in which these combinations do
not occur require no further investigation. It must,
however, be anticipated that undetected discontinuities will
be exposed as the pit is excavated and provision must be
made to deal with the resulting slope problems as they arise.

This approach to the planning of slope stability studies in
open pit mines is outlined in the chart presented in Figure 6.
and it will be seen that two distinct stages are proposed:

---

[*]The term discontinuity as used here covers faults, joints,
bedding planes or any other surfaces upon which movement
can take place.

PLANNING A SLOPE STABILITY PROGRAMME

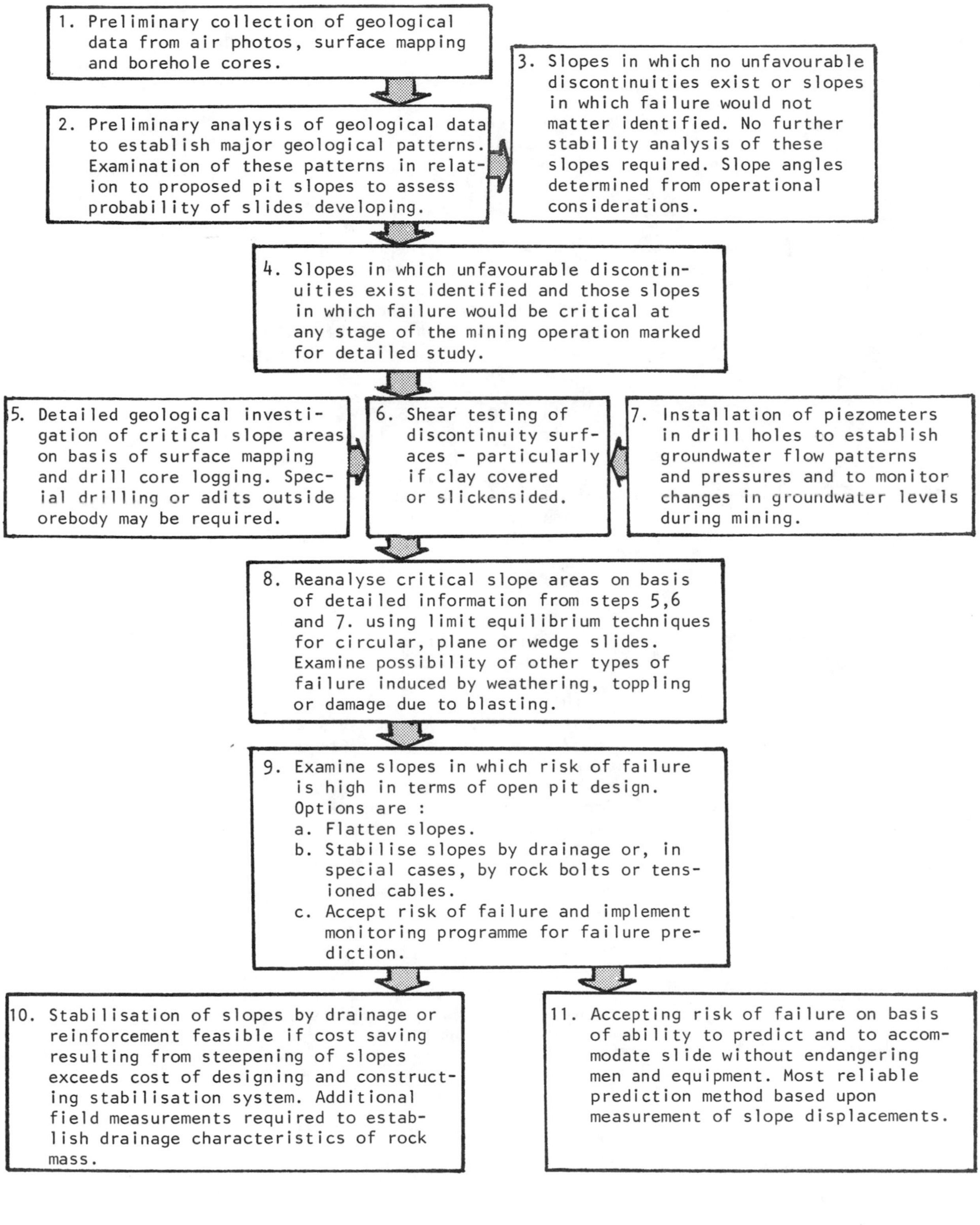

1. Preliminary collection of geological data from air photos, surface mapping and borehole cores.

2. Preliminary analysis of geological data to establish major geological patterns. Examination of these patterns in relation to proposed pit slopes to assess probability of slides developing.

3. Slopes in which no unfavourable discontinuities exist or slopes in which failure would not matter identified. No further stability analysis of these slopes required. Slope angles determined from operational considerations.

4. Slopes in which unfavourable discontinuities exist identified and those slopes in which failure would be critical at any stage of the mining operation marked for detailed study.

5. Detailed geological investigation of critical slope areas on basis of surface mapping and drill core logging. Special drilling or adits outside orebody may be required.

6. Shear testing of discontinuity surfaces - particularly if clay covered or slickensided.

7. Installation of piezometers in drill holes to establish groundwater flow patterns and pressures and to monitor changes in groundwater levels during mining.

8. Reanalyse critical slope areas on basis of detailed information from steps 5,6 and 7. using limit equilibrium techniques for circular, plane or wedge slides. Examine possibility of other types of failure induced by weathering, toppling or damage due to blasting.

9. Examine slopes in which risk of failure is high in terms of open pit design. Options are :
   a. Flatten slopes.
   b. Stabilise slopes by drainage or, in special cases, by rock bolts or tensioned cables.
   c. Accept risk of failure and implement monitoring programme for failure prediction.

10. Stabilisation of slopes by drainage or reinforcement feasible if cost saving resulting from steepening of slopes exceeds cost of designing and constructing stabilisation system. Additional field measurements required to establish drainage characteristics of rock mass.

11. Accepting risk of failure on basis of ability to predict and to accommodate slide without endangering men and equipment. Most reliable prediction method based upon measurement of slope displacements.

Figure 6 : Analysis of the stability of slopes in open pit mines.

Stage 1 involves a preliminary evaluation of the geological data available from the prospecting or exploration programme which normally includes air photo interpretation, surface mapping and diamond drilling.
Note that this data is collected for ore-reserve evaluation and it is usually necessary to re-examine it in terms of the factors which are important for stability.
The preliminary assessment of stability can be done using a number of simple techniques which will be described in the first part of this book. This preliminary study should identify those slopes in which no failure is likely, and which can therefore be designed on the basis of operational considerations, and those slopes in which the risk of failure appears to be high and which require more detailed analysis.

Stage 2, which applies only to those slopes in which potential instability could prove dangerous at some stage in the mining operation, involves a much more detailed study of the geology, the groundwater conditions and the mechanical properties of the rock mass. A detailed analysis of stability is then carried out on the basis of this information and this should provide the mine management with a set of quantative data upon which rational decisions can be based.
The second part of this book will deal with the techniques which can be used for these detailed stability studies.

At this point the reader would probably wish to ask the questions: who should do this work and how much will it cost? The following comments on these questions are offered with the warning that they represent the personal opinions of the authors and should not be regarded as general rules. Conditions will obviously vary widely from mine to mine and from country to country and the ultimate decision upon how to deal with stability problems in an open pit mine must be taken by the mine management after due consideration of the factors which are important to that particular mine.

The preliminary investigations listed above under Stage 1 should ideally be integrated into the evaluation and feasibility studies of the mine. Much of the information required for these preliminary slope studies can be obtained at minimal cost if provision is made for its collection during the exploration drilling programme. There is no reason why these preliminary studies should not be carried out by the geologists or engineers engaged in the feasibility studies on the mine. The techniques are not difficult and do not require any complex mathematical treatment. Outside assistance is only required at this stage if the company engineers or geologists feel that it is necessary to discuss their evaluation with someone with experience in slope analysis in order to check that no important points have been overlooked. There may be other situations in which the mine owners or management may consider it more efficient to contract this work out to consulting engineers or geologists and, under normal circumstances, a preliminary evaluation of slope problems would require between one and three man-months of work.

In existing open pit mines, a much more detailed knowledge of the geology and of areas of the pit which have shown signs of instability will already be available. Under these circumstances, the problems will have defined themselves and the preliminary investigations discussed above will probably not be necessary. However, some failures may have developed already to the extent that the remedial measures required to deal with them are more expensive than would have been the case if action had been taken earlier in the life of the mine.

Once the critical slopes in a mine have been defined, the more detailed studies which are then necessary will vary so much from mine to mine that general guide-lines are very difficult to define. It is unlikely that, with the exception of very large mining groups, there would be geologists or engineers in the company who would feel competent to deal with the more complex slope problems without some outside assistance. This does not mean that the techniques of slope analysis are particularly difficult but it does mean that these analyses provide only part of the answer to a slope problem. The rest of the answer comes from engineering judgement based upon experience from having dealt with a variety of slope problems.

Specialist geotechnical organisations which have dealt with a number of open pit mines studies are in a good position to provide sound advice in these cases and, because of their familiarity with similar problems, it is frequently cheaper and more efficient to use these services in preference to embarking upon a "do-it-yourself" study. If this course is adopted, it is useful if an engineer or geologist on the mine has a reasonable knowledge of techniques for stability analysis in order that an effective liaison between the management and the consultants should exist. It is hoped that this book will provide a comprehensive source of information on these techniques.

How much is a slope analysis by a specialist consultant likely to cost and what would be the cost of implementing the recommendations made as a result of this analysis? The answer to this question is subject to the same degree of uncertainty as is associated with a visit to the doctor and yet few of us would hesitate to visit the doctor if we suspected that something was wrong. Some mines can be involved in considerable expense in dealing with slope failures while others may spend virtually nothing on this problem. Experience suggests that a sum of 1% of the total mining cost may not be an unreasonable amount to spend on slope design and correction costs. Where no appropriate action on designing optimum slopes is taken, the cost of dealing with unexpected failures during the life of the mine can easily exceed this figure.

Chapter 1 references

1. MOFFIT, R.B., FRIESE-GREENE, T.W. and LILLICO, T.M. Pit slopes - their influence on the design and economics of open pit mines. *Proc. 2nd Symposium on Stability in Open Pit Mining*, Vancouver 1971. Published by A.I.M.E. New York 1972, pages 67-77.

2. STEWART, R.M. and KENNEDY, B.A. The role of slope stability in the economics, design and operation of open pit mines. *Proc. Ist Symposium on Stability in Open Pit Mining*, Vancouver 1970. Published by A.I.M.E. New York 1971, pages 5-21.

3. STEFFEN, O.K.H., HOLT,W. and SYMONS, V.R. Optimising open pit geometry and operational procedure. *Planning Open Pit Mines*, Johannesburg Symposium 1970. Published by A.A. Balkema, Amsterdam 1971, pages 9-31.

4. HALLS, J.L. The basic economics of open pit mining. *Planning Open Pit Mines*, Johannesburg Symposium 1970. Published by A.A. Balkema, Amsterdam 1971, pages 125-131.

5. SODERBERG, A. and RAUSCH, D. Pit planning and layout. *Surface Mining*. Editor Eugene P. Pfleider. Published by A.I.M.E., New York 1968. Chapter IV.

6. BLACK, R.A.L. Economic and engineering design problems in open pit mining. *Mine and Quarry Engineering*, Jan., Feb. and March 1964, 20 pages.

7. GROSZ, R.W. The changing economics of surface mining. *Proc. Intnl. Symposium on Computer Application in Mining Industry*. Published by A.I.M.E., New York 1969, pages 401-419.

8. STEWART, R.M. and SEEGMILLER, B.L. Requirements for stability in open pit mining. *Proc. 2nd Symposium on Stability in Open Pit Mining*, Vancouver 1971. Published by A.I.M.E., New York 1972, pages 1-7.

9. KENNEDY, B.A., NIERMEYER, K.E. and FAHM, B.A. A major slope failure at the Chuquicamata Mine, Chile. *Mining Engineering*. A.I.M.E., Vol.12, No.12, 1969, page 60.

10. KENNEDY, B.A., and NIERMEYER, K.E. Slope monitoring systems used in the prediction of a major slope failure at the Chuquicamata Mine, Chile. *Planning Open Pit Mines*, Johannesburg Symposium 1970. Published by A.A. Balkema, Amsterdam 1971, pages 215-225.

Chapter 2 : Basic mechanics of slope failure

Continuum mechanics approach to slope stability

A question which frequently arises in discussions on slope stability is how high and how steep can a rock slope be cut. One approach to this problem, which has been adopted by a number of investigators[11-15], is to assume that the rock mass behaves as an elastic continuum. The success which has been achieved by the application of techniques such as photoelastic stress analysis or finite element methods in the design of underground excavations has tempted many research workers to apply the same techniques to slopes. Indeed, from the research point of view, the results have been very interesting but in terms of practical rock slope engineering, these methods have limited usefulness. These limitations arise because our knowledge of the mechanical properties of rock masses is so inadequate that the choice of material properties for use in the analysis becomes a matter of pure guesswork. For example, if one attempts to calculate the limiting vertical height of a slope in a very soft limestone on the basis of its intact strength, a value in excess of 3500 feet is obtained[16]. Clearly, this height bears very little relation to reality and one would have to reduce the strength properties by a factor of at least 10 in order to arrive at a reasonable slope height.

It is appropriate to quote from a paper by Terzaghi[17] where, in discussing the problem of foundation and slope stability, he said "..... natural conditions may preclude the possibility of securing all the data required for predicting the performance of a real foundation material by analytical or any other methods. If a stability computation is required under these conditions, it is necessarily based on assumptions which have little in common with reality. Such computations do more harm than good because they divert the designer's attention from the inevitable but important gaps in his knowledge....".

Muller[18] and his co-workers in Europe have emphasised for many years the fact that a rock mass is not a continuum and that its behaviour is dominated by discontinuities such as faults, joints and bedding planes. Most practical rock slope designs are currently based upon this discontinuum approach and this will be the approach adopted in all the techniques presented in this book. However, before leaving the question of the continuum mechanics approach, the authors wish to emphasise that they are not opposed in principle to its application and indeed, when one is concerned with overall displacement or groundwater flow patterns, the results obtained from a numerical method such as the finite element technique can be very useful. Developments in numerical methods such as those reported by Goodman et al[19] and Cundall[20] show that the gap between the idealised elastic continuum and the real discontinuum is gradually being bridged and the authors are optimistic that the techniques which are currently interesting research methods will eventually become useful engineering design tools.

Maximum slope height - slope angle relationship for excavated slopes

Even if one accepts that the stability of a rock mass is dominated by geological discontinuities, there must be situations where the orientation and inclination of these discontinuities is such that simple sliding of slabs, blocks or wedges is not possible. Failure in these slopes will involve a combination of movement on discontinuities and failure of intact rock material and one would anticipate that, in such cases, higher and steeper slopes than average could be excavated. What practical evidence is there that this is a reasonable assumption?

A very important collection of data on excavated slopes was compiled by Kley and Lutton[21] and additional data has been obtained by Ross-Brown[22]. The information refers to slopes in opencast mines, quarries, dam foundation excavations and highway cuts. The slope heights and corresponding slope angles for the slopes in materials classified as hard rock have been plotted in Figure 7 which includes both stable and unstable slopes. Ignoring, for the moment, the unstable slopes, this plot shows that the highest and steepest slopes which have been successfully excavated, as far as is known from this collection of data, fall along a fairly clear line shown dashed in Figure 7. This line gives a useful practical guide to the highest and steepest slopes which can be contemplated for normal open pit mine planning. In some exceptional circumstances, higher or steeper slopes may be feasible but these could only be justified if a very comprehensive stability study had shown that there was no risk of inducing a massive slope failure.

Role of discontinuities in slope failure

Figure 7 shows that, while many slopes are stable at steep angles and at heights of several hundreds of feet, many flat slopes fail at heights of only tens of feet. This difference is due to the fact that the stability of rock slopes varies with inclination of discontinuity surfaces, such as faults, joints and bedding planes, within the rock mass. When these discontinuities are vertical or horizontal, simple sliding cannot take place and the slope failure will involve fracture of intact blocks of rock as well as movement along some of the discontinuities. On the other hand, when the rock mass contains discontinuity surfaces dipping towards the slope face at angles of between 30° and 70°, simple sliding can occur and the stability of these slopes is significantly lower than those in which only horizontal and vertical discontinuities are present.

The influence of the inclination of a failure plane on the stability of a slope is strikingly illustrated in Figure 8 in which the critical height of a dry rock slope is plotted against discontinuity angle. In deriving this curve, it has been assumed that only one set of discontinuities is present in a very hard rock mass and that one of these discontinuities "daylights" at the toe of the vertical slope as shown in the sketch in Figure 8. It will be seen that the critical vertical height H decreases from a value in excess of 200 feet, for vertical and horizontal discontinuities, to about 70 feet for a discontinuity inclination of 55°.

Clearly the presence, or absence, of discontinuities has a very important influence upon the stability of rock slopes and the detection of these geological features is one of the most critical parts of a stability investigation. Techniques for dealing with this problem are discussed in later chapters of this book.

A planar discontinuity in an open pit mine bench.

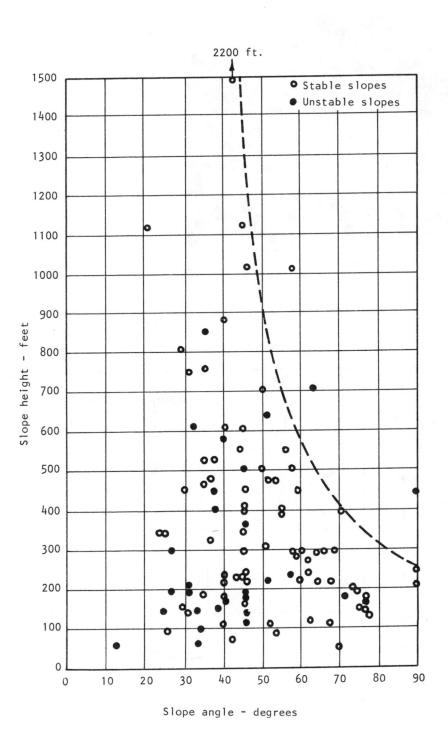

Figure 7 : Slope height versus slope angle relationships
for hard rock slopes, including data collected
by Kley and Lutton [21] and Ross-Brown [22].

Inclined planar discontinuities which daylight at the toe of a rock slope can cause instability when they are inclined at a steeper angle than the angle of friction of the rock surfaces.

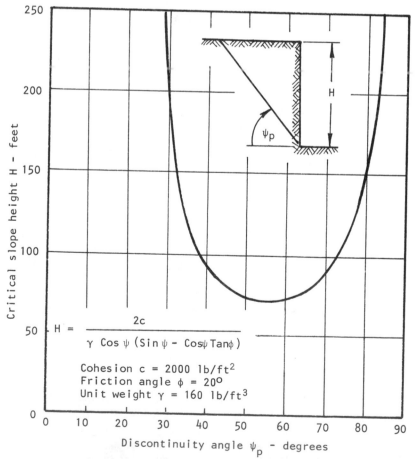

$$H = \frac{2c}{\gamma \, \cos\psi \, (\sin\psi - \cos\psi \, \tan\phi)}$$

Cohesion $c = 2000 \ lb/ft^2$
Friction angle $\phi = 20°$
Unit weight $\gamma = 160 \ lb/ft^3$

Figure 8 · Critical height of a drained vertical slope containing a planar discontinuity dipping at an angle ψ_p .

Friction, cohesion and density

The material properties which are most relevant to the
discussion on slope stability presented in this book are the
angle of friction, the cohesive strength and the density of
rock and soil masses.

Friction and cohesion are best defined in terms of the plot
of shear stress versus normal stress given in Figure 9.
This plot is a simplified version of the results which
would be obtained if a rock specimen containing a
geological discontinuity such as a joint is subjected to a
loading system which causes sliding along the discontinuity.
The shear stress τ required to cause sliding increases with
increasing normal stress σ. The slope of the line relating
shear to normal stress defines the angle of friction ϕ. If
the discontinuity surface is initially cemented or if it is
rough, a finite value of shear stress τ will be required to
cause sliding when the normal stress level is zero. This
initial value of shear strength defines the cohesive
strength c of the surface.

The relationship between shear and normal stresses for a
typical rock surface or for a soil sample can be expressed
as :

$$\tau = c + \sigma \, \text{Tan} \, \phi \qquad (1)$$

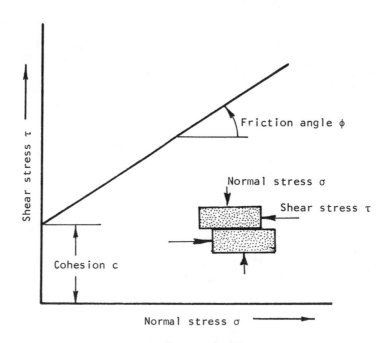

Figure 9 : Relationship between the shear stress τ required
to cause sliding along a discontinuity and the
normal stress σ acting across it.

TABLE 1 - TYPICAL SOIL AND ROCK PROPERTIES						
Description		Unit weight (Saturated/dry)		Friction angle degrees	Cohesion	
Type	Material	lb/ft³	kN/m³		lb/ft²	kPa
Cohesionless — Sand	Loose sand , uniform grain size	118/90	19/14	28-34*		
	Dense sand, uniform grain size	130/109	21/17	32-40*		
	Loose sand, mixed grain size	124/99	20/16	34-40*		
	Dense sand, mixed grain size	135/116	21/18	38-46*		
Gravel	Gravel, uniform grain size	140/130	22/20	34-37*		
	Sand and gravel, mixed grain size	120/110	19/17	48-45*		
Blasted/broken rock	Basalt	140/110	22/17	40-50*		
	Chalk	80/62	13/10	30-40*		
	Granite	125/110	20/17	45-50*		
	Limestone	120/100	19/16	35-40*		
	Sandstone	110/80	17/13	35-45*		
	Shale	125/100	20/16	30-35*		
Cohesive — Clay	Soft bentonite	80/30	13/6	7-13	200-400	10-20
	Very soft organic clay	90/40	14/6	12-16	200-600	10-30
	Soft, slightly organic clay	100/60	16/10	22-27	400-1000	20-50
	Soft glacial clay	110/76	17/12	27-32	600-1500	30-70
	Stiff glacial clay	130/105	20/17	30-32	1500-3000	70-150
	Glacial till, mixed grain size	145/130	23/20	32-35	3000-5000	150-250
Rock	Hard igneous rocks - granite, basalt, porphyry	** 160 to 190	25 to 30	35-45	720000-1150000	35000-55000
	Metamorphic rocks - quartzite, gneiss, slate	160 to 180	25 to 28	30-40	400000-800000	20000-40000
	Hard sedimentary rocks - limestone, dolomite, sandstone	150 to 180	23 to 28	35-45	200000-600000	10000-30000
	Soft sedimentary rock - sandstone, coal, chalk, shale	110 to 150	17 to 23	25-35	20000-400000	1000-20000

* Higher friction angles in cohesionless materials occur at low confining or normal stresses as discussed in Chapter 5.

** For intact rock, the density of the material does not vary significantly between saturated and dry states with the exception of some materials such as porous sandstones.

Typical values for the angle of friction and cohesion which are found in shear tests on a range of rocks and soils are listed in Table 1 together with unit weights for these materials. The values quoted in this table are intended to give the reader some idea of the magnitudes which can be expected and they should only be used for obtaining preliminary estimates of the stability of a slope.

There are many factors which cause the shear strength of a rock or soil to deviate from the simple linear dependence upon normal stress illustrated in Figure 9. These variations, together with methods of shear testing, are discussed in Chapter 5.

Sliding due to gravitational loading

Consider a block of weight W resting on a plane surface which is inclined at an angle ψ to the horizontal. The block is acted upon by gravity only and hence the weight W acts vertically downwards as shown in the margin sketch. The resolved part of W which acts down the plane and which tends to cause the block to slide is W Sin ψ. The component of W which acts across the plane and which tends to stabilise the slope is W Cos ψ.

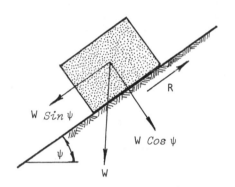

The normal stress σ which acts across the potential sliding surface is given by

$$\sigma = (W \cos \psi)/A \qquad (2)$$

where A is the base area of the block.

Assuming that the shear strength of this surface is defined by equation (1) and substituting for the normal stress from equation (2)

$$\tau = c + \frac{W \cos \psi}{A} . \tan \phi$$

or $\qquad R = cA + W \cos \psi . \tan \phi \qquad (3)$

where $R = \tau A$ is the shear *force* which resists sliding down the plane.

The block will be just on the point of sliding or in a condition of *limiting equilibrium* when the disturbing force acting down the plane is exactly equal to the resisting force :

$$W \sin \psi = cA + W \cos \psi . \tan \phi \qquad (4)$$

If the cohesion c = 0, the condition of limiting equilibrium defined by equation (4) simplifies to

$$\psi = \phi \qquad (5)$$

Influence of water pressure on shear strength

The influence of water pressure upon the shear strength of two surfaces in contact can most effectively be demonstrated by the beer can experiment.

An opened beer can filled with water rests on an inclined piece of wood as shown in the margin sketch on the next page.

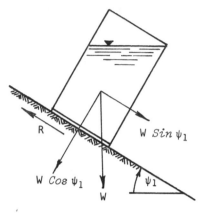

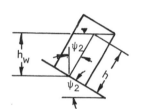

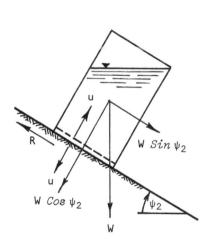

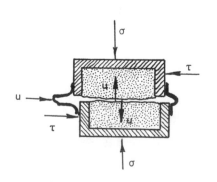

The forces which act in this case are precisely the same as those acting on the block of rock as shown in the diagram on the previous page. For simplicity the cohesion between the beer can base and the wood is assumed to be zero. According to equation (5) the can with its contents of water will slide down the plank when $\psi_1 = \phi$.

The base of the can is now punctured so that water can enter the gap between the base and the plank, giving rise to a water pressure u or to an uplift force U = uA, where A is the base area of the can.

The normal force $W \cos \psi_2$ is now reduced by this uplift force U and the resistance to sliding is now

$$R = (W \cos \psi_2 - U) \tan \phi \qquad (6)$$

If the weight per unit volume of the can plus water is defined as γ_t while the weight per unit volume of the water is γ_w, then $W = \gamma_t.h.A$ and $U = \gamma_w.h_w.A$, where h and h_w are the heights defined in the small sketch. From this sketch it will be seen that $h_w = h.\cos \psi_2$ and hence

$$U = \gamma_w/\gamma_t . W \cos \psi_2 \qquad (7)$$

Substituting in (6)

$$R = W \cos \psi_2 (1 - \gamma_w/\gamma_t) \tan \phi \qquad (8)$$

and the condition for limiting equilibrium defined in equation (4) becomes

$$\tan \psi_2 = (1 - \gamma_w/\gamma_t) \tan \phi \qquad (9)$$

Assuming the friction angle of the can/wood interface is 30°, the unpunctured can will slide when the plane is inclined at ψ_1 = 30° (from equation (5)). On the other hand, the punctured can will slide at a much smaller inclination because the uplift force U has reduced the normal force and hence reduced the frictional resistance to sliding. The total weight of the can plus water is only slightly greater than the weight of the water. Assuming γ_w/γ_t = 0.9 and ϕ = 30°, equation (9) shows that the punctured can will slide when the plane is inclined at ψ_2 = 3° 18'.

The effective stress law

The effect of water pressure on the base of the punctured beer can is the same as the influence of water pressure acting on the surfaces of a shear specimen as illustrated in the margin sketch. The normal stress σ acting across the failure surface is reduced to the *effective stress* (σ - u) by the water pressure u. The relationship between shear strength and normal strength defined by equation (1) now becomes

$$\tau = c + (\sigma - u) \tan \phi \qquad (10)$$

In most hard rocks and in many sandy soils and gravels, the cohesive and frictional properties (c and ϕ) of the materials are not significantly altered by the presence of water and hence, reduction in shear strength of these materials is due, almost entirely to the reduction of normal

stress across failure surfaces. Consequently, it is *water
pressure* rather than *moisture content* which is important in
defining the strength characteristics of hard rocks, sands
and gravels. In terms of the stability of slopes in these
materials, the presence of a small volume of water at high
pressure, trapped within the rock mass, is more important
than a large volume of water discharging from a free
draining aquifer.

In the case of soft rocks such as mudstones and shales and
also in the case of clays, both cohesion and friction can
change markedly with changes in moisture content and it is
necessary, when testing these materials, to ensure that the
moisture content of the material during test is as close as
possible to that which exists in the field. Note that the
effective stress law defined in equation (10) still applies
to these materials but that, in addition, c and ϕ change.

The effect of water pressure in a tension crack

Consider the case of the block resting on the inclined plane
but, in this instance, assume that the block is split by a
tension crack which is filled with water. The water
pressure in the tension crack increases linearly with depth
and a total force V, due to this water pressure acting on
the rear face of the block, acts down the inclined plane.
Assuming that the water pressure is transmitted across the
intersection of the tension crack and the base of the block,
the water pressure distribution illustrated in the margin
sketch occurs along the base of the block. This water
pressure distribution results in an uplift force U which
reduces the normal force acting across this surface.

The condition of limiting equilibrium for this case of a
block acted upon by water forces V and U in addition to its
own weight W is defined by

$$W \sin \psi + V = cA + (W \cos \psi - U) \tan \phi \qquad (11)$$

From this equation it will be seen that the disturbing
force tending to induce sliding down the plane is increased
and the frictional force resisting sliding is decreased and
hence, both V and U result in decreases in stability.
Although the water *pressures* involved are relatively small,
these pressures act over large areas and hence the water
forces can be very large. In many of the practical examples
considered in later chapters, the presence of water in the
slope giving rise to uplift forces and water forces in
tension cracks is found to be critical in controlling the
stability of slopes.

Reinforcement to prevent sliding.

One of the most effective means of stabilising blocks or
slabs of rock which are likely to slide down inclined
discontinuity surfaces is to install tensioned rockbolts or
cables. Consider the block resting on the inclined plane
and acted upon by the uplift force U and the force V due to
water pressure in the tension crack. A rockbolt, tensioned
to a load T is installed at an angle β to the plane as
shown. The resolved component of the bolt tension T acting

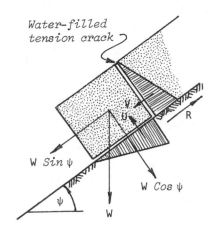

Water-filled
tension crack

W Sin ψ

W Cos ψ

ψ

W

V

U

R

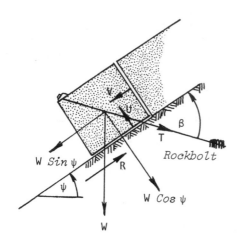

parallel to the plane is T Cos β while the component acting across the surface upon which the block rests is T Sin β. The condition of limiting equilibrium for this case is defined by

$$W \sin \psi + V - T \cos \beta = cA + (W \cos \psi - U + T \sin \beta) \tan \phi$$

(12)

This equation shows that the bolt tension reduces the disturbing force acting down the plane and increases the normal force and hence the frictional resistance between the base of the block and the plane.

The minimum bolt tension required to stabilise the block is obtained by rearranging equation 12 to give an expression for the bolt tension T and then minimising this expression with respect to the angle β , ie set dT/dβ = 0, which gives

$$\beta = \phi$$

(13)

Factor of safety of a slope

All the equations defining the stability of a block on an inclined plane have been presented for the condition of *limiting equilibrium*, i.e. the condition at which the forces tending to induce sliding are exactly balanced by those resisting sliding. In order to compare the stability of slopes under conditions other than those of limiting equilibrium, some form of index is required and the most commonly used index is the *Factor of Safety*. This can be defined as the ratio of the total force available to resist sliding to the total force tending to induce sliding. Considering the case of the block acted upon by water forces and stabilised by a tensioned rockbolt (equation 12), the factor of safety is given by

$$F = \frac{cA + (W \cos \psi - U + T \sin \beta) \tan \phi}{W \sin \psi + V - T \cos \beta}$$

(14)

When the slope is on the point of failure, a condition of limiting equilibrium exists in which the resisting and disturbing forces are equal, as defined by equation 12, and the factor of safety F = 1. When the slope is stable, the resisting forces are greater than the disturbing forces and the value of the factor of safety will be greater than unity.

Suppose that, in a practical mining situation, the observed behaviour of a slope suggests that it is on the point of failure and it is decided to attempt to stabilise the slope. Equation 14 shows that the value of the factor of safety can be increased by reducing both U and V, by drainage, or by increasing the value of T by installing rockbolts or tensioned cables. It is also possible to change the weight W of the failing mass but the influence of this change on the factor of safety must be carefully evaluated since both the disturbing and resisting forces are decreased by a decrease in W.

Practical experience suggests that, in a situation such as that described above, an increase in the factor of safety from 1.0 to 1.3 will generally be adequate for mine slopes which are not required to remain stable for long periods of time. For critical slopes adjacent to haul roads or important installations, a factor of safety of 1.5 is usually preferred.

This example has been quoted because it emphasises the fact that the factor of safety is an index which is most valuable as a design tool when used on a *comparative* basis. In this case, the mine engineers and management have decided, on the basis of the observed behaviour of the slope, that a condition of instability exists and that the value of the factor of safety is 1.0. If remedial measures are taken, their effect can be measured against the condition of slope failure by calculating the increase in the factor of safety. Hoek and Londe, in a general review of rock slope and foundation design methods[23], conclude that the information which is most useful to the design engineer is that which indicates the response of the structure to changes in significant parameters. Hence, decisions on remedial measures such as drainage can be based upon the *rate of change* of the factor of safety, even if the absolute value of the calculated factor of safety cannot be relied upon with a high degree of certainty. To quote from this general review : "The function of the design engineer is not to compute accurately but to judge soundly."

In carrying out a feasibility study for a proposed open pit mine or civil engineering project, the geotechnical engineer frequently is faced with the task of designing slopes where none have previously existed. In this case there is no background experience of slope behaviour which can be used as a basis for comparison. The engineer may compute a factor of safety of 1.3 for a particular slope design, based upon the data available to him, but he has no idea whether this value represents an adequately stable slope since he has not had the opportunity of observing the behaviour of actual slopes in this particular rock mass. Under these circumstances, the engineer is well advised to exercise caution in the choice of the parameters used in the factor of safety calculation. Conservatively low values of both cohesion and friction should be used and, if the groundwater conditions in the slope are unknown, the highest anticipated groundwater levels should be used in the calculation. Sensitivity analyses of the effects of drainage and rock-bolting can still be carried out as in the previous case but, having chosen conservative rock strength parameters, the slope designer is unlikely to be faced with unpleasant surprises when the slope is excavated.

In later chapters of this book, a number of practical examples are given to illustrate the various types of rock slope design which are likely to be encountered by the reader. The problems of obtaining rock strength values, rock structure data and groundwater conditions for use in factor of safety calculations are discussed in these examples and guidance is given on the values of the factor of safety which are appropriate for each type of design.

Slope failures for which factors of safety can be calculated

In discussing the basic mechanism of slope failure, the model of a single block of rock sliding down an inclined plane has been used. This is the simplest possible model of rock slope failure and, in most practical cases, a more complex failure process has to be considered. In some cases, the methods of calculating the factor of safety, presented in this book, cannot be used because the failure process does not involve simple gravitational sliding. These cases will be discussed later in this chapter. The method of limiting equilibrium can be used in analysing the slope failures listed below.

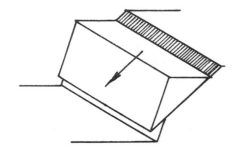

Plane failure

As shown in the margin sketch, plane failure occurs when a geological discontinuity, such as a bedding plane, strikes parallel to the slope face and dips into the excavation at an angle greater than the angle of friction. The calculation of the factor of safety follows precisely the same pattern as that used for the single block (equation 14). The base area A and the weight W of the sliding mass are calculated from the geometry of the slope and failure plane. A tension crack running parallel to the crest of the slope can also be included in the calculation.

A detailed discussion on the analysis of plane failure is given in Chapter 7.

Wedge failure

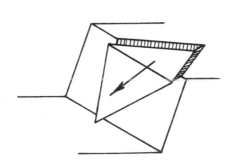

When two discontinuities strike obliquely across the slope face and their line of intersection daylights in the slope face, the wedge of rock resting on these discontinuities will slide down the line of intersection, provided that the inclination of this line is significantly greater than the angle of friction. The calculation of the factor of safety is more complicated than that for plane failure since the base areas of both failure planes as well as the normal forces on these planes must be calculated.

The analysis of wedge failures is discussed in Chapter 8.

Circular failure

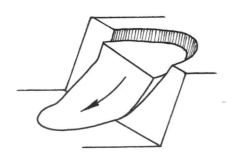

When the material is very weak, as in a soil slope, or when the rock mass is very heavily jointed or broken, as in a waste rock dump, the failure will be defined by a single discontinuity surface but will tend to follow a circular failure path. This type of failure, illustrated in the margin sketch, has been treated in exhaustive detail in many standard soil mechanics textbooks and no useful purpose would be served by repetition of these detailed discussions in this book. A set of circular failure charts is presented in Chapter 9 and a number of worked examples are included in this chapter to show how the factor of safety can be calculated for simple cases of circular failure.

Critical slope height versus slope angle relationships

One of the most useful forms in which slope design data can be presented is a graph showing the relationship between slope heights and slope angles for failure, e.g. the dashed line in Figure 7. A number of typical slope failure cases have been analysed and the relationships between critical slope heights and slope angles have been plotted in Figure 10. This figure is intended to give the reader an overall appreciation for the type of relationship which exists for various materials and for the role which groundwater plays in slope stability. The reader should not attempt to use

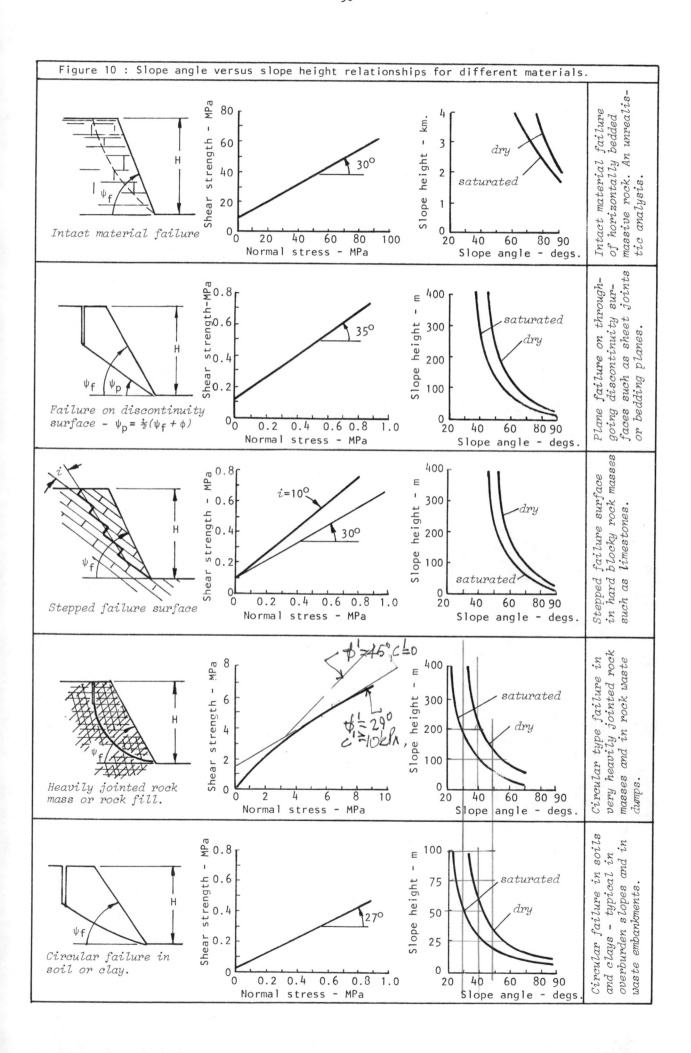

Figure 10 : Slope angle versus slope height relationships for different materials.

this figure as a basis for the design of a particular slope since the conditions may differ from those assumed in deriving the results presented in Figure 10. Individual slopes should be analysed using the methods described in Chapters 7, 8 and 9.

Slopes for which a factor of safety cannot be calculated

The failure modes which have been discussed so far have all involved the movement of a mass of material upon a failure surface. An analysis of failure or a calculation of the factor of safety for these slopes requires that the shear strength of the failure surface (defined by c and ϕ) be known. There are also a few types of slope failure which cannot be analysed by the methods already described, even if the strength parameters of the material are known, since failure does not involve simple sliding. These cases are discussed on the following pages.

Toppling failure

Consider, once again, a block of rock resting on an inclined plane as shown in Figure 11a. In this case, the dimensions of the block are defined by a height h and a base length b and it is assumed that the force resisting downward movement of the block is due to friction only , i.e. c = 0.

When the vector representing the weight W of the block falls within the base b, sliding of the block will occur if the inclination of the plane ψ is greater than the angle of friction ϕ. However, when the block is tall and slender (h > b), the weight vector W can fall outside the base b and, when this happens, the block will topple i.e. it will rotate about its lowest contact edge.

The conditions for sliding and/or toppling for this single block are defined in Figure 11b. The four regions in this diagram are defined as follows :

Region 1 : $\psi < \phi$ and b/h > Tan ψ, the block is stable and will neither slide nor topple.
Region 2 : $\psi > \phi$ and b/h > Tan ψ, the block will slide but it will not topple.
Region 3 : $\psi < \phi$ and b/h < Tan ψ, the block will topple but it will not slide.
Region 4 : $\psi > \phi$ and b/h < Tan ψ, the block can slide and topple simultaneously.

In analysing the stability of this block, the methods of limiting equilibrium can be used for regions 1 and 2 only. Failure involving toppling, i.e. regions 3 and 4 to the right of the curve in Figure 11b, cannot be analysed in this same way. Methods for dealing with toppling failure in slopes are discussed in Chapter 10.

Ravelling slopes

Travellers in mountain regions will be familiar with the accumulations of scree which occur at the base of steep slopes. These screes are generally small pieces of rock which have become detached from the rock mass and which have fallen as individual pieces into the accumulated pile. The cyclic expansion and contraction associated with the freezing and thawing of water in cracks and fissures in the rock mass is one of the principal causes of slope ravelling

Toppling failure in a slate quarry .

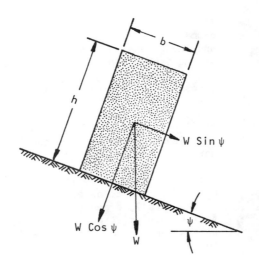

Figure 11a : Geometry of block on inclined plane.

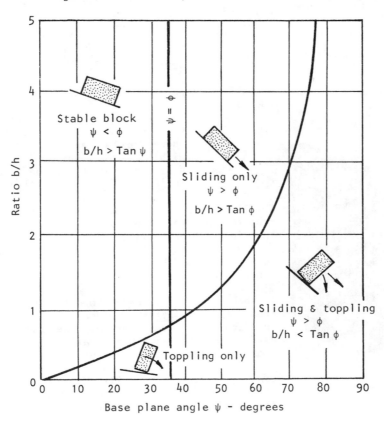

Figure 11b : Conditions for sliding and toppling of a
block on an inclined plane.

Ravelling of the weathered surface material in a slope

Slumping of columns in vertically jointed dolerite as a result of weathering in an underlying shale layer.

but a gradual deterioration of the materials which cement the individual blocks together may also play a part in this type of slope failure.

Weathering, or the deterioration of certain types of rock on exposure, will give rise also to a loosening of a rock mass and the gradual accumulation of materials on the surface and at the base of the slope. Some of the engineering implications of weathering have been reviewed by Goodman[24] who gives a selection of useful references on the subject[25-30].

Few serious attempts have been made to analyse the process of slope failure by ravelling since the fall of small individual pieces of rock does not constitute a serious hazard. When the stability of an accumulation of scree or of weathered material is likely to be altered by the excavation of a slope in this material, the stability of the excavation can be assessed by one of the methods described in Chapters 7, 8 and 9. Generally, the method of circular failure analysis, described in Chapter 9, would be used unless the size of the excavation is such that it is likely to cut back into the undisturbed rock mass.

It is important that the slope designer should recognise the influence of weathering on the nature of the materials with which he is concerned and this subject will be discussed in greater detail in Chapter 7.

Probabilistic approach to slope design

Probability theory has two distinct roles in the design of rock slopes :

a. In the analysis of populations or families of structural discontinuities to determine whether there are dominant or preferred orientations within the rock mass.

b. As a replacement for the factor of safety as an *index* of slope stability (or instability).

The first role is discussed in Chapter 3 which deals with the graphical presentation of geological data. The second role, that in which probability of failure replaces factor of safety as an index of slope stability, has been strongly advocated by McMahon[31] and has been utilised by a number of other authors[32-35].

It should clearly be understood that the use of probability theory in this latter role does not influence the other steps in a stability investigation. The collection of geological data follows the same basic pattern as that described in this book. The mechanics of failure are treated in the same way and the same limitations apply to the types of failure which can be analysed. Probability theory does not, at present, offer any particular advantages in the analysis of toppling, ravelling or buckling type failures.

The authors of this book have chosen to present all the detailed discussions on stability analysis in terms of the factor of safety. This decision has been made because it is believed that the discussion is less confusing for the non-specialist reader for whom this book is intended. The reader who feels that he has understood the basic principles of slope analysis is strongly recommended to examine the

literature on the use of probability theory to determine
for himself whether he wishes to replace the factor of
safety index by the probability of failure.

Chapter 2 references

11. BLAKE, W. Stresses and displacements surrounding an open pit in a gravity loaded rock. *U.S. Bureau of Mines Report of Investigations* 7002, Aug. 1967, 20 pages.

12. BLAKE, W. Finite element model is excellent pit tool. *Mining Engineering*, A.I.M.E., Vol. 21, No. 8, 1969, pages 79-80.

13. YU, Y.S., GYENGE, M. and COATES, D.F. Comparison of stress and displacement in a gravity loaded slope by photoelasticity and finite element analysis. *Canadian Dept. Energy, Mines and Resources Report* MR 68-24 ID, 1968.

14. WANG, F.D. and SUN, M.C. Slope stability analysis by finite element stress analysis and limiting equilibrium method. *U.S. Bureau of Mines Report of Investigations* 7341, January 1970, 16 pages.

15. STACEY, T.R. The stresses surrounding open-pit mine slopes. *Planning open pit mines.* Johannesburg Symposium, 1970. Published by A.A.Balkema, Amsterdam, 1971, pages 199-207.

16. HOEK, E. The influence of structure upon the stability of rock slopes. *Proc. 1st Symposium on Stability in Open Pit Mining,* Vancouver, 1970. Published by A.I.M.E. New York, 1971, pages 49-63.

17. TERZAGHI, K. Stability of steep slopes in hard unweathered rock. *Geotechnique*, Vol. 12, 1962, pages 251-270.

18. MULLER, L. The European approach to slope stability problems in open pit mines. Proc. 3rd Symposium on Rock Mechanics. *Colorado School of Mines Quarterly,* Vol. 54, No. 3, 1959, pages 116-133.

19. GOODMAN, R.E.., TAYLOR, R.L. and BREKKE, T.L. A model for the mechanics of jointed rock. *A.S.C.E. Proceedings, J. Soil Mech. Foundation Div.* Vol. 94, No. SM3, 1968, pages 637-659.

20. CUNDALL, P.A. A computer model for simulating progressive large-scale movements in blocky rock systems. *Symposium on Rock Fracture*, Nancy, France. October 1971, Section 2-8.

21. KLEY, R.J. and LUTTON, R.J. Engineering properties of nuclear craters : a study of selected rock excavations as related to large nuclear craters. *Report U.S. Army Engineers,* No. PNE 5010, 1967, 159 pages.

22. ROSS-BROWN, D.R. Slope design in opencast mines. *Ph.D Thesis,* Imperial College, London University, 1973, 250 pages.

23. HOEK, E. and LONDE, P. Surface workings in rock. *Advances in Rock Mechanics.* Proc. 3rd Congress of the International Society for Rock Mechanics, Denver, 1974. Published by National Academy of Sciences, Washington, D.C., 1974, Vol. 1A, pages 612-654.

24. GOODMAN, R.E. *Methods of geological engineering in discontinuous rocks.* West Publishing Co., St Paul, Minnesota, 1976, 472 pages.

25. RUXTON, B.P. and BERRY, L. Weathering of granite and associated erosional features in Hong Kong, *Bulletin Geological Society of America,* Vol. 68, 1957, page 1263.

26. DEERE, D.U. and PATTON, F.D. Slope stability in residual soils, *Proc. 4th Pan American Conference on Soil Mechanics and Foundation Engineering,* San Juan, Puerto Rico, Vol. 1, 1971, pages 87-170

27. FOOKES, P.G. and HORSWILL, P. Discussion of engineering grade zones. *Proc. Conference on In-situ testing of Soils and Rock,* Institution of Civil Engineers, London, 1970, page 53.

28. SAUNDERS, M.K. and FOOKES, P.G. A review of the relationship of rock weathering and climate and its significance to foundation engineering. *Engineering Geology,* Vol. 4, 1970, pages 289-325.

29. DEERE, D.U., MERRITT, A.H and COON, R.F. Engineering classification of in situ rock. *Technical Report No. AFWL-67-144,* Air Force Systems Command, Kirtland Air Force Base, New Mexico. 1969.

30. SPEARS, D.A. and TAYLOR, R.K. The influence of weathering on the composition and engineering properties on in-situ coal measure rocks. *International Journal of Rock Mechanics and Mining Sciences,* Vol.9, 1972, pages 729-756.

31. McMAHON, B.K. A statistical method for the design of rock slopes. *Proc. First Australia-New Zealand Conference on Geomechanics.* Melbourne, 1971, Vol.1, pages 314-321.

32. McMAHON, B.K. Design of rock slopes against sliding on pre-existing fractures. *Advances in Rock Mechanics,* Proc. 3rd Congress of the International Society for Rock Mechanics, Denver 1974. Published by National Academy of Sciences, Washington D.C., 1974, Vol 11B, pages 803-808.

33. SHUK, T. Optimisation of slopes designed in rock. *Proc. 2nd Congress of the International Society for Rock Mechanics,* Belgrade, 1970, Vol. 3, Sect. 7-2.

34. LANGEJAN, A. Some aspects of safety factors in soil mechanics considered as a problem of probability. *Proc. 6th International Conference on Soil Mechanics and Foundation Engineering,* Montreal, 1965, Vol. 2, pages 500-502.

35. SERRANO, A.A. and CASTILLO, E. A new concept about the stability of rock masses. *Advances in Rock Mechanics,* Proc. 3rd Congress of the International Society for Rock Mechanics, Denver, 1974. Published by National Academy of Sciences, Washington D.C., 1974, Vol. 11B, pages 820-826.

Chapter 3 : Graphical presentation of geological data

Introduction

The dominant role of geological discontinuities in rock slope behaviour has been emphasised already and few engineers or geologists would question the need to base stability calculations upon an adequate set of geological data. But what is an adequate set of data? What type of data and how much detailed information should be collected for a stability analysis?

This question is rather like the question of which came first - the chicken or the egg? There is little point in collecting data for slopes which are not critical but critical slopes can only be defined if sufficient information is available for their stability to be evaluated. The data gathering must, therefore, be carried out in two stages as suggested in Figure 6.

The first stage involves an examination of existing regional geology maps, air photographs, easily accessible outcrops and the core recovered during exploration drilling. A preliminary analysis of this data will indicate slopes which are likely to prove critical and which require more detailed analysis.

The second stage involves a much more detailed examination of the geological features of these critical regions and may require the drilling of special holes outside the ore body, excavation of trial pits or adits and the detailed mapping and testing of discontinuities.

An important aspect of the geological investigations, in either the first or second stages, is the presentation of the data in a form which can be understood and interpreted by others who may be involved in the stability analysis or who may be brought in to check the results of such an analysis. This means that everyone concerned must be aware of precisely what is meant by the geological terms used and must understand the system of data presentation.

The following definitions and graphical techniques are offered for the guidance of the reader who may not already be familiar with them. There is no implication that these are the best definitions or techniques available and the reader who has become familiar with different methods should certainly continue to use those. What is important is that the techniques which are used in any study should be clearly defined in documents relating to that study so that errors arising out of confusion are avoided.

Definition of geological terms

Rock Material or intact rock, in the context of this discussion, refers to the consolidated and cemented assemblage of mineral particles which form the intact blocks between discontinuities in the rock mass. In most hard igneous and metamorphic rocks, the strength of the intact rock is one or two orders of magnitude greater than that of the rock mass and failure of this intact material is not involved generally in the processes of slope failure. In softer sedimentary rocks, the intact material may be relatively weak and failure of this material may play an important part in slope failure.

An ordered structural pattern in slate

An apparently disordered discontinuity pattern in a hard rock slope

Rock mass is the *in-situ* rock which has been rendered discontinuous by systems of structural features such as joints, faults and bedding planes. Slope failure in a rock mass is generally associated with movement of these discontinuity surfaces.

Waste rock or broken rock refers to a rock mass which has been disturbed by some mechanical agency such as blasting, ripping or crushing so that the interlocking nature of the in-situ rock has been destroyed. The behaviour of this waste or broken rock is similar to that of a clean sand or gravel, the major differences being due to the angularity of the rock fragments.

Discontinuities or weakness planes are those structural features which separate intact rock blocks within a rock mass. Many engineers describe these features collectively as *joints* but this is an over-simplification since the mechanical properties of these features will vary according to the process of their formation. Hence, faults, dykes, bedding planes, cleavage, tension joints and shear joints all will exhibit distinct characteristics and will respond in different ways to applied loads. A large body of literature dealing with this subject is available and the interested reader is referred to this for further information[36],[37],[38]. For the purposes of this discussion, the term *discontinuity* will generally be used to define the structural weakness plane upon which movement can take place. The type of discontinuity will be referred to when the description provides information which assists the slope designer in deciding upon the mechanical properties which will be associated with a particular discontinuity.

Major discontinuities are continuous planar structural features such as faults which may be so weak, as compared with any other discontinuity in the rock mass, that they dominate the behaviour of a particular slope. Many of the large failures which have occurred in open pit mines have been associated with faults and particular attention should be paid to tracing these features.

Discontinuity sets refers to systems of discontinuities which have approximately the same inclination and orientation. As a result of the processes involved in their formation[36], most discontinuities occur in families which have preferred directions. In some cases, these sets are clearly defined and easy to distinguish while, in other cases, the structural pattern appears disordered.

Continuity. While major structural features such as faults may run for many tens of feet or even miles, smaller discontinuities such as joints may be very limited in their extent. Failure in a system where discontinuities terminate within the rock mass under consideration will involve failure of the intact rock bridges between these discontinuities. Continuity also has a major influence upon the permeability of a rock mass since this depends upon the extent to which discontinuities are hydraulically connected.

Gouge or infilling is the material between two faces of a structural discontinuity such as a fault. This material may be the debris resulting from the sliding of one surface upon another or it may be material which has been precipitated from solution or caused by weathering. Whatever the origin of the infilling material in a discontinuity, its presence will have an important influence upon the shear strength of that discontinuity. If the thickness of the gouge is such that the faces of the discontinuity do not come into contact, the shear strength will be equal to the shear strength of the gouge. If the gouge layer is thin so that contact between asperities on the rock surfaces can occur, it will modify the shear strength of the discontinuity but will not control it[39].

Roughness. Patton[40],[41] emphasised the importance of surface roughness on the shear strength of structural discontinuities in rock. This roughness occurs on both a small scale, involving grain boundaries and failure surfaces, and on a large scale, involving folds and flexures in the discontinuity. The mechanics of movement on rough surfaces will be discussed in the chapter dealing with shear strength.

Definition of geometrical terms

Dip is the *maximum* inclination of a structural discontinuity plane to the horizontal, defined by the angle ψ in the margin sketch. It is sometimes very difficult, when examining an exposed portion of an obliquely inclined plane, to visualise the *true dip* as opposed to the *apparent dip* which is the inclination of an arbitrary line on the plane. The apparent dip is always smaller than the true dip. One of the simplest models which can be used in visualising the dip of a plane is to consider a ball rolling down an obliquely inclined plane. The path of the ball will always lie along the line of maximum inclination which corresponds to the true dip of the plane.

Dip direction or *dip azimuth* is the direction of the horizontal trace of the line of dip, measured *clockwise* from north as indicated by the angle α in the margin sketch.

Strike is the trace of the intersection of an obliquely inclined plane with a horizontal reference plane and it is at right angles to the dip and dip direction of the oblique plane. The practical importance of the strike of a plane is that it is the visible trace of a discontinuity which is seen on the horizontal surface of a rock mass. In using strike and dip to define a plane for rock slope analysis, it is essential that the direction in which the plane dips is specified. Hence, one may define a plane as having a strike of N 45 E (or 045°) and a dip of 60° SE. Note that a plane dipping 60° NW could also have a strike of N 45 E.

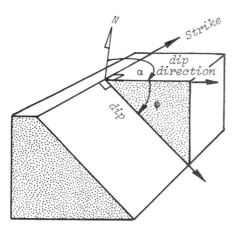

Definition of geometrical terms

Throughout this book, planes will be defined by their dip and dip direction. This convention has been chosen to avoid any possible confusion and to facilitate computation of slope geometries in later chapters. The same convention has been adopted by some geotechnical consulting organisations for stability computer programs. However, geologists are free to use strike and dip measurements for recording their field observations, if this is the convention preferred by them, and a supplementary program is used to transform these measurements into dips and dip directions before they are used as input in the slope stability programs.

Plunge is the dip of a *line*, such as the line of intersection of two planes or the axis of a borehole or a tunnel.

Trend is the direction of the horizontal projection of a line, measured clockwise from north. Hence, it corresponds to the dip direction of a plane.

In recording dip and dip direction data, many geologists use the system in which these quantities are written 35/085. Since the dip of a plane must lie between 0° and 90°, the angle defined by 35 refers to the dip. Similarly, the angle 085 refers to the dip direction which lies between 0° and 360°. The same convention can be used to define the plunge and trend of a line in space. The reader is encouraged to adopt this convention as it will help to eliminate recording errors in the field since, even if a figure is entered into an incorrect column, it will be clear that a two digit number refers to dip and a three digit number refers to dip direction.

Graphical techniques for data presentation

One of the most important aspects of rock slope analysis is the systematic collection and presentation of geological data in such a way that it can easily be evaluated and incorporated into stability analyses. Experience has shown that spherical projections provide a convenient means for the presentation of geological data. The engineer or geologist, who is not familiar with this technique, is strongly advised to study the following pages carefully. A few hours invested in such study can save many hours of frustration and confusion later when the reader becomes involved in studying designs and reading reports in which these methods have been used.

Many engineers shy away from spherical projection methods because they are unfamiliar and because they appear complex, bearing no recognisable relationship to more conventional engineering drawing methods. For many years the authors regarded these graphical methods in the same light but, faced with the need to analyse three-dimensional rock slope problems, an effort was made, with the aid of a patient geologist colleague, and the mystery associated with these techniques was rapidly dispelled. This effort has since been repaid many times by the power and flexibility which these graphical methods provide for the rock engineer.

Several types of spherical projection can be used and a comprehensive discussion on these methods has been given by Phillips[42], Turner and Weiss[38], Badgley[43], Friedman[44] and Ragan[45]. The projection which is used exclusively in this book is the *equal area projection*, sometimes called the Lambert projection or the Schmidt net.

The *equal angle* or *stereographic* projection offers certain advantages, particularly when used for geometrical construction, and is preferred by many authors. Apart from the techniques used in contouring pole populations, to be described later in this chapter, the constructions carried out on the two types of net are identical and the reader will have no difficulty in adapting the techniques, which he has learned using equal area projections, to analyses using stereographic projections.

Equal-area projection

The Lambert equal area projection will be familiar to most readers as the system used by geographers to represent the

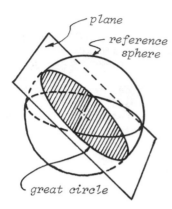

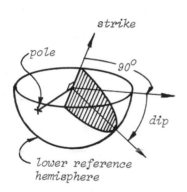

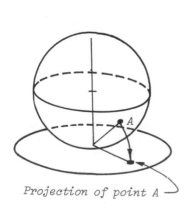

Projection of point A

spherical shape of the earth on a flat surface. In adapting this projection to structural geology, the traces of planes on the surface of a reference sphere are used to define the dips and dip directions of the planes. Imagine a reference sphere which is free to move in space but which is *not* free to rotate in any direction; hence any radial line joining a point on the surface to the centre of the sphere will have a fixed direction in space. If this sphere is now moved so that its centre lies on the plane under consideration, the great circle which is traced out by the intersection of the plane and the sphere will define uniquely the inclination and orientation of the plane in space. Since the same information is given on both upper and lower parts of the sphere, only one of these need be used and, in engineering applications, the *lower reference hemisphere* is used for the presentation of data.

In addition to the great circle, the inclination and orientation of the plane can also be defined by the *pole* of the plane. The pole is the point at which the surface of the sphere is pierced by the radial line which is *normal* to the plane.

In order to communicate the information given by the great circle and the position of the pole on the surface of the lower reference hemisphere, a two dimensional representation is obtained by projecting this information onto the horizontal or equatorial reference plane. The method of projection is illustrated in Figure 12. Polar and equatorial projections of a sphere are shown in Figure 13.

Polar and equatorial equal-area nets are presented on pages 43 and 44 for use by the reader. Good undistorted copies or photographs of these nets will be useful in following the examples given in this chapter and later in the book.

The most practical method of using the stereonet for plotting structural information is to mount it on a base-board of ¼ inch thick plywood as shown in Figure 14. A sheet of clear plastic film of the type used for drawing on for overhead projection, mounted over the net and fixed with transparent adhesive tape around its edges, will keep the stereonet in place and will also protect the net markings from damage in use. The structural data is plotted on a piece of tracing paper or film which is fixed in position over the stereonet by means of a carefully centred pin as shown. The tracing paper must be free to rotate about this pin and it is essential that it is located accurately at the centre of the net otherwise significant errors will be introduced into the subsequent analysis.

Before starting any analysis, the North point must be marked on the tracing so that a reference position is available.

Figure 12 : *Method of construction of an equal-area projection.*

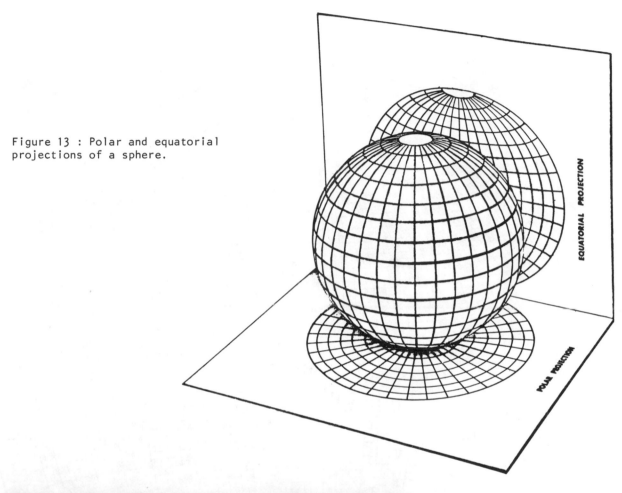

Figure 13 : Polar and equatorial projections of a sphere.

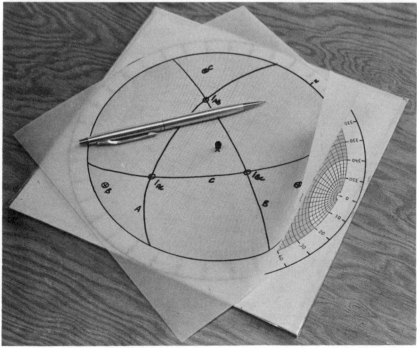

Figure 14 : Geological data is plotted and analysed on a piece of tracing paper which is located over the centre of the stereonet by means of a centre pin as shown. The net is mounted on a base-board of plywood or similar material.

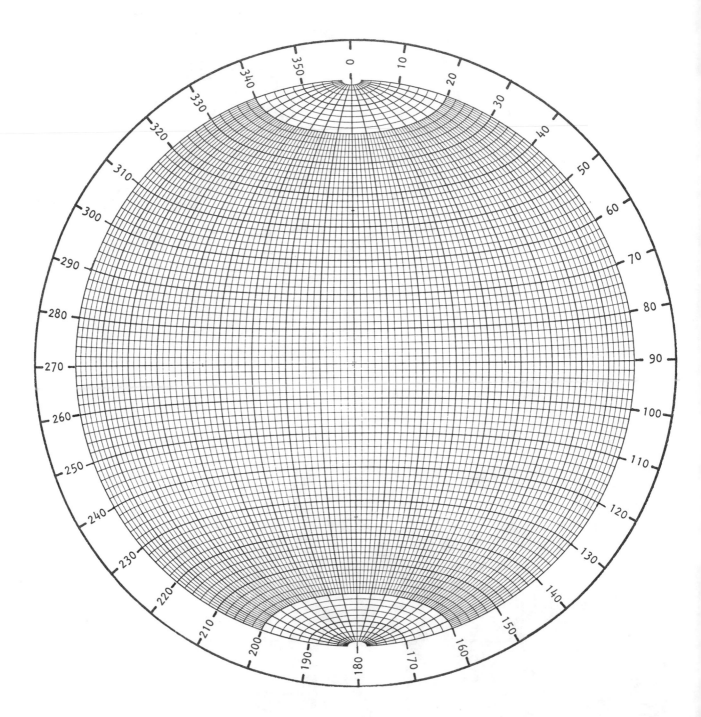

Equatorial equal-area stereonet marked in 2° intervals.

Computer drawn by Dr C.M. St John of the Royal School of Mines, Imperial College, London.

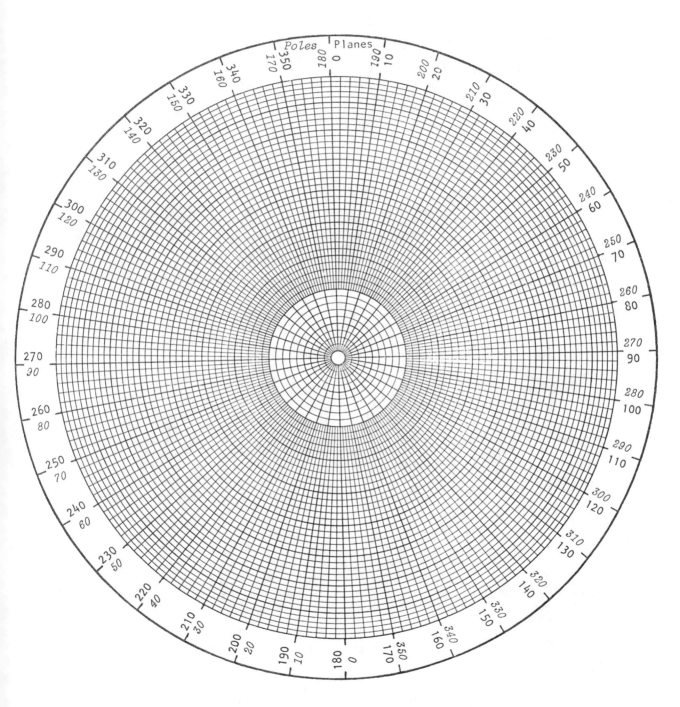

Polar equal-area stereonet marked in 2° intervals.

Computer drawn by Dr C.M. St John of the Royal School of Mines,
Imperial College, London.

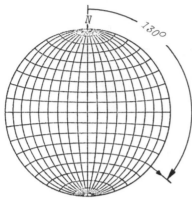

Construction of a great circle and a pole representing a plane.

Consider a plane dipping at 50° in a dip direction of 130°. The great circle and the pole representing this plane are constructed as follows:

Step 1: With the tracing paper located over the stereonet by means of the centre pin, trace the circumference of the net and mark the north point. Measure off the dip direction of 130° clockwise from north and mark this position on the circumference of the net.

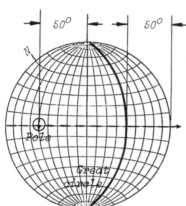

Step 2: Rotate the tracing about the centre pin until the dip direction mark lies on the W-E axis of the net, i.e. the tracing is rotated through 40°. Measure 50° from the outer circle of the net and trace the great circle which corresponds to a plane dipping at this angle.

The position of the pole, which has a dip of (90°-50°), is found by measuring 50° from the centre of the net as shown or, alternatively, 40° from the outside of the net. The pole lies on the projection of the dip direction line which, at this stage in the construction, is coincident with the W-E axis of the net.

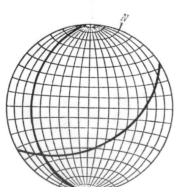

Step 3: The tracing is now rotated back to its original position so that the north mark on the tracing coincides with the north mark on the net. The final appearance of the great circle and the pole representing a plane dipping at 50° in a dip direction of 130° is as illustrated.

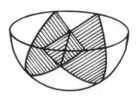

Determination of the line of intersection of two planes.

Two planes, having dips of 50° and 30° and dip directions of 130° and 250° respectively, intersect. It is required to find the plunge and the trend of the line of intersection.

Step 1: One of these planes has already been described above and the great circle defining the second plane is obtained by marking the 250° dip direction on the circumference of the net, rotating the tracing until this mark lies on the W-E axis and tracing the great circle corresponding to a dip of 30°.

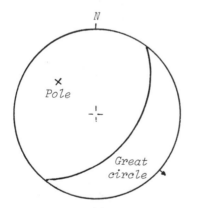

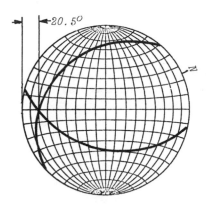

Step 2: The tracing is now rotated until the intersection of the two great circles lies along the W-E axis of the stereonet and the plunge of the line of intersection is measured as 20.5°.

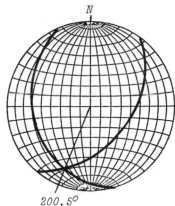

Step 3: The tracing is now rotated until the north mark coincides with the north point on the stereonet and the trend of the line of intersection is found to be 200.5°

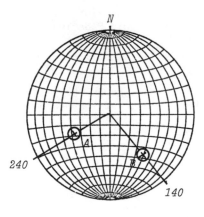

To determine the angle between two specific lines.

Two lines in space, e.g. lines of intersection or normals to planes, are specified by plunges of 54° and 40° and trends of 240° and 140° respectively. It is required to find the angle between these lines.

Step 1: The points A and B which define these lines are marked on the stereonet as described under procedure for locating the pole.

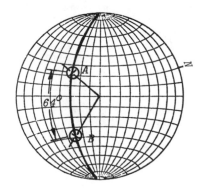

Step 2: The tracing is now rotated until these two points lie on the same great circle on the stereonet and the angle between the lines is determined by counting the small circle divisions between A and B, along the great circle. This angle is found to be 64°.

The great circle on which A and B lie defines the plane which contains these two lines and the dip and dip direction of this plane are found to be 60° and 200° respectively.

Alternative method for finding the line of intersection of two planes.

Two planes, dipping at 50° and 30° in dip directions of 130° and 250° respectively are defined by their poles A and B as shown. The line of intersection of these two planes is defined as follows:

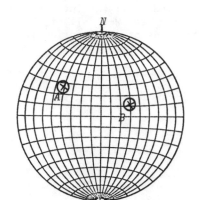

Step 1: Rotate the tracing until both poles lie on the same great circle. This great circle defines the plane which contains the two normals to the planes.

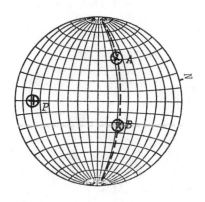

Step 2: Find the pole of this plane by measuring the dip on the W-E axis of the stereonet. This pole P defines the normal to the plane containing A and B and, since this normal is common to both planes, it is, in fact, the line of intersection of the two planes.

Hence, the pole of a plane which passes through the poles of two other planes defines the line of intersection of those planes.

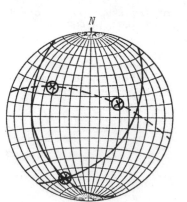

Plotting and analysis of field measurements.

In plotting field measurements of dip and dip direction, it is convenient to work with poles rather than great circles since the poles can be plotted directly on a polar stereonet such as that given on Page 44. Suppose that a plane has dip direction and dip values of 050/60, the pole is located on the stereonet by using the dip direction value of 50 given in *italics* and then measuring the dip value of 60 *from the centre of the net* along the radial line. Note that no rotation of the tracing paper, centred over the stereonet, is required for this operation and, with a little practice, the plotting can be carried out very quickly.

There is a temptation to plot the compass readings directly onto the polar stereonet, without the intermediate step of entering the measurements into a field notebook, but the authors advise against this short-cut. The reason is that

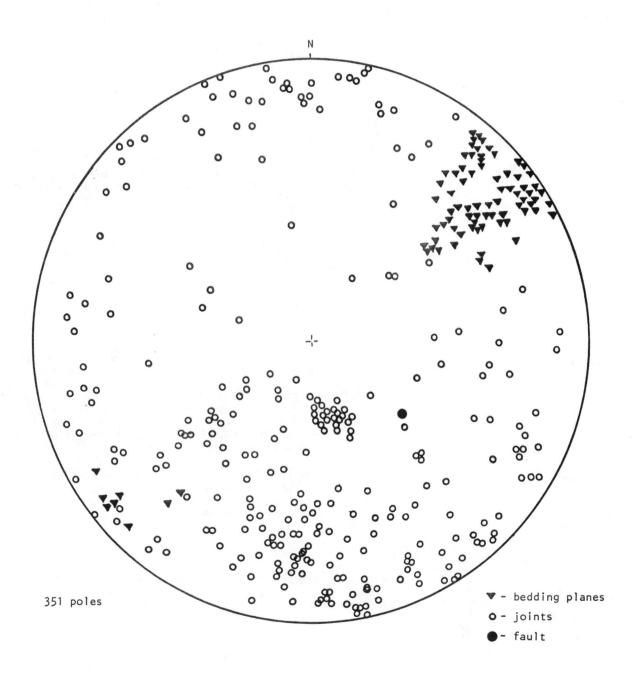

351 poles

▼ - bedding planes
○ - joints
● - fault

Figure 15 : Plot of poles of discontinuities in a hard rock mass.

the measurements may well be required for other purposes, such as a computer analysis, and it is a great deal easier to work from recorded numbers than from the pole plot. Correcting errors on a pole plot on which several hundred measurements have been recorded is also difficult and information can be lost if it has not been recorded elsewhere. Some geologists prefer to use a portable tape recorder, instead of a notebook, for the recording of field data, and the reader should not hesitate to experiment to find the method which is best suited to his own requirements. When plotting field data it is recommended that different symbols be used to represent the poles of different types of structural features. Hence, faults may be represented by heavy black dots, joints by open circles, bedding planes by triangles and so on. Since these structural features are likely to have significantly different shear strength characteristics, the interpretation of a pole plot for the purposes of a stability analysis is simplified if different types of structure can easily be identified.

A plot of 351 poles of bedding planes and joints and of one fault in a hard rock mass is given in Figure 15. Since the fault occurs at one particular location in the rock mass, its influence need only be considered when analysing the stability of the slope in that location. On the other hand, the bedding plane and joint measurements were taken over a considerable area of rock exposure and these measurements form the basis of the stability analysis of all other slopes in the proposed excavation.

Two distinct pole concentrations are obvious in Figure 15; one comprising bedding plane poles in the north-eastern portion of the stereonet and the other, representing joints, south of the centre of the net. The remainder of the poles appear to be fairly well scattered and no significant concentrations are obvious at first glance. In order to determine whether other significant pole concentrations are present, contours of pole densities are prepared.

Several methods of contouring pole plots have been suggested [42-47] but only two techniques will be described in this book. These techniques are preferred by the authors on the basis of numerous trials in which speed, convenience and accuracy of different contouring methods were evaluated.

Denness curvilinear cell counting method.

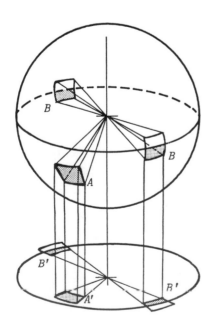

In order to overcome certain disadvantages of other contouring techniques, particularly when dealing with pole concentrations very close to the circumference of the net, Denness [46] devised a counting method in which the reference sphere is divided into 100 squares. A 1% counting square on the surface of the reference sphere, marked A in the margin sketch, projects onto the equal area stereonet as a curvilinear figure A'. When the counting cell falls across the equator of the reference sphere, only the poles falling in the lower half of the 1% cell will be shown on the stereonet since only the the lower part of the reference sphere is used in the plotting process. The counting cell marked B and its projection B' illustrate this situation. Poles which fall *above* the equator are plotted on the *opposite side* of the stereonet and hence a count of the total number of poles falling in a 1% square falling across the equator is obtained by summing the poles in the shaded

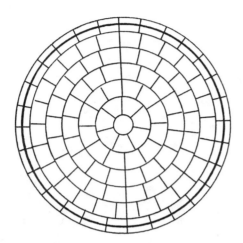

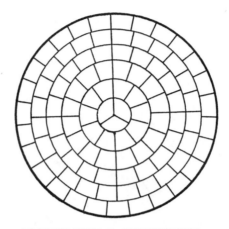

DENNESS TYPE A COUNTING NET

Cells per ring	Cell radius / Net radius	Angle
1	0.100	360.00
7	0.283	51.43
12	0.447	30.00
18	0.616	20.00
22	0.775	16.37
25	0.923	14.40
28	1.064	12.85

DENNESS TYPE B COUNTING NET

Cells per ring	Cell radius / Net radius	Angle
3	0.172	120.00
10	0.360	36.00
16	0.539	22.50
20	0.700	18.00
24	0.855	15.00
27	1.000	13.33

Figure 16 : Dimensions of Denness curvilinear cell counting nets.

portions of *both* projections marked *B'*.

Details of the two types of counting net devised by Denness are given in Figure 16. The type A net is intended for the analysis of pole plots with concentrations near the circumference of the net, representing vertically jointed strata. The type B net is more suited to the analysis of poles of inclined discontinuities and, since inclined discontinuities are of prime concern in the analysis of rock slope stability, this type of net is recommended for use by readers of this book. A type B counting net, drawn to the same scale as the stereonets on pages 43 and 44 and the pole plot in Figure 15, is reproduced in Figure 17.

In order to use this net for contouring a pole plot, a transparent copy or a tracing of the net must be prepared. Note that many photocopy machines introduce significant distortion and scale changes and care must be taken that good undistorted copies of nets with identical diameters are available before starting an analysis.

The transparent counting net is centred over the pole plot and a clean piece of tracing paper is placed over the counting net. The centre of the net and the north mark are marked on the tracing paper. The number of poles falling in each 1% counting cell is noted, in pencil, at the centre of each cell. Contours of equal pole density

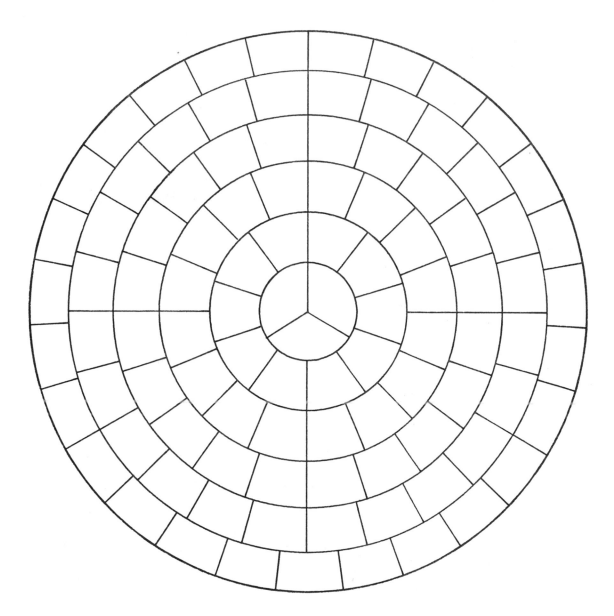

Figure 17 : Denness type B curvilinear cell counting net.

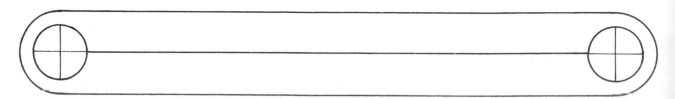

Figure 18 : Counting circles for use in contouring pole plots.

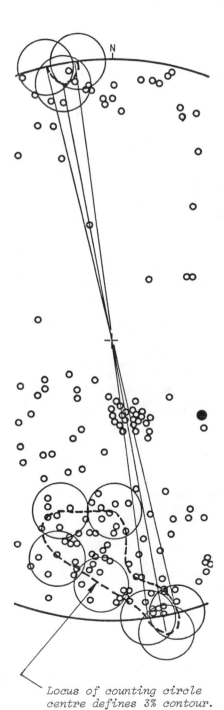

Locus of counting circle
centre defines 3% contour.

are obtained by joining the same numbers on the diagram.
If it is felt that insufficient information is available
in certain parts of the diagram, the counting net can be
rotated as indicated by the dashed lines in the margin
sketch. The new counting cell positions are used to
generate additional pole counts which are noted at the
centres of these cells. If necessary, the counting net
can be moved off centre by a small amount in order to
generate additional information in a radial direction.

Contours of equal pole densities are generally expressed
as percentages. Hence, in the case of the 351 poles plotted
in Figure 15, a 2% contour is obtained by joining pole
counts of 7 and a pole count of between 17 and 18 corresponds to a contour value of 5%.

Floating circle counting method.

One of the disadvantages of using a counting net to contour
a pole plot is that the geometry of the counting net bears no
direct relationship to the distribution of poles. When a
cluster of poles falls across the boundary between two
counting cells, a correct assessment of the pole concentration can only be obtained by allowing the cell to "float"
from its original position and to centre it on the highest
concentration of poles. In some cases, several moves of
the counting cell are needed to generate the quantity of
imformation required for the construction of meaningful
contours. Consideration of this counting procedure suggests
that an alternative, and perhaps more logical, procedure is
to use a single counting cell in a "floating" mode, its
movements being dictated by the distribution of the poles
themselves rather than by some arbitrarily fixed geometrical
pattern. This reasoning lies behind the floating or free
circle counting method[38] described below.

Figure 18 gives a pattern which can be used by the reader
for the construction of a circle counter for use with
stereonets of the diameter given on pages 43 and 44 and in
Figure 15. The diameter of the circles is one-tenth of the
diameter of the net and, therefore, the area enclosed by
these circles is 1% of the area of the stereonet. The
circles are exactly one net diameter apart and are used
together when counting poles near the circumference of the
net.

In order to construct a circle counter, trace the pattern
given in Figure 18 onto a clear plastic sheet, using drawing
instruments and ink to ensure an accurate and permanent
reproduction. The plastic sheets used for drawing on for
overhead projection, unexposed and developed photographic
film or thin sheets of clear rigid plastic are all ideal
materials for a counter. Punch or drill two small holes,
approximately 1mm in diameter at the centre of each of the
small circles.

The margin sketch illustrates the use of the circle counter
to construct a 3% contour on the pole plot given in
Figure 15. One of the small circles is moved around until
it encircles 10 or 11 poles (3% of 351 poles = 10.5) and a
pencil mark is made through the small hole at the centre of
the circle. The circle is then moved to another position
at which 10 or 11 poles fall within its circumference and
another pencil mark is made. When one of the small circles

is positioned in such a way that a part of it falls outside the stereonet, the total number of poles falling in this circle is given by adding the poles in this and in the other small circle, which must be located diametrically opposite on the stereonet as shown in the margin sketch. The locus of the small circle centre positions defines the 3% contour.

Recommended contouring procedure.

The following procedure is considered to provide an optimum compromise between speed and accuracy for contouring pole plots.

a. Use a Denness type B counting net (Figure 1) to obtain a count of the number of poles falling in each counting cell.

b. Sum these individual counts to obtain the total number of poles plotted on the net and establish the number of poles per 1% area which correspond to the different contour percentage values.

c. Draw very rough contours on the basis of the pole counts noted on the tracing paper.

d. Use the circle counter (Figure 18) to refine the contours, starting with low value contours, (say 2 or 3%) and working inwards towards the maximum pole concentrations.

The contour diagram illustrated in the margin sketch was prepared from the pole plot in Figure 15 in approximately one hour by means of this technique.

Computer analysis of structural data.

Plotting and contouring a few sets of structural geology data can be both interesting and instructive and is strongly recommended to any reader who wishes completely to understand the techniques described on the previous pages. However, faced with the need to process large volumes of such data, the task becomes very tedious and may place an unacceptably high demand on the time of staff who could be employed more effectively on other projects.

The computer is an ideal tool for processing structural geology data on a routine basis and many civil and mining engineering companies and geotechnical consulting organisations use computers for this task. A full discussion on this subject would exceed the scope of this chapter and the interested reader is referred to papers by Spencer and Clabaugh[48], Lam[49], Attewell and Woodman[50], and Mahtab *et al*[51] for details of the different approaches to the computer processing of structural geology data.

Optimum sample size.

The collection of structural geology data is time consuming and expensive and it is important that the amount of data collected should be the minimum required adequately to define the geometrical characteristics of the rock mass. In considering what constitutes an adequate definition of the geometry of the rock mass, the object of the exercise must be kept clearly in mind. In the context of this book, the purpose of attempting to define the rock mass geometry is to provide a basis for choosing the most appropriate failure mode. This is one of the most important decisions in the

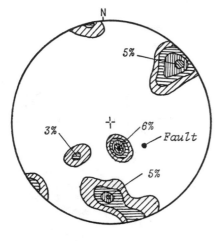

2% – 7 poles
3% – 10 poles
4% – 14 poles
5% – 17 poles
6% – 22 poles

entire process of a slope stability investigation since an incorrect choice of the failure mechanism will almost certainly invalidate the analysis. A hard rock mass, in which two or three strongly developed discontinuity sets show up as dense pole concentrations on a stereoplot, will usually fail by sliding on one or two planes or by toppling. A single through-going feature such as a fault can play a dominant role in a slope failure and it is important that such features are identified separately in order that they are not lost in the averaging which occurs during the contouring of a pole plot. A soft rock mass such as a coal deposit which may be horizontally bedded and vertically jointed or a hard rock mass in which joint orientations appear to be random may fail in a circular mode similar to that which occurs in soil.

From this brief discussion, it will be clear that the collection and interpretation of structural geology data for the purposes of slope stability analysis cannot be treated as a routine statistical exercise. The rock mass knows nothing about statistics and there are many factors, in addition to the density of pole concentrations, which have to be taken into account in assessing the most likely failure mechanism in any given slope. An appreciation of the role of these other factors, which include the strength of the rock mass and the groundwater conditions in the slope, will assist the geologist in deciding on how much structural geology data is required in order that he may make a realistic decision on the slope failure mechanism.

For the reader who has not had a great deal of experience in slope stability analyses and may find it difficult to decide when he has enough structural geology data, the following guidelines on pole plots has been adapted from a paper by Stauffer[47]:

1. First plot and contour 100 poles.
2. If no preferred orientation is apparent, plot an additional 300 poles and contour all 400. If the diagram still shows no preferred orientation, it is probably a random distribution.
3. If step 1. yields a single pole concentration with a value of 20% or higher, the structure is probably truly representative and little could be gained by plotting more data.
4. If step 1. results in a single pole concentration with a contour value of less than 20%, the following total numbers of poles should be contoured.
    12 - 20% add 100 poles and contour all 200.
    8 - 12% add 200 poles and contour all 300.
    4 - 8% add 500 to 900 poles and contour all 600 to 1000.
    less than 4% at least 1000 poles should be contoured.
5. If step 1. yields a contour diagram with several pole concentrations, it is usually best to plot at least another 100 poles and contour all 200 before attempting to determine the optimum sample size.
6. If step 5. yields 1% contours less than 15° apart and with no pole concentrations higher than say 5%, the diagram is possibly representative of a folded structure for which the poles fall within a *girdle* distribution[45].

7. If step 5. yields a diagram with smooth 1% contours
   about 20° apart with several 3-6% pole concentrations,
   then an additional 200 poles should be added and all
   400 poles contoured.
8. If step 7. results in a decrease in the value of the
   maximum pole concentrations and a change in the
   position of these concentrations, the apparent pole
   concentrations on the original plot were probably due
   to the manner in which the data were sampled and it
   is advisable to collect new data and carry out a new
   analysis.
9. If step 7. gives pole concentrations in the same
   positions as those given by step 5, add a further
   200 poles and contour all 600 to ensure that the pole
   concentrations are real and not a function of the
   sampling process.
10. If step 5. yields several pole concentrations of
    between 3 and 6% but with very irregular 1% contours,
    at least another 400 poles should be added.
11. If step 5. yields several pole concentrations of less
    than 3% which are very scattered and if the 1%
    contour is very irregular, at least 1000 and possibly
    2000 poles will be required and any pole concentration
    of less than 2% should be ignored.

Stauffer's work involved a very detailed study of the
statistical significance of pole concentrations and his paper
was not written with any particular application in mind.
Consequently, the guidelines given above should be used for
general guidance and should not be developed into a set of
rules.

The following caution is quoted from Stauffer's paper:
   'A practised eye can identify point clusters, cell
   groupings and gross symmetry even for small samples
   of weak preferred orientations. It is probably true,
   however, that geologists are more prone to call a
   diagram preferred than to dismiss it as being random.
   This is understandable; most geologists examine a
   diagram with the intent of finding something significant,
   and are loath to admit their measurements are not
   meaningful. The result is a general tendency to make
   interpretations more detailed than the nature of the
   data actually warrants.'

The authors feel that it is necessary to add their own words
of caution in emphasising that a contoured pole diagram is a
necessary but not a sufficient aid in slope stability studies.
It must always be used in conjunction with intelligent field
observations and a final decision on the method of analysis
to be used on a particular slope must be based upon a
balanced assessment of all the available facts.

Evaluation of potential slope problems.

Different types of slope failure are associated with different
geological structures and it is important that the slope
designer should be able to recognise the potential stability
problems during the early stages of a project. Some of the
structural patterns which should be watched for when examining
pole plots are outlined on the following pages.

Figure 19 shows the four main types of failure considered in
this book and gives the appearance of typical pole plots of

geological conditions likely to lead to such failures. Note
that in assessing stability, the cut face of the slope must
be included in the stereoplot since sliding can only occur
as a result of movement towards the free face created by the
cut.

The diagrams given in Figure 19 have been simplified for the
sake of clarity. In an actual rock slope, combinations of
several types of geological structures may be present and
this may give rise to additional types of failure. For
example, presence of discontinuities which can lead to
toppling as well as planes upon which wedge sliding can
occur could lead to the sliding of a wedge which is
separated from the rock mass by a "tension crack".

In a typical field study in which structural data has been
plotted on stereonets, a number of significant pole
concentrations may be present. It is useful to be able to
identify those which represent potential failure planes
and to eliminate those which represent structures which are
unlikely to be involved in slope failures. John[52], Panet[53]
and McMahon[32] have discussed methods for identifying
important pole concentrations but the authors prefer a method
developed by Markland[54].

Markland's test is designed to establish the possibility of
a wedge failure in which sliding takes place along the line
of intersection of two planar discontinuities as illustrated
in Figure 19. Plane failure, Figure 19b, is also covered
by this test since it is a special case of wedge failure.
If contact is maintained on both planes, sliding can only
occur along the line of intersection and hence this line of
intersection must "daylight" in the slope face. In other
words, the plunge of the line of intersection must be less
than the dip of the slope face, measured in the direction of
the line of intersection as shown in Figure 20.

As will be shown in the chapter dealing with wedge failure,
the factor of safety of the slope depends upon the plunge of
the line of intersection, the shear strength of the
discontinuity surfaces and the geometry of the wedge. The
limiting case occurs when the wedge degenerates to a plane,
i.e. the dips and dip directions of the two planes are the
same, and when the shear strength of this plane is due to
friction only. As already discussed, sliding under these
conditions occurs when the dip of the plane exceeds the
angle of friction ϕ and hence, a first approximation of
wedge stability is obtained by considering whether the plunge
of the line of intersection exceeds the friction angle for
the rock surfaces. Figure 20b shows that the slope is
potentially unstable when the point defining the line of
intersection of the two planes falls within the area
included between the great circle defining the slope face
and the circle defined by the angle of friction ϕ.

The reader who is familiar with wedge analysis will argue
that this area can be reduced further by allowing for the
influence of "wedging" between the two discontinuity planes.
On the other hand, the stability may be decreased if water
is present in the slope. Experience suggests that these
two factors will tend to cancel one another in typical wedge
problems and that the crude assumption used in deriving
Figure 20b is adequate for most practical problems. It
should be remembered that this test is designed to identify

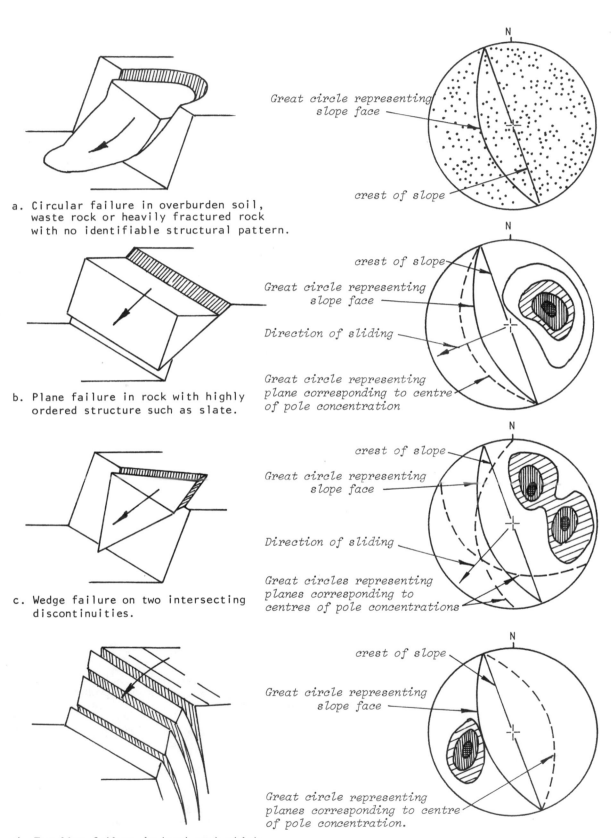

a. Circular failure in overburden soil, waste rock or heavily fractured rock with no identifiable structural pattern.

Great circle representing slope face

crest of slope

b. Plane failure in rock with highly ordered structure such as slate.

crest of slope

Great circle representing slope face

Direction of sliding

Great circle representing plane corresponding to centre of pole concentration

c. Wedge failure on two intersecting discontinuities.

crest of slope

Great circle representing slope face

Direction of sliding

Great circles representing planes corresponding to centres of pole concentrations

d. Toppling failure in hard rock which can form columnar structure separated by steeply dipping discontinuities.

crest of slope

Great circle representing slope face

Great circle representing planes corresponding to centre of pole concentration.

Figure 19 : Main types of slope failure and stereoplots of structural conditions likely to give rise to these failures.

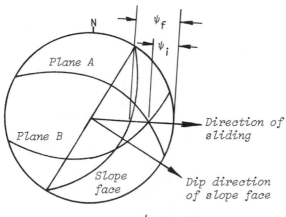

Figure 20a: Sliding along the line of intersection of planes A and B is possible when the plunge of this line is less than the dip of the slope face, measured in the direction of sliding, ie

$$\psi_f > \psi_i$$

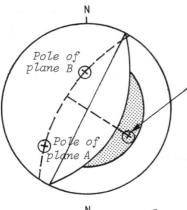

Figure 20b : Sliding is assumed to occur when the plunge of the line of intersection exceeds the angle of friction, ie

$$\psi_f > \psi_i > \phi$$

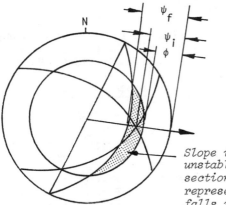

Figure 20c : Representation of planes by their poles and determination of the line of intersection of the planes by the pole of the great circle which passes through their poles.

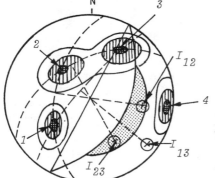

Figure 20d : Preliminary evaluation of the stability of a 50° slope in a rock mass with 4 sets of structural discontinuities.

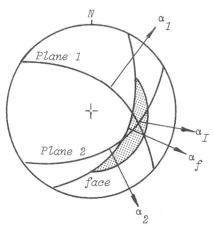

Wedge failure along α_I

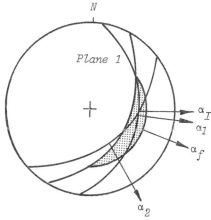

Sliding on plane 2 only
(Direction of maximum dip)

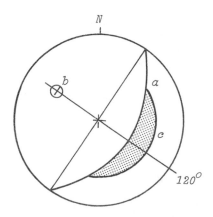

Overlay for checking wedge
failure potential

critical discontinuities and, having identified them, a more detailed analysis would normally be necessary in order to define the factor of safety of the slope.

A refinement to Markland's test has been discussed by Hocking[55] and this refinement has been introduced to permit the user to differentiate between the sliding of a wedge along the line of intersection or along one of the planes forming the base of the wedge. If the conditions for Markland's test are satisfied, i.e. the line of intersection of two planes falls within the shaded crescent shown in the margin sketch, and if the dip direction of either of the planes falls *between* the dip direction of the slope face and the trend of the line of intersection, sliding will occur on the steeper of the two planes rather than along the line of intersection. This additional test is illustrated in the margin sketches on this page.

Figures 20a and 20b show the discontinuity planes as great circles but, as has been discussed on the previous pages, field data on these structures is normally plotted in terms of poles. In Figure 20c the two discontinuity planes are represented by their poles and, in order to find the line of intersection of these planes, the method described on page 47 is used. The tracing on which the poles are plotted is rotated until both poles lie on the same great circle. The pole of this great circle defines the line of intersection of the two planes.

As an example of the use of Markland's test consider the contoured stereoplot of poles given in Figure 20d. It is required to examine the stability of a slope face with a dip of 50° and dip direction of 120°. A friction angle of 30° is assumed for this analysis. An overlay is prepared on which the following information is included:

a. The great circle representing the slope face

b. The pole representing the slope face

c. The friction circle

This overlay is placed over the contoured stereoplot and the two are rotated together over the stereonet to find great circles passing through pole concentrations. The lines of intersection are defined by the poles of these great circles as shown in Figure 20d. From this figure it will be seen that the most dangerous combinations of discontinuities are those represented by the pole concentrations numbered 1,2 and 3. The intersection I_{13} falls outside the critical area and is unlikely to give rise to instability. The pole concentration numbered 4 will not be involved in sliding but, as shown in Figure 19d, it could give rise to toppling or the opening of tension cracks. The poles of planes 1 and 2 lie outside the angle included between the dip direction of the slope face and the line of intersection I_{12} and hence failure of this wedge will be by sliding along the line of intersection I_{12}. However, in the case of planes 2 and 3, the pole representing plane 2 falls within the angle between the dip direction of the slope face and the line of intersection I_{23} and hence failure will be by sliding on plane 2. This will be the most critical instability condition and will control behaviour of the slope.

Suggested method of data presentation and analysis for open pit planning.

During the early feasibility studies on a proposed open pit mine, an estimate of the safe slope angles is required for the calculation of ore to waste ratios and for the preliminary pit layout. The only structural information which may be available at this stage is that which has been obtained from the logging of diamond drill cores, drilled for mineral evaluation purposes, and from the mapping of whatever surface outcrops are available. Scanty as this information is, it does provide a basis for a first estimate of potential slope problems and the authors suggest that this data should be treated in the manner illustrated in Figure 21.

A contour plan of a proposed open pit mine is presented in Figure 21 and contoured stereoplots of available structural data are superimposed on this plan. Two distinct structural regions, denoted by A and B, have been identified and the boundary between these regions has been marked on the plan. For the sake of simplicity, major faults have not been shown but it is essential that any information on faults should be included on large scale plans of this sort and that the potential stability problems associated with these faults should be evaluated.

An overlay is prepared as has been described on previous pages, and is aligned in the dip direction of each portion of the slope as indicated in the drawing. In preparing the overlay included in Figure 21 it has been assumed that the overall slope angle is 45° and that the average friction angle of the discontinuity surfaces in the rock mass is 30°.

The stability evaluation given in Figure 21 shows that the western and southern portion of the pit is likely to be stable at the proposed slope of 45°. This suggests that, if the rock is strong and free of major faults, these slopes could probably be steepened or, alternatively, this portion of the pit wall could be used as a haul road location with steep faces above and below the haul road.

On the other hand, the north-eastern portion of the pit contains a number of potential slope problems. The northern face is likely to suffer from plane sliding on the discontinuity surface A_1. Note that the pole A_1 is almost coincident with the pole of the slope face, an indication of potentially critical plane failure problems. Wedge failures on planes A_1 and A_3 are possible in the north-eastern corner of the pit and toppling failure due to A_2 planes may occur in the eastern slopes. Indications of potential instability such as those included in Figure 21 would suggest that serious consideration should be given to flattening the slopes in the north-eastern part of the proposed pit.

It is interesting to note that three types of structurally controlled slope failure can occur in the same structural region , depending upon the orientation of the slope face. This suggests that, where it is possible, realignment of the slope can be used to eliminate or minimise slope problems.

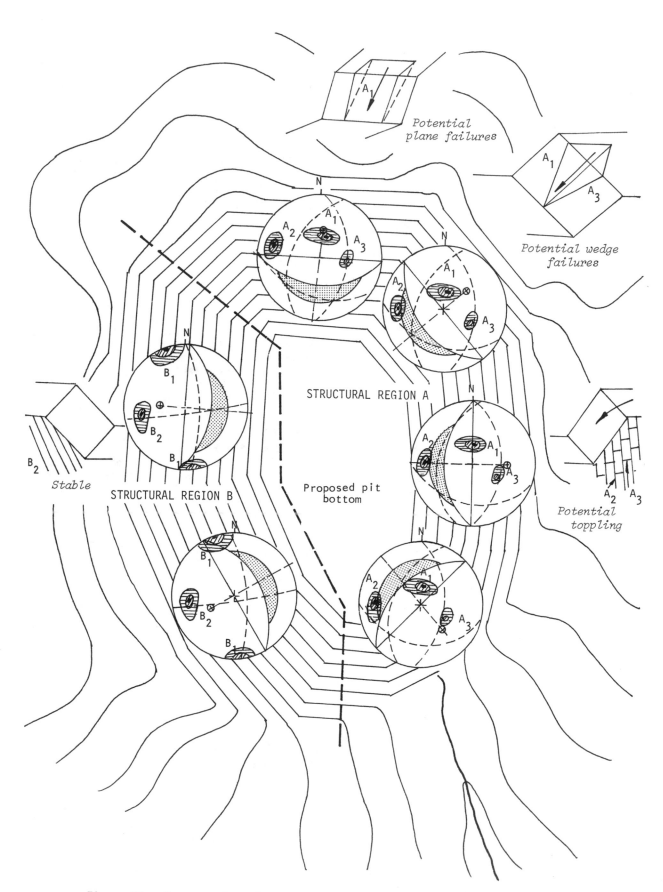

Figure 21 : Presentation of structural geology information and preliminary evaluation of slope stability of a proposed open pit mine.

Chapter 3 references

36. PRICE, N.J. *Fault and joint development in brittle and semi-brittle rock.* Pergamon Press, London, 1966. 176 pages.

37. LOUDERBACK, G.D. Faults and engineering geology. In *Application of Geology to Engineering practice (Berkey volume).* Geological Society of America, 1950, 327 pages.

38. TURNER, F.J. and WEISS, L.E. *Structural analysis of metamorphic tectonites.* McGraw-Hill Book Co., New York, 1963, 545 pages.

39. GOODMAN, R.E. The deformability of joints. In *Determination of the in-situ modulus of deformation of rock.* American Society for Testing and Materials Special Technical Publication, Number 477, 1970. Pages 174-196.

40. PATTON, F.D. Multiple modes of shear failure in rock. *Proc. 1st International Congress of Rock Mechanics.* Lisbon 1966, Vol. 1, pages 509-513.

41. PATTON, F.D. and DEERE, D.U. Significant geological factors in rock slope stability. *Planning open pit mines,* Johannesburg Symposium 1970. Published by A.A. Balkema, Amsterdam 1971, pages 143-151.

42. PHILLIPS, F.C. *The use of stereographic projections in structural geology.* Edward Arnold, London. Third edition (paperback), 1971, 90 pages.

43. BADGLEY, P.C. *Structural methods for the exploration geologist.* Harper Brothers, New York, 1959, 280 pages.

44. FRIEDMAN, M. Petrofabric techniques for the determination of principal stress directions in rock. *Proc. Conference State of Stress in the Earth's crust.* Santa Monica, 1963. Elsevier, New York, 1964, pages 451-550.

45. RAGAN, D.M. *Structural geology - an introduction to geometrical techniques.* John Wiley & Sons, New York, 2nd Edition 1973, 220 pages.

46. DENNESS, B. A revised method of contouring stereograms using variable curvilinear cells. *Geol. Mag.* Vol 109, Number 2, 1972, pages 157-163.

47. STAUFFER, M.R. An empirical-statistical study of three-dimensional fabric diagrams as used in structural analysis. *Canadian Journ. Earth Sciences.* Vol. 3, 1966, pages 473-498.

48. SPENCER, A.B. and CLABAUGH, P.S. Computer program for fabric diagrams. *American Journal of Science.* Vol.265, 1967, pages 166-172.

49. LAM, P.W.H. Computer methods for plotting beta diagrams. *American Journal of Science.* Volume 267, 1969, pages 1114-1117.

50. ATTEWELL P.B. and WOODMAN J.P. Stability of discontinuous rock masses under polyaxial stress systems. *Proc. 13th Symposium on Rock Mechanics.* University of Illinois, Urbana. 1971, pages 665 - 683.

51. MAHTAB, M.A., BOLSTAD,D.D., ALLDREDGE, J.R. and SHANLEY, R.J. Analysis of fracture orientations for input to structural models of discontinuous rock. *U.S. Bureau of Mines Report of Investigations.* No. 669, 1972.

52. JOHN, K.W. Graphical stability analysis of slopes in jointed rock. *Journal Soil Mechanics and Foundation Div.* ASCE. Vol. 94,No. SM2, 1968, pages 497-526 with discussion and closure in Vol. 95, No. SM6, 1969, pages 1541-1545.

53. PANET, M. Discussion on graphical stability analysis of slopes in jointed rock by K.W.John. *Journal Soil Mechanics and Foundation Div.* ASCE Vol.95, No.SM2, 1969, pages 685 - 686.

54. MARKLAND, J.T. A useful technique for estimating the stability of rock slopes when the rigid wedge sliding type of failure is expected. *Imperial College Rock Mechanics Research Report* No. 19, 1972, 10 pages.

55. HOCKING, G. A method for distinguishing between single and double plane sliding of tetrahedral wedges. *Intnl. J. Rock Mechanics and Mining Sciences.* Vol. 13, 1976, pages 225-226.

Chapter 4 : Geological data collection

Introduction

If one examines the amount of time spent on each phase of a rock slope stability investigation, by far the greatest proportion of time is devoted to the collection and interpretation of geological data. However, a search through slope stability literature reveals that the number of publications dealing with this topic is insignificant as compared with theoretical papers on the mechanics of slope failure (idealised slopes, of course). On first appearances it may be concluded that the engineer has displayed a remarkable tendencey to put the proverbial cart before the horse. Deeper examination of the problem reveals that this emphasis on theoretical studies has probably been necessary in order that the engineer (and geologist) should be in a better position to identify *relevant* geological information and, as a consequence, be in a position to deal with this phase of the investigation more efficiently.

As engineers, the authors would not attempt to instruct the geologists on how to go about collecting and interpreting geological data. In fact, experience suggests that attempts by engineers to set up elaborate rock classification systems and standard core logging forms have been remarkably unsuccessful because geologists tend to be highly individualistic and prefer to work from their own point of view rather than that decreed by someone else. What has been attempted in this text, is to present an engineer's view of rock slope stability in such a way that the geologist can decide for himself what geological data is relevant and how he should go about collecting it.

One of the factors which normally plays an important part in the geological investigations for an open pit mine, as compared with site investigations for a dam foundation or a highway cutting, is that the initial work is directed towards mineral exploration. Structural information on faults, joints etc. is normally only collected if this information is likely to assist in defining the ore body. Consequently, when an assessment of safe slope angles is called for during the feasibility studies, there may be insufficient information available for such an assessment to be made. Management and mine planning engineers are usually surprised when a consultant, called in to advise on slope problems, requests more geological information and it is frequently concluded that this is merely a device for increasing the amount of work on a job. This misunderstanding can be eliminated if it is realised that different types of geological information are required for mineral evaluation and for slope design.

Obviously a great deal of time and effort can be saved if the geologist who is concerned with mineral exploration is aware of the needs of the slope engineer and can include this data in his investigations. The benefits of this process are increasingly being recognised and the authors have been involved in a number of projects in which excellent collections of structural geology data have been available at an early stage of the feasibility studies. It is suggested that this must become a general practice if rock slope design is to become an integral part of rational mine planning.

On the following pages, a review is given of techniques which have been found useful in the geological data collection phase of slope stability investigations.

*The Lake Edgar fault which
stretches for many miles across
Tasmania*

*Mapping exposed structures in a
rock mass. An aluminium plate
is being used for projecting
planes.*

Regional geological investigations

A frequent mistake in rock engineering is to start an
investigation with a detailed examination of drill cores.
While these cores provide essential information, it is
necessary to see this information in the context of the
overall geological environment in which the proposed mine or
road or dam site is to exist. Structural discontinuities,
upon which local failure of a bench can occur, are related
to the regional structural pattern of the area and it is
therefore useful to start an investigation by building up a
picture of the regional geology.

Air photographs and topographic maps are readily available
for most land areas of the world and these provide an
important source of information. In some countries, detailed
regional geology maps are available and these should certainly
be obtained as early as possible in the investigation.

Stereoscopic examination of adjacent pairs of air photographs
is particularly useful and even an inexperienced observer
can detect linear surface features which usually signify the
presence of underlying geological structures. The
experienced observer can provide a surprising amount of
relevant information from an examination of air photographs
and many consultants provide routine air-photo interpretation
services. The interested reader is referred to the excellent
text by Miller[56] on this subject.

In addition to air photographs, full use should be made of
any exposures available on site. Adjacent mines or quarries,
road cuttings and exposures in river or stream beds are all
excellent sources of structural information and access to
such exposures can normally be arranged through local land-
owners.

Mapping of exposed structures.

Mapping of visible structural features on outcrops or
excavated faces is a slow and tedious process but there are
unfortunately few alternatives to the traditional techniques
used by the geologist. The most important tool for use in
mapping is obviously the geological compass and many different
types are available. An instrument which has been developed
specifically for the type of mapping required for stability
analyses is illustrated in Figure 22 and the use of this type
of compass, reading directly in terms of dip and dip direction,
can save a great deal of time. It must be emphasised that
there are several other types of compass available and that
all of them will do a perfectly adequate job of structural
mapping. Which instrument is chosen, from those illustrated
in Figure 22, is a matter of personal preference.

Several authors, including Broadbent and Rippere[57], have
discussed the question of the sampling of areas to be mapped
and a line sampling method. This involves stretching a
100 foot tape at approximately waist height along a face or
a tunnel wall and recording every structural feature which
intersects the tape line. Weaver and Call[58] and Halstead
et al[59] also used a technique, which they call fracture set
mapping, which involves mapping all structures occuring in
20 ft. x 6 ft. bands spaced at 100 ft. intervals along a
face. Da Silveira et al[60] mapped all structures exposed on
the face of a rectangular tunnel.

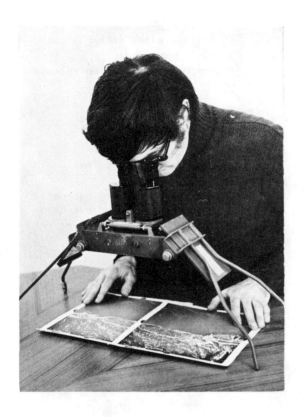

A stereoviewer being used for the examination of air photographs.

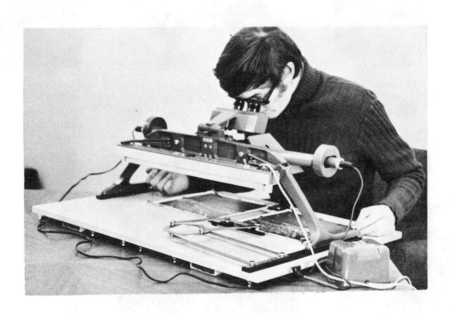

A more elaborate stereoviewer which can be used for the measurement of differences in elevation from air photographs. The instrument illustrated is a model SB180 folding mirror stereoscope manufactured by Rank Precision Industries Ltd., P.O.Box 36, Leicester LEi 9JB, England.

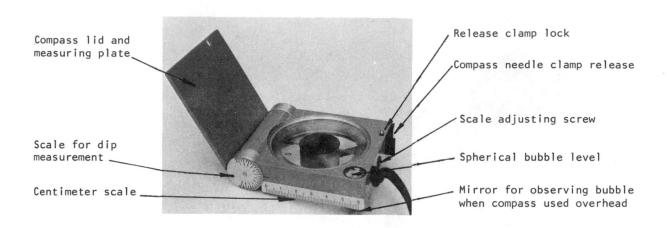

Compass lid and measuring plate

Release clamp lock

Compass needle clamp release

Scale for dip measurement

Scale adjusting screw

Centimeter scale

Spherical bubble level

Mirror for observing bubble when compass used overhead

Figure 22 : Geological compass designed by Professor Clar and manufactured by F.W.Breithaupt & Sohn, Adolfstrasse 13, Kassel 3500, West Germany.

Level compass, release clamp and turn compass until needle points north. Clamp needle.

Use coin to turn adjusting screw to correct magnetic deviation. Zero on scale now reads true north.

Place measuring plate against rock face and level compass, release needle clamp and re-clamp after needle has settled.

Read dip of plane. In this example, dip is 35 degrees.

Read dip direction of plane. In this case, dip direction is 61°.

The difference in magnetic declination in different hemispheres results in the needle jamming if a northern hemisphere compass is used in the southern hemisphere. Manufacturers will supply appropriate instrument if hemisphere is specified.

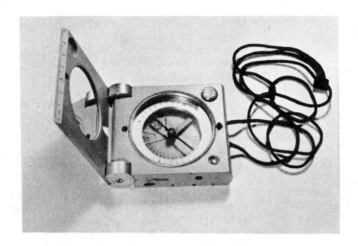

Figure 22a : A Clar type compass manufactured in East Germany
and distributed by Carl Zeiss Jena Ltd., 2 Elstree
Way, Boreham Wood, Hertfordshire WD6 1NH, England.
This instrument has sighting marks and a mirror
fitted in the lid so that it can be used to take
bearings in the same way as a Brunton compass. It
also has a suspended pointer for measuring the inclination of the line of sight. The needle is un-
damped but it can be clamped in the same way as a
Clar compass.

Figure 22b: A Brunton compass or pocket transit which is one of the
most versatile field instruments for geologists although
it is not as convenient as the Clar compass for the
direct measurement of dip and dip direction of planes.
This instrument is particularly convenient for taking
bearings and for measuring the inclination of the line
of sight.

Using a high pressure water jet to clean a rock face

A Wild P 30 phototheodolite set up for photography in an open pit mine

All these methods can be used when an excavated slope or a tunnel is available but, during early exploration studies the geologist may simply have to make do with whatever exposures are available and use his ingenuity to compile as much relevant data as possible. The geologist concerned with mapping surface outcrops, as opposed to excavated faces, must contend with the problem of weathering and of surface coverings of soil and vegetation. A novel solution to this problem was used on the site investigation for the abutments of the Gordon arch dam in Tasmania where the vegetation and soil covering on the rock slopes were washed off by means of high pressure water jets. This process exposed the underlying rock and also accentuated structural features by washing out shallow infillings of weathering products. Obviously, this technique is only applicable in special cases but it is worth considering when critical slope problems are anticipated.

An increasingly common practice in site investigations for large civil engineering structures such as dams is the excavation of exploration adits into the rock mass. These adits provide excellent access to freshly exposed rock and may also be used for the study of groundwater conditions. Relatively few mining companies have used adits for bulk sampling during early feasibility studies and the authors would encourage others to consider seriously the merits of this approach. In spite of the relatively high cost of excavating the adits, the benefits gained in terms of the quality of the information obtained, both mineralogical and structural, probably compensates for this cost and, in some cases, provides the cheapest means of collecting the information required.

All structural mapping techniques suffer from some form of bias since structures parallel or nearly parallel to an exposed face will not daylight as frequently as those perpendicular to the face. This problem has been discussed by Terzaghi[61] and most geologists apply Terzaghi's corrections to structural data obtained from surface mapping and borehole cores. However, Broadbent and Rippere[57] argue that these corrections are excessive when mapping on a typical open pit mine slope face which has been created by normal blasting. In this case the face will be highly irregular as a result of fracturing controlled by discontinuities and the assumptions made by Terzaghi in deriving these corrections are no longer valid. Under such circumstances, Broadbent and Rippere suggest that the data should be presented without correction.

Photographic mapping of exposed structures

Before leaving the question of the mapping surface exposures, mention must be made of photogrammetric techniques which have been considered for use in structural mapping. Although not yet in wide use, photogrammetric methods offer considerable advantages and the authors believe that they will increasingly be used in rock engineering.

The equipment required consists of a *phototheodolite* such as that illustrated in the margin photograph which is simply a theodolite with a suitable camera located between the upper and lower circles. The field set-up is illustrated in Figure 23 and a rock face with targets painted on it for photogrammetric measurement is shown in Figure 24.

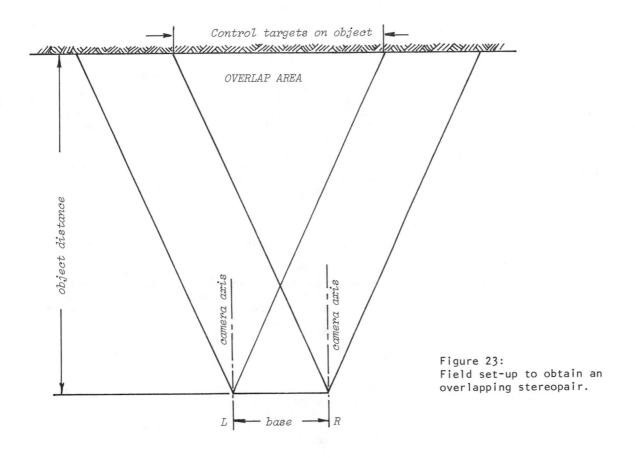

Control targets on object

OVERLAP AREA

object distance

camera axis

camera axis

L base R

Figure 23:
Field set-up to obtain an
overlapping stereopair.

Figure 24:
Rock face with targets painted
on it as controls for photogrammetry. On high, steep faces
targets painted on metal plates
can be lowered on ropes.

The two plates taken at the left and right hand camera stations (Figure 23) are then viewed in a *stereocomparator* or similar instrument which produces a stereoscopic model of the overlapping region on the two plates. Measurements of the x, y and z co-ordinates of points in this three-dimensional model can be made to an accuracy of about 1 part in 5000 of the mean object distance. Hence, a point on a face photographed from 5000 feet can be located to an accuracy of 1.2 inches.

Photogrammetric techniques can be used for both structural mapping and for quantity surveys of excavation volumes and a number of companies offer these services on a commercial basis. It would exceed the scope of this book to discuss details of photogrammetry and the interested reader is referred to the publications listed at the end of this chapter for further information[62,63,64].

Most people think of photogrammetry in terms of expensive equipment and specialist operators and, while this is true for precise measurements, many of the principles can be used to quantify photographs taken with a normal hand-held camera. These principles are described in a useful text-book by Williams[65].

Measurement of surface roughness

Patton[40] emphasised the importance of surface roughness on the shear strength of rock surfaces and his concepts are now widely accepted. The influence of roughness on strength is discussed in the next chapter and the following remarks are restricted to the measurement of roughness in the field.

A variety of techniques have been used to measure roughness and, in the authors' opinion, the most practical method is that suggested by Fecker and Rengers[66] and illustrated in Figure 25 which is self-explanatory.

Diamond drilling for structural purposes

All mining engineers are familiar with diamond drilling for mineral exploration and many would assume that the techniques used in exploration drilling are adequate for structural investigations. This is unfortunately not the case and special techniques are required to ensure that continuous cores which are as nearly undisturbed as possible are obtained. It should be remembered that it is the discontinuities and not the intact rock which control the stability of a rock slope and the nature, infilling, inclination and orientation of these discontinuities are of vital importance to the slope designer.

Contracts for mineral exploration drilling are normally negotiated on a fixed rate of payment per foot or meter drilled. In order to keep these rates low drilling contractors encourage their drillers to aim for the maximum length of hole per shift. In addition, machines are run frequently beyond their effective life in order to reduce capital costs. These practices result in relatively poor core recovery in fractured ground and it can seldom be claimed that the core is undisturbed.

To negotiate a geotechnical drilling contract on the same basis as a mineral exploration programme is to invite trouble.

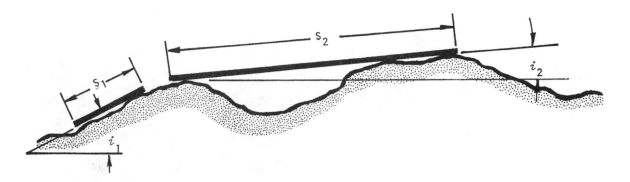

Figure 25a : Measurement of surface roughness with different
base lengths. Short base length gives high values
for the effective roughness angle while long bases
give smaller angles.

Figure 25b :
5.5cm diameter measuring plate fitted to
a Breithaupt geological compass.

Photographs reproduced with permission of
Dr. N. Rengers from a paper by Fecker and
Rengers[66].

Figure 25c :
42 cm diameter measuring plate fitted to
a Breithaupt geological compass for surface
roughness measurement.

73

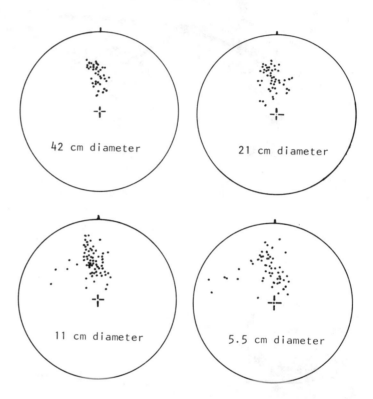

Figure 25 d : Stereoplots of poles from measurements on
rough rock surface using measuring plates of different
diameters. The average dip of the plane is 35° and its
dip direction is 170°.

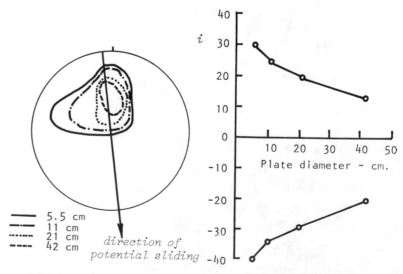

Figure 25 e : Contours of maximum scatter for different
base diameters and plot of effective roughness angle i
along direction of potential sliding. Example adapted
from paper by Fecker and Rengers[66].

Good quality undisturbed core can only be achieved if the driller has time to "feel" his way through the rock and if he is using first class equipment. Consequently, many geotechnical engineers attempt to negotiate contracts in terms of payment for core recovery rather than the length of hole drilled. Alternatively, payment on an hourly rate rather than a drilling rate removes the pressure from the driller and permits him to aim at quality rather than quantity in his drilling.

In addition to providing financial incentives for the driller, it is essential to ensure that his equipment is designed for the job, in good working order and that it is correctly used.

Drilling machines

Good core recovery in fractured ground depends upon the application of the correct thrust onto the rotating drill bit. The fixed rate of advance provided by a screw-feed machine will mean high bit pressures in hard formations. In soft formations, the bit pressure will be very low but the slow progress of the bit will allow the soft material to be eroded by the flushing water. In contrast, a hydraulic feed machine will maintain the same thrust and will allow the drill to move rapidly through soft formations, thereby minimising the erosion.

Machines such as that illustrated in Figure 26a are widely used for exploration drilling and are ideal for structural drilling. Their one disadvantage is that they tend to be bulky and it is difficult to set them up on very rough sites which may be of particular interest to the rock engineer. On the other hand, the machine illustrated in Figure 26b can easily be rigged in very difficult locations and it can be operated as much as 100 feet away from the prime mover and hydraulic pump unit. Hydraulic chucks on this machine allow easy rod changing and permit one man operations once the machine has been set up. Although this machine will not drill to the same depths as larger machines, it does provide adequate capacity for structural drilling outside the orebody where holes of 100-200 feet are generally required.

Core barrels

The aim of structural drilling is to recover undisturbed core upon which measurements of structural features can be made. This can be achieved either by the use of multiple-tube core barrels or by the use of large diameter barrels.

In a multiple-tube core barrel, the inner tube or tubes are mounted on a bearing so that they remain stationary while the outer barrel, which carries the diamond bit, rotates. The core, cut out by the bit, is transferred into the non-rotating inner barrel where it remains undisturbed until the barrel is removed from the hole.

Removing the core from the barrel is the most critical part of the operation. More than once the authors have seen core removed from an expensive double-tube core barrel by thumping the outer barrel with a 4 lb. hammer - a process guaranteed to disturb any undisturbed core which may be in the barrel. By far the most satisfactory system is to use a split inner-barrel which is removed from the core barrel

Figure 26a:
A typical hydraulic thrust diamond drilling machine being used
for exploration drilling. High
quality core recovery can be
achieved with such a machine.
(Photograph reproduced with
permission of Atlas Copco, Sweden)

Figure 26b :
Atlas Copco Diamec 250 drilling machine
set up for structural drilling in a
difficult location on a quarry bench.
The machine illustrated has been modified
to allow drilling with core barrels of up
to 3 inches (76 mm) outer diameter.

assembly with the core inside it and then split to reveal
the undisturbed core. Sometimes a thin plastic or metal
barrel is fitted inside the non-rotating barrel in order to
provide the support for the core when it is transferred into
the core-box.

Detailed literature on double and triple tube core barrels
is available from a number of manufacturers and the reader
who is unfamiliar with structural drilling is advised to
consult this literature. More general information on
drilling can be obtained from some of the references listed
at the end of this chapter[67-70].

Experience has shown that core recovery increases with
increasing core diameter and there is a tendency to use
larger diameter core barrels for structural drilling. The
most common size used at present is NX (2 1/8 inch or 56 mm)
but cores of 3 inch, 4 inch and 6 inch are favoured by some.
Rosengren[67] describes the use of large diameter thin walled
drilling equipment for underground hard rock drilling at
Mt. Isa, Australia. The National Coal Board in England
frequently uses 4½ inch diameter double-barrel drills with
air-flushing for exploration of potential opencast sites.

An extreme example, illustrated opposite, is a 48 inch
diameter Calyx core from a dam-site investigation in
Tasmania. This type of drilling would obviously only be
justified in very special circumstances which the authors
could not visualise occurring on open pit sites.

Core orientation

It should have become obvious, from previous chapters, that
the dip and dip direction of discontinuities are most
important in slope stability evaluations. Consequently,
however successful a drilling programme has been in terms of
core recovery, the most valuable information of all will
have been lost if no effort has been made to orient the core.

One approach to this problem is to use inclined boreholes
to check or to deduce the orientation of structural features.
For example, if surface mapping suggests a strong
concentration of planes dipping at 30° in a dip direction
of 130°, a hole drilled in the direction of the normal to
these planes, i.e. dipping at 60° in a dip direction of 310°,
will intersect these planes at right angles and the accuracy
of the surface mapping prediction can be checked. This
approach is useful for checking the dip and dip direction
of critical planes such as those in the slates on the
western side of the hypothetical open pit shown in Figure 21.

Alternatively, if two or more non-parallel boreholes have
been drilled in a rock mass in which there are recognisable
marker horizons, the orientation and inclination of these
horizons can be deduced using graphical techniques. This
approach is extensively discussed in published literature
and is usefully summarised by Phillips[42]. Where no
recognisable marker horizons are present, as is frequently
the case in metalliferous ore bodies, this technique is of
little value.

A second approach is to attempt to orient the core itself
and, while the techniques available abound with practical
difficulties which are the despair of many drillers, these

*A 48 inch diameter Calyx core
recovered by the Hydroelectric
Commission of Tasmania during
site investigations for a dam.*

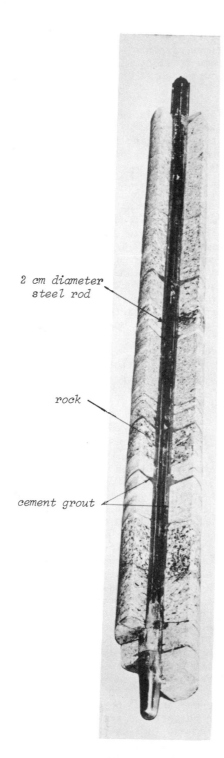

2 cm diameter
steel rod

rock

cement grout

A reinforced core, sectioned to show the rod grouted into the coaxial pilot hole. Note that the top end of the rod is square in order that it can be oriented. (Photograph reproduced with permission of Dr. M. Rocha.)

methods do provide some of the best results currently obtainable. In fact the greatest possible service which could be rendered to the rock engineer by the manufacturers of drilling equipment would be the production of *simple* core orientation systems.

One of the best core orientation devices is that manufactured by Craelius and this consists of a metal holder, of the same diameter as the core, which contains six movable prongs. The device is clamped in the front of the core barrel and lowered into the hole - its orientation fixed by the orientation of the rigid core-barrel or by a simple marker within the device. When the six prongs come into contact with the end of the hole, where the stub usually will have broken off or parted on an inclined surface, these prongs take up the profile of the core stub. The prongs are locked in place and the device released to move up the core-barrel as drilling proceeds. When the core is removed, the core orientation tool is matched to the upper end of the core and the first piece of core is oriented with respect to the known orientation of the device at the time of the fixing of the prongs. Provided that good core recovery has been achieved, it should then be possible to reconstruct the core which is then oriented with respect to the first piece. The use of the Craelius orientator is illustrated on page 78.

An alternative method, used in the Christensen-Hugel core barrel, is to scribe a reference mark on the core. The reference mark is oriented by a magnetic borehole survey instrument mounted in the core-barrel[71].

Rosengren[67] describes a simple device for core orientation in inclined holes. A short dummy barrel holding a marking pen and a mercury orienting switch is lowered down the hole on light-weight rods. The device is rotated until the mercury switch operates at a known orientation and the marker pen is then pushed onto the bottom of the hole, marking the core stub in this known position. Another system used by Rosengren for core orientation in inclined holes is to break a small container of paint against the end of the hole. The paint will run down the face of the core stub, thereby marking its orientation with repect to the vertical.

The most elaborate system of core orientation is to drill a small diameter hole at the end of the parent hole and to bond a compass or an oriented rod into this hole[69]. This scheme has been taken further by Rocha[72] in order to recover intact an oriented core. Rocha describes drilling a pilot hole along the axis of a core and grouting an oriented rod into this hole. Overcoring this reinforced material gives an intact stick of oriented core in the worst types of material.

Walton of the National Coal Board in England has used a similar technique in which a wireline tool containing a bomb of polyester resin is lowered down a hole in which a pilot hole has been drilled. The resin charge is released and flows into the hole, carrying with it a floating compass. Resin reinforced cores have been successfully recovered from depths of 125 m (410 feet) in coal measures but there are many practical difficulties associated with keeping the pilot hole in position and preventing caving in poor quality rock.

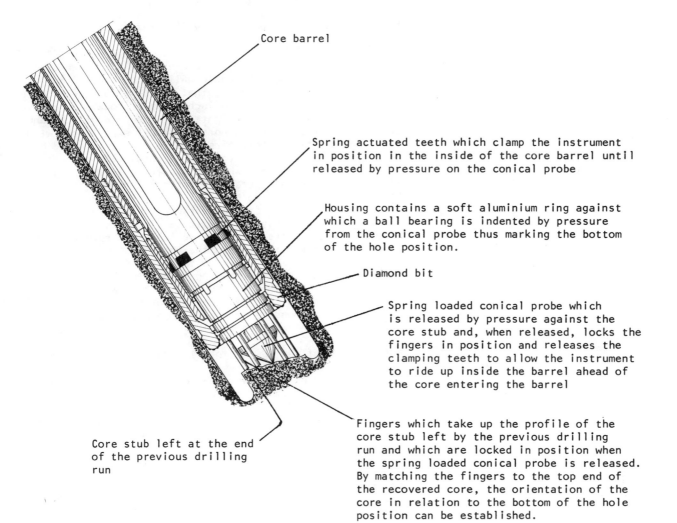

Core barrel

Spring actuated teeth which clamp the instrument
in position in the inside of the core barrel until
released by pressure on the conical probe

Housing contains a soft aluminium ring against
which a ball bearing is indented by pressure
from the conical probe thus marking the bottom
of the hole position.

Diamond bit

Spring loaded conical probe which
is released by pressure against the
core stub and, when released, locks the
fingers in position and releases the
clamping teeth to allow the instrument
to ride up inside the barrel ahead of
the core entering the barrel

Core stub left at the end
of the previous drilling
run

Fingers which take up the profile of the
core stub left by the previous drilling
run and which are locked in position when
the spring loaded conical probe is released.
By matching the fingers to the top end of
the recovered core, the orientation of the
core in relation to the bottom of the hole
position can be established.

Details and method of operation of the Craelius core orienter .
(Drawing from Craelius technical literature)

Examination of borehole walls

Because of the practical problems involved in orientation of core, another approach is to examine the walls of the borehole in an effort to map traces of structural features.

A borehole periscope, consisting of a rigid tube which supports a system of lenses and prisms, is probably the most successful instrument for borehole examination. A major advantage of this device is that it is oriented from outside the hole but a disadvantage is that it is only effective to borehole depths of approximately 100 feet[73].

Various types of borehole cameras have been developed[74,75] and small diameter television cameras have also been used for the examination of borehole walls[59]. Broadbent and Rippere[57] report rather sadly that out of 1116 feet of core recovered from a 2201 foot 8 inch diameter borehole, only 37 feet was successfully oriented with the aid of a borehole television camera. The authors' personal experiences with these devices have almost persuaded them that it would have been more profitable to invest in the provision of geology courses for leprechauns (who are reputedly small enough to fit down a borehole of reasonable size). However, it must be admitted that, in the hands of specialist operators, these instruments can provide very valuable information. It seems more than likely that, with developments in the field of electronics, better and more reliable instruments of this type will become available in the years to come.

The mining and civil engineering industries have a great deal to learn from the oil industry in this area of borehole interpretation and well logging devices such as the Televiewer are bound to find greater applications in site investigation in years to come[76,77,78].

Presentation of geological information

The collection of structural geological data is a difficult enough problem. Communication of this data to everyone concerned in the design of an open pit mine is even more difficult. In the previous chapter it was suggested that the dips and dip directions of discontinuities are most conveniently presented on equal-area stereoplots. This information, in itself, is not adequate for the design of a rock slope since the strength of the rock mass is also required.

Ideally, the following information is required for each significant discontinuity.

    a)   Location in relation to map references or pit plan

    b)   Depth below reference datum

    c)   Dip

    d)   Dip direction

    e)   Frequency or spacing between adjacent discontinuities

    f)   Continuity or extent of discontinuity

    g)   Width or opening of discontinuity.

h)   Gouge or infilling between faces of discontinuity

i)   Surface roughness of faces of discontinuity

j)   Waviness or curvature of discontinuity surface

k)   Description and properties of intact rock between
     discontinuities.

Much of this information cannot be used quantitatively in a
stability calculation but it all assists the engineer or
geologist in deciding upon the most probable failure mode
and in assigning reasonable strength properties to the rock
mass. Consequently it is important that it should be
recorded and presented in such a way that the maximum amount
of relevant information is conveyed to those who were not
involved in the logging itself.

Although standard methods of data presentation have been
suggested [79],[80], many geologists prefer to work out their
own systems to suit their particular requirements. In the
authors' opinion, it does not matter what system is used
provided that it conveys the required information. If it
does not serve this purpose then the geologist should have
the courage to change it.

A final word to engineers - we sometimes believe ourselves
capable of quantifying a subject to a much greater extent
than is actually possible. The geologist is frequently
blamed for poor quality data and yet - if he did provide us
with precise information on all the parameters listed above,
would we really know what to do with it? How do you decide
on the mechanical properties of a rock mass? The real
answer is that we do not know but it is very convenient
to have someone to blame for our lack of knowledge. In the
final analysis, the best we can do is make an informed guess
and the more information we have available at that time the
better. This information must include a personal assessment
of the rock conditions so that the geologist's reports can
be read against a background appreciation of the actual site
conditions. This imposes an obligation on the rock engineer
to devote a little less time to his calculations and a
little more to field observations. This obligation was
summed up by Londe in a lecture on the design of rock
foundations when he said "The time has come for us to
consult not only the experts but the rock as well"[81].

Chapter 4 references

Selected references on geological data collection

56. MILLER, V.C. *Photogeology*. International Series on Earth Sciences. McGraw Hill Book Co., New York, 1961. 248 pages.

57. BROADBENT, C.D. and RIPPERE, K.H. Fracture studies at the Kimberley Pit. Proc. Symposium on *Planning Open Pit Mines*, Johannesburg, 1970. Published by A.A. Balkema, Amsterdam, 1971, pages 171-179.

58. WEAVER, R. and CALL, R.D. Computer estimation of oriented fracture set intensity. *Symposium on Computers in Mining and Exploration*. University of Arizona, Tucson, 1965. 17 pages. (Unpublished paper available from University of Arizona.)

59. HALSTEAD, P.N., CALL, R.D. and RIPPERE, K.H. Geological structural analysis for open pit slope design, Kimberley pit, Ely, Nevada. *A.I.M.E. Preprint 68AM-65*, 1968. Unpublished paper available from A.I.M.E., New York.

60. DA SILVEIRA, A.F., RODRIGUES, F.P., GROSSMAN, N.F. and MENDES, F. Qualitative characterization of the geometric parameters of jointing in rock masses. *Proc. 1st Congress of the International Society of Rock Mechanics*, Lisbon 1966, Vol. 1, pages 225-233.

61. TERZAGHI, R.D. Sources of error in joint surveys. *Geotechnique*, Vol. 15, 1965, pages 287-304.

62. MOFFIT, F.R. *Photogrammetry*. International Textbook Company, Scranton, Pennsylvania, 1967.

63. CALDER, P.N., BAUER, A. and MACDOUGALL, A.R. Stereo-photography and open pit mine design. *Canadian Inst. Min. Metall. Bulletin*, Vol. 63, No. 695, 1970, page 285.

64. ROSS-BROWN, D.M. and ATKINSON, K.B. Terrestial photo-grammetry in open pits. *Proc. Inst. Min. Metall. London*, Vol. 81, 1972, pages A205 - A214.

65. WILLIAMS, J.C.C. *Simple photogrammetry*. Academic Press, London, 1969, 211 pages.

66. FECKER, E. and RENGERS, N. Measurement of large scale roughness of rock planes by means of profilograph and geological compass. *Proc. Symposium on Rock Fracture*, Nancy, France, 1971, Paper 1-18.

67. ROSENGREN, K.J. Diamond drilling for structural purposes at Mount Isa. *Industrial Diamond Review*. Vol. 30, No. 359, 1970, pages 388-395.

68. JEFFERS, J.P. Core barrels designed for maximum core recovery and drilling performance. *Proc. Diamond Drilling Symposium*, Adelaide, August, 1966.

69. HUGHES, M.D. Diamond drilling for rock mechanics investigations. *Rock Mechanics Symposium*, University Sydney, Australia, 1969, pages 135-139.

70. MOYE, D.G. Diamond drilling for foundation exploration. *Civil Engineering Transactions, Institution of Engineers of Australia.* Vol. CE9, 1967, pages 95-100.

71. KEMPE, W.F. Core orientation. *Proc. 12th Exploration Drilling Symposium,* University of Minnesota, 1967.

72. ROCHA, M. A method of integral sampling of rock masses. *Rock Mechanics,* Vol. 3, No. 1, 1967, pages 1-12.

73. KREBS, E. Optical surveying with a borehole periscope. *Mining Magazine,* Vol. 116, 1967, pages 390-399.

74. BURWELL, E.B. and NESBITT, R.H. The NX borehole camera. *Transactions American Inst. Mining Engineers,* Vol. 194. 1954, pages 805-808.

75. HOEK, E. and PENTZ, D.L. Review of the role of rock mechanics research in the design of opencast mines. *Proc. 9th Commonwealth Mining and Metallurgical Congress,* London, 1969, paper No. 4.

76. BALTOSSER, R.W. and LAWRENCE, H.W. Application of well logging techniques in metallic mineral mining. *Geophysics,* Vol. 35, 1970, pages 143-152.

77. TIXIER, M.P. and ALGER, R.P. Log evaluation of non-metallic mineral deposits. *Geophysics,* Vol. 35, 1970, pages 124-142.

78. ZEMANEK, J. The borehole televiewer - a new logging concept for fracture location and other types of borehole inspection. *Society of Petroleum Engineers,* Houston, Texas, September 1968, 14 pages.

79. GEOLOGICAL SOCIETY, LONDON ENGINEERING GROUP. The logging of rock cores for engineering purposes. *Quarterly Journal of Engineering Geology,* Vol. 3, No.1, 1970, 25 pages.

80. ROYAL DUTCH SHELL COMPANY. *Standard Legend,* Internal Shell Report, The Hague, 1958. 200 pages. (A very comprehensive report on logging and geological data presentation).

81. LONDE, P. The role of rock mechanics in the reconnaissance of rock foundations, water seepage in rock slopes and the analysis of the stability of rock slopes. *Quarterly Journal of Engineering Geology.* Vol. 5, 1973, pages 57-127.

Chapter 5 : Shear strength of rock

Introduction

In analysing the stability of a rock slope, the most important factor to be considered is the *geometry* of the rock mass behind the slope face. As discussed in Chapter 3, the geometrical relationship between the discontinuities in the rock mass and the slope and orientation of the excavated face will determine whether parts of the rock mass are free to slide or fall.

The next most important factor is the *shear strength* of the potential failure surface which may consist of a single discontinuity plane or a complex path following several discontinuities and involving some fracture of the intact rock material. Determination of reliable shear strength values is a critical part of a slope design because, as will be shown in later chapters, relatively small changes in shear strength can result in significant changes in the safe height or angle of a slope. The choice of appropriate shear strength values depends not only upon the availability of test data but also upon a careful interpretation of these data in the light of the behaviour of the rock mass which makes up the full scale slope. While it may be possible to use the test results obtained from a shear test on a rock joint in designing a slope in which failure is likely to occur along a single joint surface , similar to the one tested, these shear test results could not be used directly in designing a slope in which a complex failure process involving several joints and some intact rock failure is anticipated. In the latter case, some modification would have to be made to the shear strength data to account for the difference between the shearing process in the test and that anticipated in the rock mass. In addition, differences in the shear strength of rock surfaces can occur because of the influence of weathering, surface roughness, the presence of water under pressure and because of differences in scale between the surface tested and that upon which slope failure is likely to occur.

From this discussion it will be clear that the choice of appropriate shear strength values for use in a rock slope design depends upon a sound understanding of the basic mechanics of shear failure and of the influence of various factors which can alter the shear strength characteristics of a rock mass. It is the aim of this chapter to provide this understanding and to encourage the reader to explore further in the literature on this subject.

Shear strength of planar discontinuities

Supposing that one were to obtain a number of samples of rock, each of which had been cored from the same block of rock which contains a through-going discontinuity such as a bedding plane. This bedding plane is still cemented, in other words, a tensile force would have to be applied to the two halves of the specimen on either side of the discontinuity in order to separate them. The bedding plane is absolutely planar, having no surface undulations or roughness. Each specimen is subjected to a normal stress σ, applied across the discontinuity surface as illustrated in the margin sketch, and the shear stress τ required to cause a displacement u is measured.

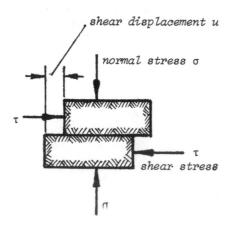

shear displacement u

normal stress σ

τ

τ

shear stress

σ

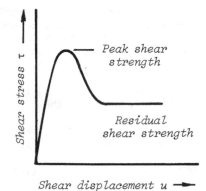

Peak shear strength

Residual shear strength

Shear stress τ

Shear displacement u ⟶

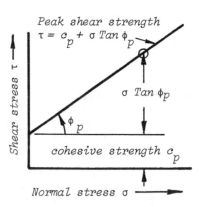

Peak shear strength
$\tau = c_p + \sigma \, Tan \, \phi_p$

$\sigma \, Tan \, \phi_p$

ϕ_p

cohesive strength c_p

Shear stress τ

Normal stress σ ⟶

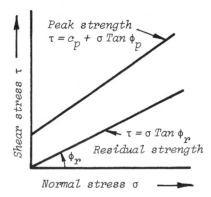

Peak strength
$\tau = c_p + \sigma \, Tan \, \phi_p$

$\tau = \sigma \, Tan \, \phi_r$
Residual strength

ϕ_r

Shear stress τ

Normal stress σ ⟶

Plotting the shear stress level at various shear displacements, for one of the tests carried out at a constant normal stress level, results in the type of curve illustrated in the upper margin sketch. At very small displacements, the specimen behaves elastically and the shear stress increases linearly with displacement. As the forces resisting movement are overcome, the curve becomes non-linear and then reaches a peak at which the shear stress reaches its maximum value. Thereafter the shear stress required to cause further shear displacement drops rapidly and then levels out at a constant value called the residual shear strength.

If the peak shear strength values obtained from tests carried out at different normal stress levels are plotted, a curve such as that illustrated in the centre margin sketch results. This curve will be approximately linear, within the accuracy of the experimental results, with a slope equal to the peak friction angle ϕ_p and an intercept on the shear stress axis of c_p, the cohesive strength of the cementing material. This cohesive component of the total shear strength is independent of the normal stress but the frictional component increases with increasing normal stress as shown in the sketch. The peak shear strength is defined by the equation

$$\tau = c_p + \sigma \, Tan \, \phi_p \qquad (15)$$

which, with the exception of the subscripts, is identical to equation 1 on page 22.

Plotting the residual shear strength against the normal stress gives a linear relationship defined by the equation

$$\tau = \sigma \, Tan \, \phi_r \qquad (16)$$

which shows that all the cohesive strength of the cementing material has been lost. The residual friction angle ϕ_r is usually lower than the peak friction angle ϕ_p.

Influence of water on shear strength of planar discontinuities

The most important influence of the presence of water in a discontinuity in rock is a reduction of shear strength due to a reduction of the *effective* normal stress as a result of water *pressure*. Equation 10 on page 25 shows that this normal stress reduction can be incorporated into the shear strength equation in the following manner :

$$\tau = c + (\sigma - u) Tan \, \phi \qquad (10)$$

where u is the water pressure within the discontinuity and c is either equal to c_p or zero and ϕ either ϕ_p or ϕ_r , depending upon whether one is concerned with peak or residual strength.

As discussed on pages 25 and 26, the influence of water upon the cohesive and frictional properties of the rock discontinuity depends upon the nature of the filling or cementing material. In most hard rocks and in many sandy soils and gravels, these properties are not significantly altered by water but many clays, shales, mudstones and similar materials will exhibit significant changes as a result of changes in moisture content. It is important, therefore, that shear tests should be carried out on samples which are as close as possible to the in situ moisture content of the rock.

Shearing on an inclined plane

In the previous discussion it has been assumed that the discontinuity surface along which shearing occurs is exactly parallel to the direction of the shear stress τ. Let us now consider the case where the discontinuity surface is inclined at an angle i to the shear stress direction, as illustrated in the margin sketch.

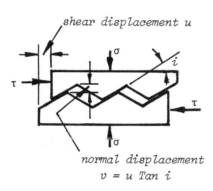

In this case, the shear and normal stresses acting *on the failure surface* are not τ and σ but are given by the following equations :

$$\tau_i = \tau \cos^2 i - \sigma \sin i \cos i \qquad (17)$$

$$\sigma_i = \sigma \cos^2 i + \tau \sin i \cos i \qquad (18)$$

If it is assumed that the discontinuity surface has zero cohesive strength and that its shear strength is given by

$$\tau_i = \sigma_i \tan \phi \qquad (19)$$

then equations 17 and 18 can be substituted into equation 19 to give the relationship between the *applied* shear and normal stresses as :

$$\tau = \sigma \tan (\phi + i) \qquad (20)$$

shear displacement u

normal displacement
$v = u \tan i$

Patton's experiments on shearing regular projections.

This equation was confirmed in a series of tests on models with regular surface projections carried out by Patton[40] who must be credited with having emphasised the importance of this simple relationship in the analysis of rock slope stability.

Patton convincingly demonstrated the practical significance of this relationship by measurement of the average value of the angle i from photographs of bedding plane traces in unstable limestone slopes. Three of these traces are reproduced in the margin sketch and it will be seen that the rougher the bedding plane trace, the steeper the angle of the slope. Patton found that the inclination of the bedding plane trace was approximately equal to the sum of the average angle i and the basic friction angle ϕ found from laboratory tests on planar surfaces.

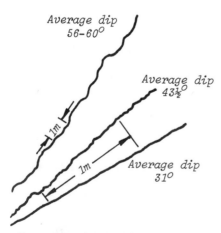

Average dip
56-60°

Average dip
43½°

1m

1m

Average dip
31°

Patton's observations on bedding plane traces in unstable limestone slopes.

An extremely important aspect of shearing on discontinuities which are inclined to the direction of the applied shear stress τ is that any shear displacement u must be accompanied by a normal displacement v. In the case of a specimen with several projections, such as that tested by Patton, this means that the overall volume of the specimen will increase or that the specimen *dilates*. This dilation plays a very important part in the shearing behaviour of actual rock surfaces as will be shown in subsequent discussions.

Note that, up to this point, the discussion has been limited to the problem of shearing along a single discontinuity or along a family of parallel discontinuities and that the question of *fracture* of the material on either side of the discontinuities has not been considered. As will be shown on the following pages, fracturing of interlocking surface projections on rock discontinuities is an important factor which has to be considered when attempting to understand the behaviour of actual rock surfaces.

Surface roughness

The discussion on the previous page has been simplified because Patton found that, in order to obtain reasonable agreement between his field observations on the dip of unstable bedding planes and the sum of the roughness angle i and the basic friction angle ϕ, it was necessary to measure only the first order roughness of the surfaces. This is defined in the margin sketch which shows that the first order projections are those which correspond to the major undulations on the bedding surfaces. The small bumps and ripples on the surface have much higher i values and Patton called these the second order projections.

Later studies by Barton[82] show that Patton's results were related to the normal stress acting across the bedding planes in the slopes which he observed. At very low normal stresses, the second order projections come into play and Barton quotes a number of values of $(\phi + i)$ which were measured at extremely low normal stresses. These values are summarised in the following table :

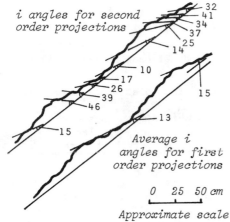

i angles for second order projections

Average i angles for first order projections

0   25   50 cm

Approximate scale

Patton's measurement of i angles for first and second order projections on rough rock surfaces.

Type of surface	Normal stress σ kg/cm^2	$(\phi + i)$	Tested by
Limestone, slightly rough bedding surfaces	1.57	77°	Goodman[39]
	2.09	73°	
	6.00	71°	
Limestone, rough bedding surfaces	3.05	66°	Goodman[39]
	6.80	72°	
Shale, closely jointed seam in limestone	0.21	71°	Goodman[39]
	0.21	70°	
Quartzite, gneiss and amphibolite discontinuities beneath natural slopes	-	80°	Paulding[83]
Beneath excavated slopes	-	75°	
Granite, rough undulating artificial tension fractures	1.5	72°	Rengers[84]
	3.5	69°	

Assuming a basic friction angle of 30°, these results show that the effective roughness angle i varies between 40° and 50° for these very low normal stress levels. In fact, one can assume that almost no fracturing of the very small second order projections takes place at these low normal stress levels and that these steep-sided projections control the shearing process. As the normal stress increases, the second order projections are sheared off and the first order projections take over as the controlling factor. One can imagine that, as the normal stress increases even further, the first order projections will be sheared off and a situation will eventually be reached where shearing takes place through the intact rock material which makes up the projections and the effective roughness angle i is reduced to zero.

The transition from dilation to shearing was studied theoretically and experimentally by Ladanyi and Archambault[85,86] who proposed the following equation for peak shear strength:

$$\tau = \frac{\sigma(1 - a_s)(\dot{v} + \text{Tan}\,\phi) + a_s.\tau_r}{1 - (1 - a_s)\dot{v}\,\text{Tan}\,\phi} \qquad (21)$$

where a_s is the proportion of the discontinuity surface which is sheared through projections of intact rock material,

$\dot{v}$ is the dilation rate dv/du at peak shear strength and

τ_r is the shear strength of the intact rock material.

At very low normal stress levels when almost no shearing through projections takes place, $a_s \rightarrow 0$ and $\dot{v} \rightarrow \text{Tan}\,i$, equation 21 reduces to equation 20. At very high normal stresses when $a_s \rightarrow 1$, $\tau \rightarrow \tau_r$.

Ladanyi and Archambault suggested that τ_r, the shear strength of the material adjacent to the discontinuity surfaces, can be represented by the equation of a parabola in accordance with a proposal by Fairhurst[87]:

$$\tau_r = \sigma_J \frac{\sqrt{1+n} - 1}{n}(1 + n\frac{\sigma}{\sigma_J})^{\frac{1}{2}} \qquad (22)$$

where σ_J is the uniaxial compressive strength of the rock material adjacent to the discontinuity which, due to weathering or loosening of the surface, may be lower than the uniaxial compressive strength of the rock material within the body of an intact block,

n is the ratio of uniaxial compressive to uniaxial tensile strength of the rock material.

Hoek[88] has suggested that, for most hard rocks, n is approximately equal to 10.

Note that, in using Ladanyi and Archambault's equation, it is not necessary to use the definition of τ_r given by equation 22. Any other appropriate intact rock material shear strength criterion, such as $\tau_r = c_J + \sigma\text{Tan}\phi_J$, can be used in place of equation 22 if the user feels that such a criterion gives a more accurate representation of the behaviour of the rock with which he is dealing[89].

The quantity a_s in equation 21 is not easy to measure, even under laboratory conditions. The dilation rate $\dot{v}$ can be measured during a shear test but such measurements have not usually been carried out in the past and hence it is only possible to obtain values for $\dot{v}$ from a small proportion of the shear strength data which has been published. In order to overcome this problem and to make their equation more generally useful, Ladanyi and Archambault carried out a large number of shear tests on prepared rough surfaces and, on the basis of these tests, proposed the following empirical relationships:

$$\dot{v} = \left(1 - \frac{\sigma}{\sigma_J}\right)^K \text{Tan}\,i \qquad (23)$$

$$a_s = 1 - (1 - \frac{\sigma}{\sigma_J})^L \qquad (24)$$

where, for rough rock surfaces, $K = 4$ and $L = 1.5$.

Substituting equations 22, 23 and 24 with $n = 10$, $K = 4$ and $L = 1.5$ into equation 21, and dividing through by σ_J, one obtains the following equation :

$$\frac{\tau}{\sigma_J} = \frac{\frac{\sigma}{\sigma_J}(1 - \frac{\sigma}{\sigma_J})^{1.5}\left((1 - \frac{\sigma}{\sigma_J})^4 \text{Tan } i + \text{Tan}\phi\right) + 0.232\left(1 - (1 - \frac{\sigma}{\sigma_J})^{1.5}\right)(1 + 10\frac{\sigma}{\sigma_J})^{0.5}}{1 - \left((1 - \frac{\sigma}{\sigma_J})^{5.5} \text{Tan } i \text{ Tan } \phi\right)}$$

$$(25)$$

While this equation may appear complex, it will be noted that it relates the two dimensionless groups τ/σ_J and σ/σ_J and that the only unknowns are the roughness angle i and the basic friction angle ϕ.

Figure 27, below, shows that Ladanyi and Archambault's equation 25 gives a smooth transition between Patton's equation 20 for dilation of a rough surface and Fairhurst's equation 22 for the shear strength of the rock material adjacent to the joints.

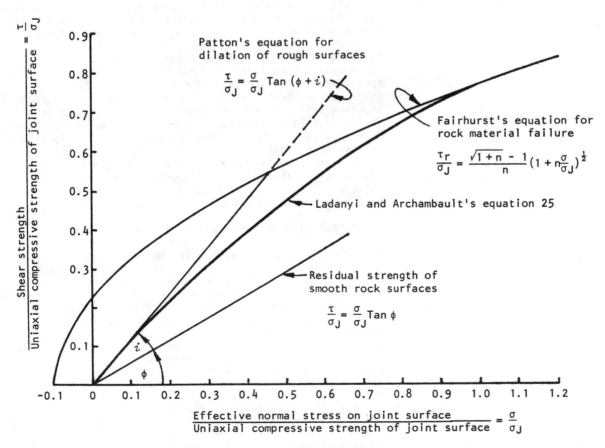

Figure 27 : Transition from dilation to shearing predicted by Ladanyi and Archambault's equation. Plotted for $i = 20^o$ and $\phi = 30^o$.

An alternative approach to the problem of predicting the shear strength of rough joints was proposed by Barton[82]. Based upon careful tests and observations carried out on artificially produced rough 'joints' in material used for model studies of slope behaviour[90,91], Barton derived the following empirical equation :

$$\tau = \sigma \, Tan \, (\phi + JRC . Log_{10} \frac{\sigma_J}{\sigma})$$

(26)

where JRC is a Joint Roughness Coefficient which is defined in Figure 28. The roughness angle i in equation 20 has been replaced by the normal stress dependent term containing JRC.

Barton's equation has been plotted in Figure 29, for JRC values of 20, 10 and 5. For comparison the residual strength of a smooth joint with $\phi = 30°$ and Ladanyi and Archambault's equation for $i = 20$ and $\phi = 30$ are included in the same figure.

Note that, while Barton's equation is in close agreement with Ladanyi and Archambault's (for $i = JRC = 20$) at very low normal stress levels, the equations diverge as the normal stress level increases. This is because Barton's equation reduces to $\tau = \sigma \, Tan \, \phi$ as $\sigma/\sigma_J \to 1$ whereas Lanadyi and Archambault's equation reduces to $\tau = \tau_r$, the shear strength of the rock material adjacent to the joint surface. Barton's equation tends, therefore, to be more conservative than Ladanyi and Archambault's at higher normal stress levels.

Barton's original studies were carried out at extremely low normal stress levels and his equation is probably most applicable in the range $0.01 < \sigma/\sigma_J < 0.3$. Since the normal stress levels which occur in most rock slope stability problems fall within this range, the equation is a very useful tool in rock slope engineering and the authors have no hesitation in recommending its use within the specified stress range. Note that, as $\sigma/\sigma_J \to 0$, the logarithmic term in equation 26 tends to infinity and the equation ceases to be valid. Barton[82] suggests that the maximum value of the term in the brackets in equation 26 should be $70°$ as shown in Figure 29.

Shear testing of discontinuities in rock

From the discussion presented in the preceding pages it will be evident that, in order to obtain shear strength values for use in rock slope design, some form of testing is required. This may take the form of a very sophisticated laboratory or in situ test in which all the characteristics of the in situ behaviour of the rock discontinuity are reproduced as accurately as possible. Alternatively, the test may involve a very simple determination or even estimate of the joint compressive strength σ_J, the roughness angle i and the basic friction angle ϕ for use in Barton's or Ladanyi and Archambault's equation. The choice of the most appropriate method depends upon the nature of the problem being investigated, the facilities which are available and the amount of time and money which has been allocated to the solution of the problem. In carrying out a detailed design for a critical slope such as that adjacent to a major item of plant or in the abutment of an arch dam, no expense and effort would be spared in attempting to

EXAMPLES OF ROUGHNESS PROFILES

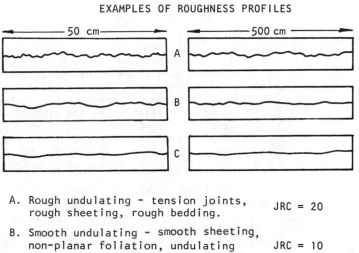

A. Rough undulating - tension joints, rough sheeting, rough bedding. JRC = 20

B. Smooth undulating - smooth sheeting, non-planar foliation, undulating bedding. JRC = 10

C. Smooth nearly planar - planar shear joints, planar foliation, planar bedding. JRC = 5

Figure 28 : Barton's definition of Joint Roughness Coefficient JRC.

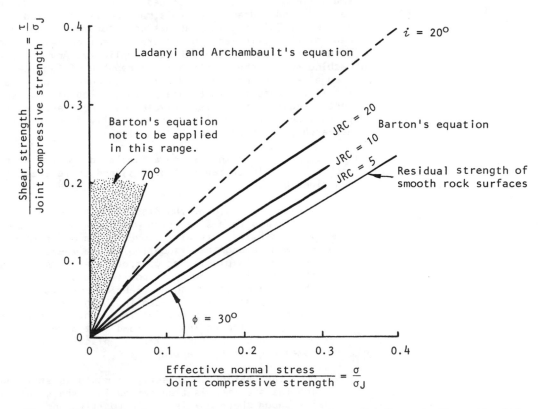

Figure 29 : Barton's prediction for the shear strength of rough discontinuities.

obtain reliable shear strength values for critical discontinuities encountered in the rock mass. On the other hand, preliminary stability calculations carried out during the feasibility study on an open pit mine slope are generally restricted in terms of access to the rock mass and also time and money available for the study, hence elaborate and expensive testing is not justified. Under these circumstances, realistic estimates of the shear strength on the basis of the approaches proposed by Barton and by Ladanyi and Archambault normally have to be used.

Figure 30 illustrates the arrangements for a large scale in situ shear test to be carried out in an underground adit. This type of test costs several thousands of pounds or dollars and would only be justified under the most critical conditions. Alternative in situ shear test arrangements have been discussed by Serafim and Lopes[92], Haverland and Slebir[93], Ruiz and Camargo[94] and Brawner, Pentz and Sharp[95].

A laboratory shear machine designed and built at the Imperial College of Science and Technology in London is illustrated in Figure 31. This machine accepts samples of approximately 12" x 16" and has a capacity of 100 tons in both normal and shear directions. The loading rate is variable over a very wide range and normal and shear displacements can be monitored continuously during the test. Individual tests on this machine are relatively expensive and its use is normally only justified on major projects.

A portable shear machine for testing rock discontinuities in small field samples has been described by Ross-Brown and Walton[96] and a drawing of this machine is presented in Figure 32. This machine was designed for field use and many of the refinements which are present on larger machines were sacrificed for the sake of simplicity. Any competent machine shop technician should be capable of fabricating such a machine and the reader is encouraged to utilise the ideas presented in Figure 32 to develop his own shear testing equipment. Alternatively, machines manufactured to this design are available commercially from Robertson Research Ltd., Llandudno, North Wales.

The steps involved in testing a sample in the portable shear machine are illustrated in Figure 33 and these steps are described below :

   a. A sample containing the discontinuity to be tested is trimmed to a size which will fit into the mould. The two halves are wired together in order to prevent movement along the discontinuity and the sample is then cast in plaster or concrete. Care must be taken that the discontinuity is positioned accurately so that it lies in the shear plane of the machine. A bed of clean gravel, placed in the bottom of the first mould, is sometimes helpful in supporting the specimen during setting up and, provided that a wet mix is used, the gravel can be left in place so the it becomes part of the casting.

   b. Once the castings have set, the mould is stripped and the specimen is transferred into the shear machine. The upper shear box is set in position and a small normal load is applied in order to prevent movement of the specimen. The wires binding the two halves of

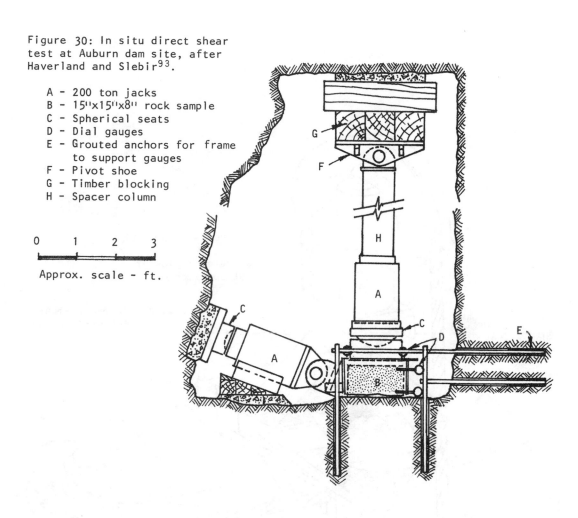

Figure 30: In situ direct shear test at Auburn dam site, after Haverland and Slebir[93].

A - 200 ton jacks
B - 15"x15"x8" rock sample
C - Spherical seats
D - Dial gauges
E - Grouted anchors for frame to support gauges
F - Pivot shoe
G - Timber blocking
H - Spacer column

0   1   2   3

Approx. scale - ft.

Figure 31 : Large scale laboratory shear machine at the Imperial College of Science and Technology, London.

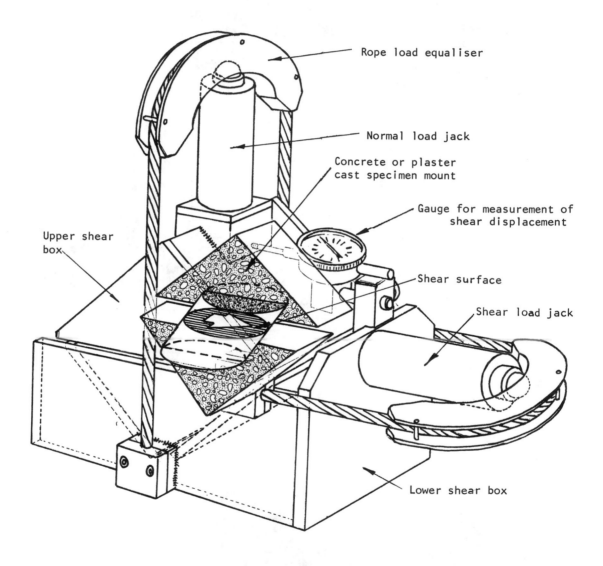

Figure 32: Drawing of a portable shear machine showing the position of
the specimen and the shear surface. Drawing adapted from
one by Robertson Research Ltd. A typical machine is 20 inches
(51 cm) long and 18 inches (48 cm) high and weighs 85 lb.(39 kg.)

the specimen together are then cut and the shear load cable is placed in position.

c. The specimen is now ready for testing and the normal load is increased to the value chosen for the test. This normal load is maintained constant while the shear load is increased. A note is kept of displacements during the application of the shear load.

d. Once the peak strength has been exceeded, usually after a shear displacement of a few millimeters, the displacement is allowed to continue and it will be found that a lower shear load is required in order to sustain movement.

e. The machine illustrated is limited to a displacement of approximately 1 in. (2.5 cm) and, in order to determine the residual shear strength, a displacement in excess of this value is normally required. This can only be achieved if the normal load is released and the upper half of the specimen is moved back to its starting position. In another version of the machine, manufactured by Robertson Research Limited, a second jack acting in opposition to the shear load jack has been added to allow for shear reversal under constant normal load.

Which of these systems is more representative of the shearing process in the rock mass is uncertain since, in one case, the detrital material is disturbed when the normal load is released while, in the other, the direction of rolling of the particles is reversed. It is possible that different rock structures behave differently under these conditions. The authors tend to prefer the single jack system since they feel that it is important that the direction of shearing under load should be kept constant.

f. In this, as in most shear machines, the *loads* applied to the specimen are measured and these have to be divided by the surface area of the discontinuity surface in order to obtain normal and shear *stresses*. The initial area should be determined by direct measurement and the reduction in surface area with displacement should be calculated.

g. The shear strength of rock is not generally sensitive to the loading rate and no difficulty should arise if the loads are applied at a rate which will permit measurements of loads and displacements to be carried out at regular intervals. A typical test would take between 15 and 30 minutes.

Portable shear machine with two shear load jacks which allow shear reversal under constant normal load.

The specimen size which can be accommodated in the portable shear machine illustrated in Figure 32 is limited to about 4'' x 4'' (10cm x 10cm) and this means that it is very difficult to test joints with surface roughness which is representative of the in situ conditions. Consequently, it is recommended that the use of this machine should be restricted to the measurement of the basic friction angle ϕ. This can be done by testing sawn surfaces or by testing field samples and by subtracting the average roughness angle i, measured on the specimen surface before testing, from the measured angle $(\phi + i)$ as determined in the test.

Figure 33a : Sample , wired together to prevent premature movement along the discontinuity, is aligned in a mould and the lower half is cast in concrete, plaster or similar material. When the lower half is set, the upper half of the mould is fitted and the entire mould plus sample is turned upside-down in order that the second half of the casting can be poured.

Figure 33b : Specimen, still wired together, is fitted into the lower shear box and the upper shear box is then fitted in place. Note that the load cables can be bent out of the way for easy access.

Figure 33c : The wires binding the specimen halves together are cut and the normal and shear loads are applied.

Estimating joint compressive strength and friction angle

When it is impossible to carry out any form of shear test, the shear strength characteristics of a rock surface can be approximated from Ladanyi and Archambault's equation or from Barton's equation. In order to solve either of these equations it is necessary to determine or to estimate values for σ_J, the joint material compressive strength, ϕ, the basic angle of friction of smooth surfaces of this rock type and i, the average roughness angle of the surface or JRC, Barton's Joint Roughness Coefficient.

The uniaxial compressive strength of the joint wall material can be obtained by coring through the joint surface and then testing specimens prepared from this core. This is a complex process and , if facilities and time are available to carry out such tests, they would almost certainly be available for a direct shear test. Consequently, uniaxial compressive strength tests on material samples would seldom be a logical way in which to obtain the value of σ_J.

A simpler alternative which can be used in either the field or the laboratory is the Point Load Index test[97]. This simple and inexpensive test can be carried out on unprepared core and the loading arrangement is illustrated in the margin sketch. Two types of commercially available point load testing machine are illustrated in Figures 34 and 35. A reasonable correlation exists between the Point Load Index and the uniaxial compressive strength of the material[98], as shown in Figure 36, and is given by :

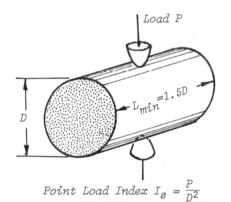

Load P

$L_{min} = 1.5D$

D

Point Load Index $I_s = \dfrac{P}{D^2}$

$$\sigma_c = 24 \; I_s{}^* \qquad\qquad (27)$$

where σ_c is the uniaxial compressive strength and
I_s is the point load strength index.

Note that σ_J, the compressive strength of the rock material adjacent to the joint surface, may be lower than σ_c as a result of weathering or loosening of the surface.

In order to judge whether a point load test is valid, the fractured pieces of core should be examined. If a clean fracture runs from one loading point indentation to the other, the test results can be accepted. However, if the fracture runs across some other plane, as may happen when testing schistose rocks, or if the points sink into the rock surface causing excessive crushing or deformation, the test should be rejected.

Figure 34 : Point Load Index test equipment manufactured by Robertson Research Ltd., Llandudno, North Wales.

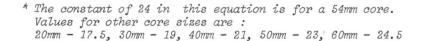

* *The constant of 24 in this equation is for a 54mm core. Values for other core sizes are :*
20mm – 17.5, 30mm – 19, 40mm – 21, 50mm – 23, 60mm – 24.5

Figure 35 : Point load test equipment
manufactured by Engineering Laboratory
Equipment Limited, Hemel Hempstead,
Hertfordshire, England.

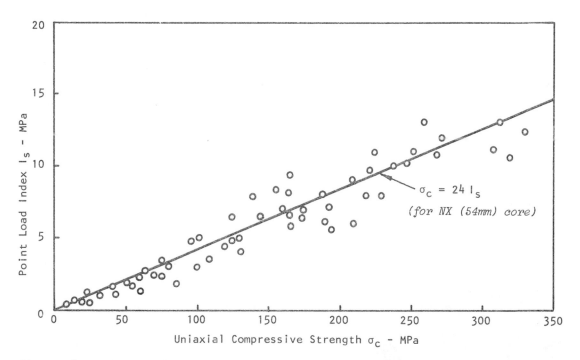

Figure 36 : Relationship between point load strength index and uniaxial compressive
strength.   1 MPa = 10.2 kg/cm^2 = 145 lb/in^2.

A less reliable but simpler alternative to the point load
test for determining the uniaxial compressive strength of
rock is the use of the Schmidt hammer [89,99]. An advantage
of this method is that it can be applied directly to an
unprepared joint surface and can be used to obtain a direct
estimate of the joint compressive strength σ_J.

The relationship between compressive strength and Schmidt
hardness is given in Figure 37. Suppose that a horizontally
held type L hammer gives a reading of 48 on a rock with a
density of 27 kN/m^3, the uniaxial compressive strength σ_c

is given by the graph as 125 ±50 MPa. Note that the hammer
should always be perpendicular to the rock surface.

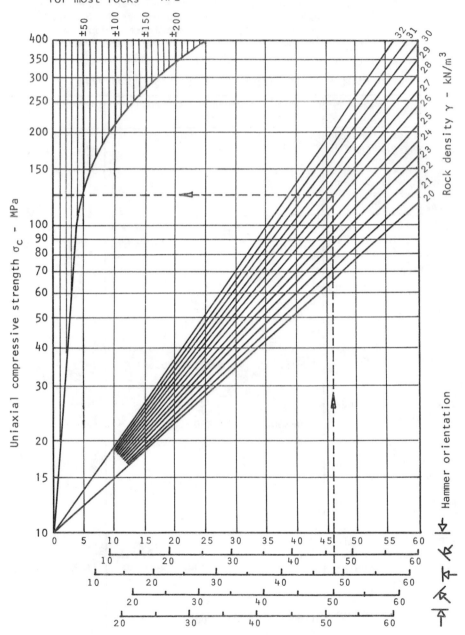

Figure 37 : Relationship between Schmidt hardness and the uniaxial
compressive strength of rock, after Deere and Miller[100].

$$1 \text{ MPa} = 1 \text{ MN/m}^2 = 10.2 \text{ kg/cm}^2 = 145 \text{ lb/in}^2$$
$$1 \text{ kN/m}^3 = 102 \text{ kg/m}^3 = 6.37 \text{ lb/ft}^3.$$

When the slope designer finds himself in a situation where
no facilities at all are available, for example on a proposed
mine site during the earliest stages of exploration, he has
to resort to a method of determining the compressive strength
of the rock by a method which is best termed " kicking the
rock". In order to assist such an adventurer, a very approximate set of guidelines have been tabulated below, based
upon papers by Deere and Miller[100], Piteau[101], Robertson[102]
and upon consulting experience. Cohesive soils have been
included in this table since these are important as joint
filling materials, to be discussed in the next section.

TABLE II - APPROXIMATE CLASSIFICATION OF COHESIVE SOIL AND ROCK

No.	Description	Uniaxial compressive strength lb/in²	kg/cm²	MPa	Examples
S1	VERY SOFT SOIL - easily moulded with fingers, shows distinct heel marks.	<5	<0.4	<0.04	
S2	SOFT SOIL - moulds with strong pressure from fingers, shows faint heel marks.	5-10	0.4-0.8	0.04-0.08	
S3	FIRM SOIL - very difficult to mould with fingers, indented with finger nail, difficult to cut with hand spade.	10-20	0.8-1.5	0.08-0.15	
S4	STIFF SOIL - cannot be moulded with fingers, cannot be cut with hand spade, requires hand picking for excavation .	20-80	1.5-6.0	0.15-0.60	
S5	VERY STIFF SOIL - very tough , difficult to move with hand pick, pneumatic spade required for excavation.	80-150	6-10	0.6-1.0	
R1	VERY WEAK ROCK - crumbles under sharp blows with geological pick point, can be cut with pocket knife.	150-3500	10-250	1-25	Chalk, rocksalt
R2	MODERATELY WEAK ROCK - shallow cuts or scraping with pocket knife with difficulty, pick point indents deeply with firm blow.	3500-7500	250-500	25-50	Coal, schist, siltstone
R3	MODERATELY STRONG ROCK - knife cannot be used to scrape or peel surface, shallow indentations under firm blow from pick point.	7500-15000	500-1000	50-100	Sandstone, slate, shale
R4	STRONG ROCK - hand-held sample breaks with one firm blow from hammer end of geological pick.	15000-30000	1000-2000	100-200	Marble, granite, gneiss
R5	VERY STRONG ROCK - requires many blows from geological pick to break intact sample.	> 30000	> 2000	> 200	Quartzite, dolerite, gabbro, basalt

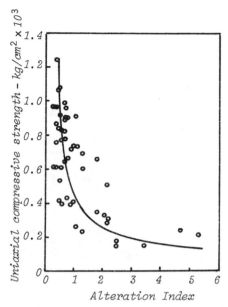

Variation of compressive strength with degree of alteration for a granite . After Serafim[103].

Approximate values for the basic friction angle φ for different rocks.

Rock	φ-degrees
Amphibolite	32
Basalt	31 - 38
Conglomorate	35
Chalk	30
Dolomite	27 - 31
Gneiss (schistose)	23 - 29
Granite (fine grain)	29 - 35
Granite (coarse grain)	31 - 35
Limestone	33 - 40
Porphyry	31
Sandstone	25 - 35
Shale	27
Siltstone	27 - 31
Slate	25 - 30

Lower value is generally given by tests on wet rock surfaces.

After Barton[82].

A final question which must be considered when estimating the compressive strength of the material adjacent to the discontinuity surface relates to the weathering or alteration of the material. This question has been discussed by Barton[82] who suggests that weathering can reduce the strength of the near surface material to as low as one quarter of the uniaxial compressive strength of the intact unweathered material. This process will vary according to the rock type since very dense and impervious rocks such as basalt would gradually acquire a thin skin of weathered material, granites would weather more deeply and, porous rock such as sandstone could weather more or less uniformly to considerable depth.

The graph reproduced in the margin is from Serafim[103] and shows the significant strength reduction with increasing alteration. The alteration index , described by Hamrol[104], is the weight of water absorbed by the rock in a quick absorption test, divided by the dry weight of rock, expressed as a percentage. Rocha[105] has also shown the rapid fall in shear strength of rock which is associated with the first few percent increase in alteration index. Useful discussions on weathering have also been given by Fookes, Dearman and Franklin[106] and Franklin and Chandra[107].

Because of the wide range of conditions which can be encountered in the field, the authors have not attempted to give specific guidelines on the allowance which should be made for weathering when considering the uniaxial compressive strength of rock in order to determine the joint wall compressive strength σJ. The reader should be aware of this problem and should give consideration to the allowance which he should make under each particular set of circumstances.

Turning to the question of the basic friction angle φ for use in Ladanyi and Archambault's and in Barton's equations, ideally, this quantity should be determined by direct shear testing on smooth rock surfaces which have been prepared by means of a clean, smooth diamond saw cut. Alternatively, residual shear strength values, obtained from shear tests in which the specimen has been subjected to considerable displacement, can be used to obtain the value of φ. Note that either of these tests should be carried out over a range of normal stress levels to ensure that a linear relationship between shear strength and normal stress with zero cohesion is obtained. This precaution is necessary because the shear strength at very low normal stresses can be influenced by extremely small surface roughness on the specimen. Tilting tests, in which the angle of inclination of the specimen required to cause sliding is measured, are not reliable for the determination of the basic friction angle because of the influence of very small scale surface roughness.

If a shear test is carried out on a field sample with rough surfaces, the surface profile can be measured, before testing, and the average roughness angle i subtracted from the angle of inclination of the line relating shear strength to normal stress. This correction should only be used when the shear test results fall reasonably close to a straight line which passes through the origin.

When no test results are available, the tabulation given in the margin can be used to obtain an estimate of the basic angle of friction.

The roughness angle i which is required for an evaluation of Ladanyi and Archambault's equation can be obtained from measurements such as those described in Figure 25 on pages 72 and 73. Care should be taken to ensure that the scale of measurement is appropriate to the scale of the problem. In very large slopes features such as folds in the bedding planes may contribute to the effective roughness of the potential sliding surface. The average i value for such surfaces can be measured off photographs, as was done by Patton[40], or by measuring the dips, with a geological compass, along a line marked on the plane . This line should be in the direction of potential sliding and should be long enough to ensure that several roughness 'wavelengths' are included in the measurement.

Barton's Joint Roughness Coefficient JRC is only approximately related to the roughness angle i and he suggests[82] that the value of JRC should be estimated by simple visual comparison with Figure 28. Note that two scales are given in this figure and that the user would use the scale most appropriate to the problem which he is considering.

Shear strength of filled discontinuities

Up to this point the discussion has been restricted to the shear strength of surfaces in which rock-to-rock contact occurs along the entire length of the surface. A common problem which is encountered in rock slope design is a discontinuity which is filled with some form of soft material. This filling may be detrital material or gouge from previous shear movements, typical in faults, or it may be material which has been deposited in open joints as a result of the movement of water through the rock mass. In either case, the presence of a significant thickness of soft, weak filling material can have a major influence on the stability of the rock mass.

Goodman [39] demonstrated the importance of joint fillings in a series of tests in which artificially created saw-tooth joint surfaces were coated with crushed mica. The decrease in shear strength with increasing filling thickness is illustrated in Figure 38 which shows that, once the filling thickness exceeds the amplitude of the surface projections, the strength of the joint is controlled by the strength of the filling material.

A very comprehensive review of the shear strength of filled discontinuities has been prepared by Barton[108] and this paper is highly recommended to any reader who wishes to study this subject in greater detail. A list of shear strength values for filled joints, based upon one compiled by Barton, is given in Table III.

When a major discontinuity with a significant thickness of filling is encountered in a rock mass in which a slope is to be excavated, it is prudent to assume that shear failure will occur through the filling material. Consequently, at least for the preliminary analysis, the influence of surface roughness should be ignored and the shear strength of the discontinuity should be taken as that for the filling material. Determination of the shear strength of this filling material should be carried out in accordance with well

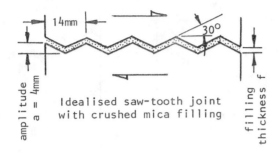

Idealised saw-tooth joint with crushed mica filling

Figure 38 : Influence of joint filling thickness on the shear strength of an idealised saw-tooth joint. After Goodman [39].

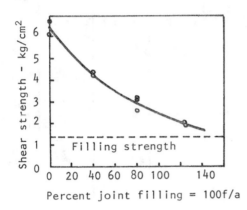

Percent joint filling = 100f/a

established soil mechanics principles.

Another major factor which must be considered in relation to filled joints is the influence which they have on the permeability of the rock mass. The permeability of clay gouge and similar joint filling material may be three or four orders of magnitude lower than that of the surrounding rock mass and this can give rise to damming of groundwater into compartments within the rock mass. When water pressure is allowed to build up behind a clay-filled discontinuity such as a fault, the overall stability of the slope can be jeopardised and the situation is made worse by the fact that this filling material has a very low shear strength and that failure of the slope may be initiated along this discontinuity.

From this discussion it will be clear that an extremely important aspect of a site investigation programme for a rock slope design is the detection of major discontinuities which are filled with clay or other filling materials. If the presence of such discontinuities is suspected, a special effort should be made during the site investigation programme to check whether they do exist. This may involve the drilling of holes in critical locations as well as the careful tracing of outcrops and intersections with any existing excavations. Determination of the orientation and inclination of such discontinuities for subsequent inclusion in stability analyses is important as is the sampling of the filling material for shear strength testing.

TABLE III - SHEAR STRENGTH OF FILLED DISCONTINUITIES

Rock	Description	Peak strength		Residual strength		Tested by
		$c'kg/cm^2$	$\phi°$	$c'kg/cm^2$	$\phi°$	
Basalt	Clayey basaltic breccia, wide variation from clay to basalt content.	2.4	42			Ruiz, Camargo Midea and Nieble[109].
Bentonite	Bentonite seam in chalk	0.15	7.5			Link[110]
	Thin layers	0.9-1.2	12-17			
	Triaxial tests	0.6-1.0	9-13			Sinclair and Brooker[111]
Bentonitic shale	Triaxial tests	0 - 2.7	8.5-29			Sinclair and Brooker[111]
	Direct shear tests			0.3	8.5	
Clays	Over-consolidated, slips, joints and minor shears	0 - 1.8	12-18.5	0 - 0.03	10.5-16	Skempton and Petley[112]
Clay shale	Triaxial tests	0.6	32			Sinclair and Brooker[111]
Clay shale	Stratification surfaces			0	19-25	Leussink and Muller-Kirchenbauer[113]
Coal measure rocks	Clay mylonite seams, 1.0 to 2.5cm thick	0.11-0.13	16	0	11-11.5	Stimpson and Walton[114]
Dolomite	Altered shale bed, approximately 15 cm thick.	0.41	14.5	0.22	17	Pigot and Mackenzie[115]
Diorite, grano-diorite and porphyry	Clay gouge (2% clay, PI = 17%)	0	26.5			Brawner[116]
Granite	Clay filled faults	0 - 1.0	24 - 45			Rocha[105]
	Weakened with sandy-loam fault filling	0.5	40			Nose[117]
	Tectonic shear zone, schistose and broken granites, disintegrated rock and gouge.	2.42	42			Evdokimov and Sapegin[118]
Greywacke	1-2mm clay in bedding planes			0	21	Drozd[119]
Limestone	6mm clay layer			0	13	Krsmanovic et al[120]
	1-2cm clay fillings	1.0	13-14			Krsmanovic & Popovic[121]
	<1mm clay fillings	0.5-2.0	17-21			
Limestone, marl and lignites	Interbedded lignite layers	0.8	38			Salas and Uriel[122]
	Lignite/marl contact	1.0	10			
Limestone	Marlaceous joints, 2cm thick	0	25	0	15-24	Bernaix[123]
Lignite	Layer between lignite and underlying clay	0.14-0.3	15-17.5			Schultze[124]
Montmorillonite clay		3.6	14	0.8	11	Eurenius[125]
	8 cm seams of bentonite (montmorillonite) clay in chalk.	0.16-0.2	7.5-11.5			Underwood[126]
Schists, quartzites and siliceous schists.	10-15cm thick clay filling	0.3-0.8	32			Serafim and Guerreiro[127]
	Stratification with thin clay	6.1-7.4	41			
	Stratification with thick clay	3.8	31			
Slates	Finely laminated and altered	0.5	33			Coates, McRorie and Stubbins[128]
Quartz/kaolin/pyrolusite	Remoulded triaxial tests	0.42-0.9	36-38			

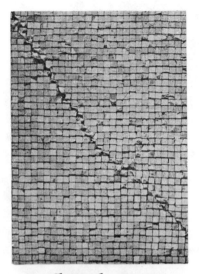

Shear plane

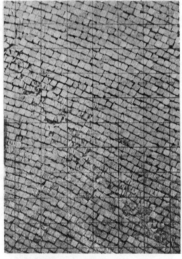

Shear zone

Kink band

Shear strength of closely jointed rock masses

When a hard rock mass contains a number of joint sets and when the joint spacing is very close, in relation to the size of slope being considered, the behaviour of the rock mass may differ significantly from that of the single discontinuity surface considered in the first part of this chapter. The loosened state of the rock mass, resulting from the close jointing, permits individual blocks within the mass to translate and to rotate to a far greater degree than can occur in more intact rock and this gives rise to an overall strength reduction.

The determination of the shear strength of closely jointed rock masses has long been recognised as an important engineering problem and a number of excellent papers have been published on this subject [85,86,129-136]. Closely related research has also been carried out on the shear strength characteristics of rockfill [137-139] and many similarities can be found between the results of this work and that on closely jointed rock. It would not be practical to attempt a detailed review of all of this work in this book and the discussion will be limited to the relationships proposed by Ladanyi and Archambault [85,86] which the authors consider to be a good summary of current thinking on this topic.

Ladanyi and Archambault carried out a large number of model studies using small blocks of commercially compressed concrete. Each model contained 1800 blocks measuring $\frac{1}{2}$"x$\frac{1}{2}$"x2.5" packed tightly together to form a 2.5" thick model slab. Biaxial loads were applied in the plane of the model slab, the loading direction being varied in relationship to the "joint" orientation.

Three distinct types of failure occurred and these are illustrated in the photographs reproduced in the margin.
  Case 1 - Shear along a well defined plane inclined to both discontinuity sets.
  Case 2 - Formation of a narrow failure zone in which block rotation has occurred in addition to the sliding and material failure of Case 1.
  Case 3 - Formation of a kink band of rotated and separated columns of 3, 4 or 5 blocks.

On the basis of these model studies, Ladanyi and Archambault proposed modified forms of equations 22, 23 and 24 (pages 87 and 88) which could be used in equation 21 to predict the shear strength of closely jointed rock masses. After very careful consideration of these modifications and after discussion with Ladanyi, the authors have introduced a further slight modification which removes an anomaly which can occur when using Ladanyi and Archambault's equations. The final equations resulting from these discussions and modifications are as follows :

$$\tau = \frac{\sigma(1 - a_s)(\dot{v} + \mathrm{Tan}\,\phi) + a_s n \sigma_b \frac{\sqrt{1+n}-1}{n}(1 + n\frac{\sigma}{n\sigma_b})^{\frac{1}{2}}}{1 - (1 - a_s)\dot{v}\,\mathrm{Tan}\,\phi} \qquad (28)$$

where

$$\dot{v} = (1 - \frac{\sigma}{\sigma_b})^K \tan i \qquad (29)$$

$$a_s = 1 - (1 - \frac{\sigma}{\sigma_b})^L \qquad (30)$$

and σ_b is the uniaxial compressive strength of the individual blocks within the rock mass,

η is the degree of interlocking which defines the freedom of the blocks to translate and to rotate before being sheared or fractured *.

The suggested values for K, L and η for the three types of failure are as follows :

Case 1 - shear plane formation, K = 4 and L = 1.5 as for single rough discontinuity surfaces (page 88) with η = 0.7 to allow for loosening of the rock mass as a result of close jointing.

Case 2 - shear zone formation, K = 5 to allow for increased freedom of block to rotate, L = Tan i and η = 0.6 which allows for a looser rock mass than in case 1.

Case 3 - kink band formation, K = 5 as for case 2 and L = $(2/n_r)^3$ Tan i where n_r is the number of rows of blocks in the kink band which is normally 3 to 5. In this case η = 0.5 to allow for the very loose condition of the rock mass.

Figure 39 gives a set of shear strength curves, calculated by means of equations 28, 29 and 30 using ϕ = 30°, i = 20° and, for case 3, n_r = 4 in addition to the values suggested above. For comparison, the shear strengths for intact rock and for failure on a single rough discontinuity and the residual strength of a smooth plane are shown as dashed curves. These curves are reproduced from Figure 27, assuming $\sigma_J = \sigma_b$.

The curve for case 1, the formation of a single shear plane, coincides with that for a rough joint at very low normal stresses when the behaviour is strongly dilatant. As the normal stress increases, the curve for the rock mass falls below that for the intact rock as a result of the weakening effect of the close jointing. Rosengren and Jaeger[133] noted this type of behaviour on tests on marble which had been heated to break the grain boundary material, resulting in a very tightly interlocking model rock mass. In spite of the fact that the grains were still in their original positions, the fact that the tensile strength of the grain boundaries had been reduced to zero gave rise to a strength reduction of about 20% at high normal stresses.

---

* In Ladanyi and Archambault's original equations, the term $\eta\sigma_b$ appears in equations 28, 29 and 30. The authors have omitted η from equations 29 and 30 because they consider that it applies to the interlocking of the rock mass and not to the dilation rate $\dot{v}$ and the sheared area a_s which are functions of the shape and orientations of the individual blocks within the rock mass.

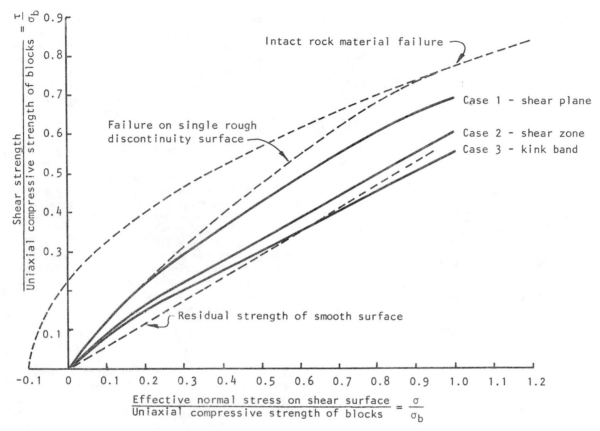

Figure 39 : Shear strength curves for three types of failure in closely jointed interlocking rock masses. Plotted for $i = 20^{\circ}$ and $\phi = 30^{\circ}$.

It is interesting to note that the shear strength curve for case 3, the formation of a kink band, falls below the residual strength of a smooth surface at high normal stresses. This phenomenon has been observed by Ashby[140] and also by Nascimento and Teixeira[141] and it has been discussed by Goodman[24] who, on page 199 of his book, states " .. our appreciation and understanding of these phenomena (block rotation and kink band formation) is just beginning". Clearly, if the formation of a kink band can give rise to shear strengths which are lower than the residual strength of smooth surfaces, the implications can be important in terms of slope design in closely jointed rock masses. Since this strength reduction appears to be relatively small, the authors do not consider these findings alarming but they do feel that the subject merits further study.

In evaluating equations 28, 29 and 30, the choice of the uniaxial compressive strength of the blocks σ_b is very important. Analysis of rock test data suggests that σ_b may be significantly lower than the uniaxial compressive strength of the rock material making up the blocks and it is presumed that this strength reduction is due to weathering of the block surfaces. It may be that the behaviour of the blocks is dominated by the strength of their surfaces, in which case $\sigma_b = \sigma_J$, the joint compressive strength defined on page 87. Until this matter has been clarified by further research, it is recommended that the reader should assume that the value of σ_b is approximately one quarter of the uniaxial compressive strength of unweathered intact rock material. An over-estimate of the value of σ_b will result in an optimistic shear strength curve at low normal stresses and this, in turn, will give rise to optimistic slope designs.

Testing closely jointed rock masses

Determination of the shear strength of a closely jointed rock mass presents formidable experimental problems and relatively few attempts have been made to carry out direct shear or any other type of in situ test. These experimental difficulties explain why most of the studies on closely jointed rock masses have been carried out on models made up of blocks of artificial "rock".

Jaeger[142] has described one of the most elaborate tests ever attempted on closely jointed materials and his paper is recommended reading for anyone who is faced with this type of problem. The rock mass tested by Jaeger was an andesite with several sets of joints having an average spacing of approximately one inch. The joints were free of infilling but the surfaces had been weathered as a result of high water flows.

Six inch (153 mm) cores of undisturbed material were recovered by very careful triple-tube diamond drilling and these cores were transported in the inner core tubes to Jaeger's laboratory in Canberra. In the laboratory, the core tubes were carefully cut along two diametrically opposite axial lines so that one half of the tube could be lifted off, leaving the undisturbed core resting in the remaining half of the tube. The core was then wrapped in thin copper sheeting and carefully rolled over to transfer the core from the core tube into the copper sheet which was then soldered to form a sealed sleeve. The specimen ends were prepared by careful diamond sawing through the copper sheeted core.

Cores prepared in the manner described were then tested triaxially as illustrated in the margin sketch. The results obtained by Jaeger are reproduced in Figure 40 where curve A represents the maximum and curve B the minimum shear strengths given by the range of specimens tested. Also shown in this figure is the residual strength of smooth andesite joints and it will be noted that the interlocking of the angular rock particles gives rise to a very important increase in shear strength.

Many civil engineering laboratories have large triaxial cells designed for testing rockfill and some of these cells are suitable for tests similar to those by Jaeger. One such cell was used by the Snowy Mountains Authority laboratories and is illustrated in the photograph reproduced in the margin. Marsal[137] has described the design of a very large cell to accommodate 113cm diameter by 250cm high specimens. These large cells are very expensive to manufacture and to operate and most laboratories use smaller cells for 6 inch (153 mm) cylindrical samples. Testing procedures should follow the guidelines set out by Bishop and Henkel[143] for the triaxial testing of soils.

Water pressure

Confining pressure

Axial load

Triaxial testing of granular materials.

Large triaxial cell for rock-fill testing in the laboratory of the Snowy Mountains Authority in Australia.

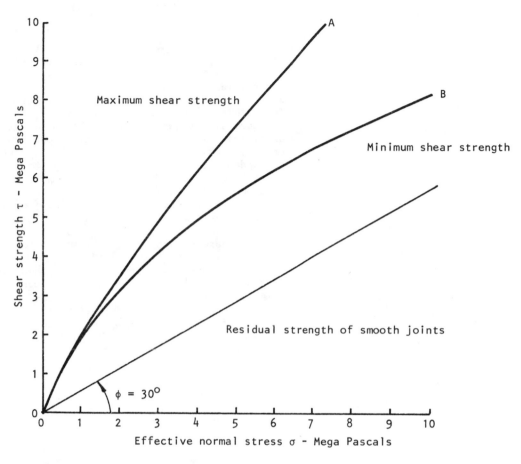

Figure 40 : Shear strength of closely jointed Andesite determined by Jaeger[142] by means of triaxial tests on core samples.

The recovery of undisturbed samples of closely jointed rock is a very difficult process and, in many cases, may not be possible because of the lack of adequate facilities. Under such circumstances, the authors feel that a reasonable alternative is to approach the problem as one would that of testing compacted rockfill. The procedure for testing rockfill has been fully described by Marsal[137] and by Marachi, Chan and Seed[138] and is briefly outlined below.

A representative sample of material is collected from an exposure of the rock mass. No attempt is made to prevent the particles from falling apart but care should be taken to ensure that all the individual pieces within the chosen sample volume are collected. A grading curve to the sample is obtained by running the material through a set of sieves and weighing the amount retained on each sieve. The maximum particle size which should be included in the test specimen is approximately one sixteenth of the diameter of the test specimen and, if the grading curve of the field sample gives larger sizes than this, for the triaxial cell available, the grading curve should be scaled down. This is done by removing all the particles larger than the permitted size and by adjusting the amount retained on each smaller sieve size until the grading curve is parallel to that of the original

109

sample. Very fine material (less than 200 mesh US standard sieve size) is removed from the sample since this could enter voids between larger particles and give rise to strength characteristics similar to those of filled joints. The sample is then compacted by placing it in layers into a confined sleeve and subjecting it to vibration and/or direct axial load. If a significant axial load has to be used in order to achieve a density equivalent to the in situ density, some particle breakage will probably occur and the grading curves should be checked for the compacted material. The specimens are now ready for triaxial testing which would normally be carried out in the drained condition, in other words, any water pressure which could build up in the specimen is allowed to dissipate by loading at a suffiently slow rate and by providing drainage in the loading platens.

Figure 41 gives a set of results obtained from such a series of tests which were carried out on 6 inch (153mm) diameter samples of graded closely-jointed andesite, similar to that tested by Jaeger. It is interesting to examine these results in some detail since they provide a good example which can be used in a discussion on the interpretation of triaxial test data.

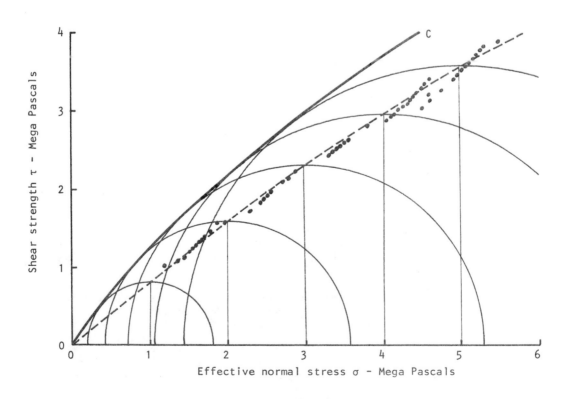

Figure 41 : Results of triaxial tests on recompacted samples of closely jointed andesite.

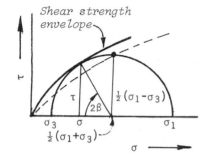

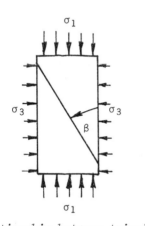

Relationship between triaxial test results and shear strength curve

Interpretation of triaxial test data

A convenient method for determining a shear strength curve from a set of triaxial test data is illustrated in Figure 41 and is as follows:

From the measured values of the axial stress at failure σ_1 and the lateral confining pressure σ_3, calculate the values of $\frac{1}{2}(\sigma_1 + \sigma_3)$ and $\frac{1}{2}(\sigma_1 - \sigma_3)$ which correspond to the normal and shear stresses on a 45° plane through the specimen. These values are plotted as points on the shear strength versus effective normal stress graph and a best fit curve is drawn through these points. This curve represents the locus of the tops of Mohr circles as illustrated in the margin sketch. A set of idealised Mohr circles is now drawn at regular normal stress increments and the envelope to these circles is fitted by eye. This envelope defines the shear strength of the material as tested. Note that the envelope touches the Mohr circle at a point defined by the angle β, the theoretical failure plane inclination. Unfortunately this angle seldom coincides with the actual failure plane inclination in the specimen because of end effects due to the contact between the specimen ends and the platens.

If it is intended to fit a theoretical shear strength relationship, such as that proposed by Ladanyi and Archambault, to the experimentally derived shear strength curve, the following procedure can be followed :

From an examination of the angularity of the individual blocks making up the specimen, decide upon a value for the roughness angle i. This value may range from 40° for very angular blocks to 10° for relatively rounded pieces. The basic friction angle ϕ is determined from tests on smooth planes of the rock material or is estimated from the table of values given in the margin on page 100. In general, the value of ϕ will be close to 30°. From the appearance of the failed specimen, a decision is made on which of the three cases (defined on page 105) is most appropriate and the values of the parameters K, L and η are chosen accordingly. Now calculate the values of τ/σ_b for a range of σ/σ_b values and plot the theoretical shear strength curve as was done in Figure 39 on page 106. A value for σ_b, the uniaxial compressive strength of the block material, is now chosen to give coincidence between the experimental and theoretical curves. Check that the value of σ_b chosen bears some relationship to the uniaxial compressive strength of the unweathered rock material. Generally σ_b should be about $\frac{1}{4}$ to $\frac{1}{2}$ the value of the uniaxial compressive strength of fresh material.

This process was followed in the analysis of the shear strength results given in Figures 40 and 41 and the following values gave close agreement between the shear strength predicted by equation 28 and that determined experimentally:

	Figure 40 curve A	Figure 40 curve B	Figure 41 curve C
i°	35	35	25
ϕ°	30	30	30
K	5	5	5
L	0.7 (Case 2)	0.7 (Case 2)	0.0584*
η	0.6	0.6	0.5
σ_b MPa	52	30	30

* Case 3 assuming $n_r = 4$.

The uniaxial compressive strength of fresh andesite is estimated to be in excess of 100 MPa (15000 lb/in^2).

The purpose of this curve fitting exercise was to attempt to arrive at a rational explanation for the behaviour of the different types of samples tested in order that the probable shear strength of the in situ rock mass could be judged with a reasonable degree of confidence. The sceptical reader may argue that, with all the variables available in Ladanyi and Archambault's equation, a theoretical shear strength curve could be fitted to *any* set of experimental data. If the equation is used simply as a curve generator and the variables are all adjusted until the desired result is obtained, this argument is certainly true. However, if an honest attempt is made to decide upon the values of the variables as outlined on the previous page, the process can be of considerable assistance to someone faced with a critical slope design problem.

Implications of a curvilinear shear strength curve in rock slope design

The importance of a non-linear shear strength relationship in slope design is most conveniently demonstrated by means of a practical example. Figure 42a illustrates a curvilinear failure curve for a closely jointed rock material. It is required to design a series of slopes of different heights to be excavated in this material. A factor of safety of 1.5 is required and the effect of groundwater in the slope is to be allowed for in the design. It is assumed that the material will fail along a circular failure path.

Several methods of circular failure analysis are available which allow the effective normal stress at different points along the failure surface to be calculated and one of these methods could be used for the slope design. Having determined the effective normal stress at a point, the corresponding shear strength can be obtained directly from the shear strength curve given in Figure 42a . The values of shear strength along the failure surface are then used to calculate the resistance of this surface to sliding and, hence, the factor of safety of the slope.

An alternative method involves the use of circular failure charts (discussed in Chapter 9) and this method has been used in deriving the slope design given in Figure 42b. Tangents, numbered 1, 2 and 3 in Figure 42a, are drawn to the curvilinear shear strength curve and the *apparent* cohesion and *apparent* friction angle are determined for each tangent. These. values are used to calculate the slope height versus slope angle relationship for the assumed groundwater conditions and required factor of safety. These relationships, numbered to correspond to the tangents from which they were derived, are plotted in Figure 42b. The *envelope* to these individual curves defines the overall slope design.

This example shows that it is dangerous to draw a straight line through a limited set of shear strength data and to extrapolate this line beyond the range of normal stresses used in the tests. Suppose that tangent number 2 had been drawn through a set of results obtained at normal stresses of between 1000 and 2000 lb/in^2 and that the apparent cohesion and friction angle defined by this tangent had been used for the entire slope design. Figure 42b shows that both

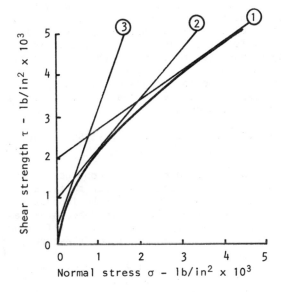

Figure 42a : Non-linear relationship between shear strength and normal stress for a closely jointed rock mass with tangents at various normal stress levels.

Figure 42b : Relationship between slope height and slope face angle for a factor of safety of 1.5. The envelope to the individual curves, corresponding to the tangents in Figure 42a, gives the overall design curve.

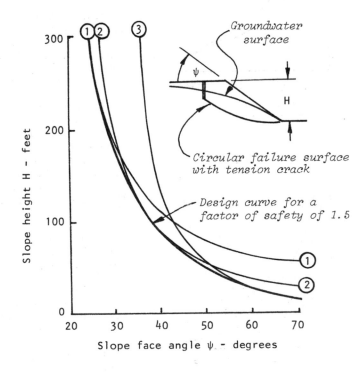

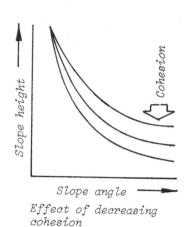

Effect of decreasing
cohesion

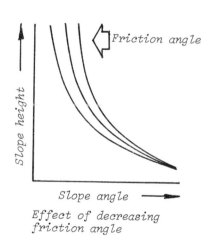

Effect of decreasing
friction angle

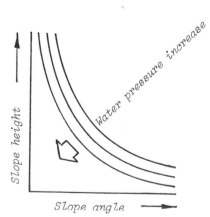

Effect of increasing
water pressure

the height of steep slopes and the angle of high slopes
would have been over-estimated if tangent number 2 had been
the only basis for the slope design.

The sketches in the margin show the influence of variations
in cohesive strength, angle of friction and in groundwater
conditions on the relationship between slope height and
slope angle for typical rock slopes. A reduction in the
cohesive strength causes a reduction in the safe height of
steep slopes. A decrease in the angle of friction gives
rise to a reduction in the safe angle of high slopes. An
increase in groundwater pressure causes a general decrease
in the stability of the slopes.

Shear strength determination by back analysis of slope
failures.

By this time the reader should have been convinced that the
determination of the shear strength of rock surfaces or of
closely jointed rock masses by in situ or by laboratory
testing is not a simple matter. Even when successful tests
have been carried out, the slope designer is still faced
with the problem of relating the results of these tests to
the full scale design. In view of these difficulties , it
is tempting to consider the possibility of back-analysing
existing slope failures in order to determine the shear
strengths which must have been mobilised in the full scale
rock mass at the time of failure. This technique has been
used successfully in soil mechanics for many decades and,
with care, it can be used to obtain useful data for rock
slope design.

Because of the influence of surface roughness on individual
discontinuities and the interlocking of blocks in closely
jointed rock masses, most shear strength curves for rock
tend to be curvilinear. This means that the values of the
cohesive strength and angle of friction determined from the
back analysis of a failure in a particular slope are valid
only for the normal stress level in that slope at the time
of failure. This is equivalent to determining one of the
tangents in Figure 42a and the dangers of using such a
tangent as a basis for general slope design have been
emphasised in the preceding discussion. It is very rare
to find a set of slope failures in slopes of different
heights in the same material and hence there is little
possibility of determining a number of tangents and, from
them, the complete shear strength curve.

In spite of these limitations, back analysis of slope
failures can be a very important source of shear strength
data. At the very least, it can provide the basis for the
design of slopes *having similar dimensions* to those in which
the failure has occurred. If these field data are supplemented by laboratory tests and theoretical analysis, using the
concepts set out earlier in this Chapter, it may be possible
to extend the normal stress range over which the shear
strength results can be used

Table IV on page 114 lists a number of slope failures which
have been back-analysed and the values of cohesive strengths
and friction angles obtained in these analyses are plotted
in Figure 43. These results can be used as a very rough
guide to the shear strength values which may be found in
different geological materials.

TABLE IV - SOURCES OF SHEAR STRENGTH DATA PLOTTED IN FIGURE 43					
Point number	Material	Location	Slope height feet	Analysed by	Ref.
1	Disturbed slates and quartzites	Knob Lake, Canada		Coates, Gyenge and Stubbins	144
2	Soil			Whitman and Bailey	145
3	Jointed porphyry	Rio Tinto, Spain	150-360	Hoek	146
4	Ore body hanging wall in granitic rocks	Grangesberg, Sweden	200-800	Hoek	147
5	Maximum height and angle of excavated slopes - see Figure 7 on page 20				
6	Bedding planes in limestone	Somerset, England	200	Roberts and Hoek	148
7	London clay	England		Skempton and Hutchinson	149
8	Gravelly alluvium	Pima, Arizona		Hamel	150
9	Faulted rhyolite	Ruth, Nevada		Hamel	151
10	Sedimentary series	Pittsburgh, Pennsylvania		Hamel	152
11	Koalinised granite	Cornwall, England	250	Ley	153
12	Clay shale	Fort Peck Dam, Montana		Middlebrooks	154
13	Clay shale	Gardiner Dam, Canada		Fleming et al	155
14	Chalk	Chalk cliffs, England	50	Hutchinson	156
15	Bentonite/clay	Oahe Dam, South Dakota		Fleming et al	155
16	Clay	Garrison Dam, North Dakota		Fleming et al	155
17	Weathered granites	Hong Kong	40-100	Hoek and Richards	157
18	Weathered volcanics	Hong Kong	100-300	Hoek and Richards	157

Sample collection and preparation

Before leaving the subject of shear strength, the authors feel that some comments must be made on the question of sample collection and preparation for laboratory testing. From the discussion in this Chapter it will be clear that samples for laboratory shear testing should be disturbed as little as possible during collection. The importance of the strength of the joint surface material has been emphasised and it is important that this material be retained on the joints to be tested. When the joints are infilled with clay or gouge materials, it is essential that this material should not be disturbed and that the two halves of the specimen should not be displaced relative to one another. When the materials to be tested are sensitive to changes in moisture content, the specimens must be sealed after collection in order to retain the in situ moisture content.

Because of difficulties of access, most samples for shear testing are collected by geologists whose only equipment consists of a geological pick and a rucksack. While this may do a great deal to enhance the rugged pioneering spirit

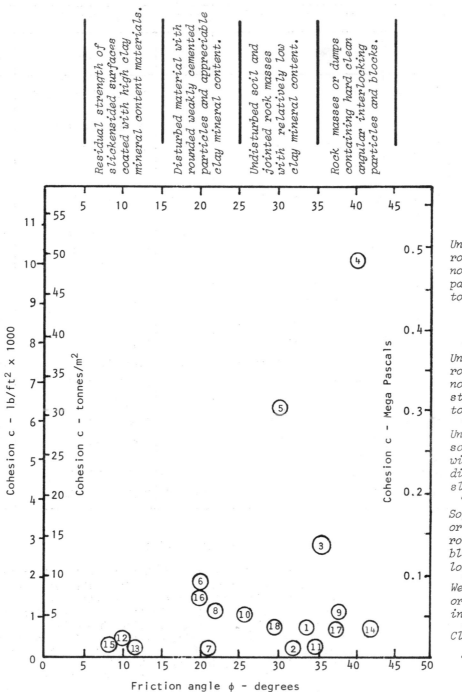

Figure 43 : Relationship between the friction angles and
cohesive strengths mobilised at failure for
the slope failure analyses listed in Table IV.

of geologists, it does little for the quality of the results
obtained from tests on such samples. The fault seldom lies
with the geologists in such cases since they are simply
doing the best possible job with the resources available.
The fault generally lies with those responsible for planning
the investigation and with the low priority allocated to
shear strength testing.

The authors feel very strongly that poor quality shear tests
are worse than no shear tests at all since the results of
such tests can be extremely misleading and can give rise
to serious errors in slope design. When the funding or
the time available on a project is limited or when correct
sample collection is impossible for some practical reason,
it is recommended that one of the following alternative
courses should be followed.

a. Having established the geological conditions on site
   as accurately as possible, the slope designer should
   return to his office and devote the time which he
   would have spent in sample collection and testing to
   a thorough study of published papers and reports on
   slope stability in similar geological environments.
   It is frequently possible to find cases which have
   been studied which are very similar to that under
   consideration and the results presented in such studies
   may well provide an adequate basis for slope design.
   When using the results of a published study, extreme
   care should be taken in checking that the shear strength
   values were not derived from poor quality tests. It is
   also important that a sensitivity study should be
   carried out to determine the influence of shear strength
   variations on the proposed slope design. If this
   influence is significant, it may be necessary to insist
   that a few high quality shear tests be carried out in
   order to narrow the range of possible variation.

b. When hand samples are available, shear test samples
   can be prepared by diamond saw cutting and then lapping
   the surfaces with a coarse grit . This removes surface
   roughness effects and reduces the test to one in which
   the basic angle of friction ϕ is determined. This value
   is then used, together with surface roughness and
   joint compressive strength measurements, observations
   or estimates, to calculate the shear strength of the
   rock joint by means of Barton's equation (26 on page
   89) or Ladanyi and Archambault's equation (25 on page
   88).

When both time and money are available for high quality
shear strength testing, the slope designer has to decide
upon the type of test which will give the results required.
In the case of a critical slope, eg. the abutment of an
arch dam, it may be justifiable to carry out in situ tests
similar to that illustrated in Figure 30 on page 92. In
many cases it is more economical and more convenient to
collect samples in the field for subsequent testing in the
laboratory.

The obvious source of samples for shear strength testing
is the collection of diamond drill cores which usually are
available on project sites. Provided that these cores have
been obtained by careful drilling, using double or triple-
tube equipment, and provided that the core size is in excess
of about 2 inch (50 mm), reasonable samples can be obtained

from these cores. The main drawback to such samples is
their size and this means that the influence of surface
roughness must be considered separately. Ideally, small
samples should only be used to test infilled or planar
joints which do not suffer from roughness effects. This
may not always be possible and, when a rough joint is tested,
the effect of surface roughness must be assessed and allowed
for in the interpretation of the test results.

In order to overcome the size limitation of normal diamond
drill core, some slope stability engineers have used large
diamond core bits (8 to 10 inches, 200 to 250 mm) mounted
on concrete coring rigs which are bolted to the rock face.
Obviously, the length of core which can be obtained using
such a system is limited to the length of the core barrel
which can be supported by the rig and, in general, this is
of the order of 1½ to 3 feet (0.5 to 1m). In spite of this
limitation, some excellent samples have been obtained with
equipment of this type.

Londe[81] has described the use of a wire saw for specimen
recovery and this technique is illustrated in Figures 44
and 45. Four holes, marked A,B,C and D in Figure 44, are
drilled along the top and bottom edges of the specimen.
These holes are large enough to accommodate small wheels
which change the direction of the wire. Wire cuts are made
between holes A-B, B-C and C-D and then the rear face of
the specimen is cut free by passing the wire through the
side and top cuts and cutting between holes A and D as
illustrated in Figure 44. Finally, when the rear face has
been cut free, the rods carrying the deflection wheels
are withdrawn so that the base of the specimen, between
holes A and D, is cut free. While this process may appear
complex, it has a great deal of merit since wire-sawing
is one of the gentlest rock cutting processes and, with
care, the specimen can be recovered with very little disturbance.

When sampling rock cores which deteriorate very rapidly on
exposure, a technique described by Stimpson, Metcalfe and
Walton[158] can be used to protect the core during shipment
and to prevent loss of in situ moisture. This technique
is illustrated in Figure 46 and the steps are described
below :

Step 1 - The mould, consisting of a plastic pipe split
         along its axis, is prepared to receive the core.
         The core is supported on thin metal rods held in
         drilled spacers.

Step 2 - The core, wrapped in aluminium foil, is placed on
         the supporting rods and the end spacers are
         adjusted to the length of the core.

Step 3 - The upper half of the mould is fixed in place
         with adjustable steel bands and polyurethane foam
         is poured through the access slot. Suitable material
         is Bibbithane RM 114 or RM 118 manufactured by
         Bibby Chemicals, Paragon Works, Baxenden , Accring-
         ton, Lancashire BB5 2SL, England.

Step 4 - The foam sets in approximately 30 minutes after
         which the mould is stripped and the rods withdrawn.

Step 5 - The core is removed just before testing by cutting
         the foam with a coarse-toothed wood saw.

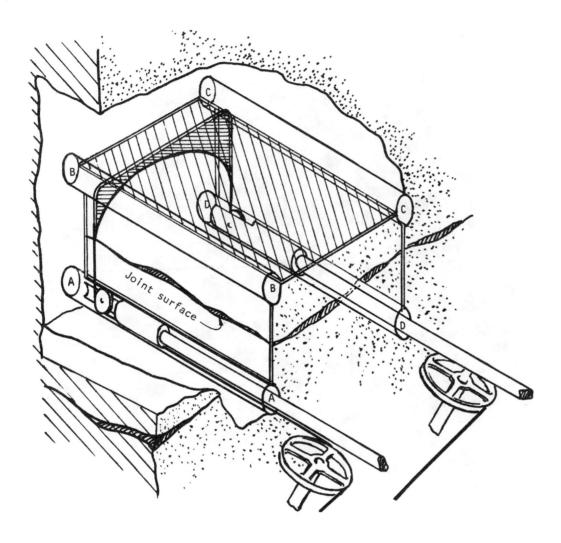

Figure 44 : Wire-sawing technique for sampling rock specimens for
shear testing, described by Londe[81]. See text for details.

Figure 45 : Wire-sawing a shear specimen from the sidewall of
a tunnel as described by Londe[81]. Photograph reproduced
with permission of Pierre Londe, Coyne and Bellier, France.

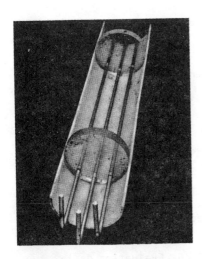

Step 1

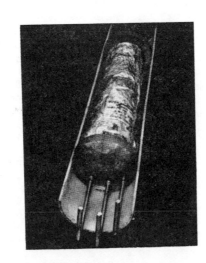

Step 2

Step 3

Step 4

Step 5

Figure 46 : Preservation of fragile
rock core in foamed plastic as
described by Stimpson, Metcalfe
and Walton[158].
Steps are described in the text on
page 117.

*Note that rubber gloves should be
worn when working with component
liquids and with unset plastic foam.*

Chapter 5 references

82. BARTON, N.R. Review of a new shear strength criterion for rock joints. *Engineering Geology*, Elsevier. Vol.7, 1973, pages 287-332.

83. PAULDING, B.W.Jr. Coefficient of friction of natural rock surfaces. *Proc. ASCE J. Soil Mech. Foundation Div.* Vol. 96 (SM2), 1970, pages 385-394.

84. RENGERS, N. Roughness and friction properties of separation planes in rock. *Thesis*. Tech. Hochschule Fredericiana, Karlsruhe, Inst. Bodenmech. Felsmech. Veröff, 47, 1971, 129 pages.

85. LANDANYI, B. and ARCHAMBAULT, G. Simulation of shear behaviour of a jointed rock mass. *Proc. 11th Symposium on Rock Mechanics*, published by AIME, New York, 1970, pages 105-125.

86. LADANYI, B. and ARCHAMBAULT, G. Evaluation de la résistance au cisaillement d'un massif rocheux fragmenté. *Proc. 24th International Geological Congress*, Montreal, 1972, Section 13D, pages 249-260.

87. FAIRHURST, C. On the validity of the Brazilian test for brittle materials. *Intnl. J. Rock Mechanics and Mining Sciences*. Vol. 1, 1964, pages 535-546.

88. HOEK, E. Brittle failure of rock. *Rock Mechanics in Engineering Practice*. Edited by K.G.Stagg and O.C.Zienkiewicz, Published by J.Wiley, London, 1968, pages 99-124.

89. MARTIN, G.R. and MILLER, P.J. Joint strength characteristics of a weathered rock. *Proc. Third Congress International Society for Rock Mechanics*, Denver, Vol.2A, 1974, pages 263-270.

90. BARTON, N.R. A relationship between joint roughness and joint shear strength. *Proc. International Symposium on Rock Fracture*. Nancy, France, 1971, Paper 1-8.

91. BARTON, N.R. A model study of the behaviour of excavated slopes. *Ph.D. Thesis*, University of London, Imperial College of Science and Technology. 1971, 520 pages.

92. SERAFIM, J.L. and LOPES, J.B. In-situ shear tests and triaxial tests on foundation rocks of concrete dams. *Proc. 5th International Conference on Soil Mechanics and Foundation Engineering*. Paris, Vol. VI, 1961, page 533.

93. HAVERLAND, M.L. and SLEBIR, E.J. Methods of performing and interpreting in-situ shear tests. *Proc. 13th Symposium on Rock Mechanics*. ASCE, 1972, pages 107-137.

94. RUIZ, M. and CAMARGO, F. A large scale field test on rock. *Proc. First Congress of the International Society for Rock Mechanics*. Lisbon, Vol. VI., 1966, page 257.

95. BRAWNER, C.O., PENTZ, D.L. and SHARP, J.C. Stability studies of a footwall slope in layered coal deposit. *Proc. 13th Symposium on Rock Mechanics*. ASCE, 1972, pages 329-365.

96. ROSS-BROWN, D.M. and WALTON, G. A portable shear box for testing rock joints. *Rock Mechanics*, Vol.7, No.3, 1975, pages 129-153.

97. BROCH, E. and FRANKLIN, J.A. The point-load strength test. *Intnl. J. Rock Mechanics and Mining Sciences*. Vol.9, 1972, pages 669-697.

98. BIENIAWSKI, Z.T. Estimating the strength of rock materials. *J. South African Institute of Mining and Metallurgy*, Vol. 74, 1974, pages 312-320.

99. BARTON, N.R and CHOUBEY, V. The shear strength of rock joints in theory and practice. *Rock Mechanics*, 1977, In press.

100. DEERE, D.U., and MILLER, R.P. Engineering classification and index properties for intact rock. *Technical Report No. AFNL-TR-65-116 Air Force Weapons Laboratory*. New Mexico, 1966.

101. PITEAU, D.R. Geological factors significant to the stability of slopes cut in rock. *Proc. Symposium on Planning Open Pit Mines*. Johannesburg. Published by A.A. Balkema, Amsterdam, 1971, pages 33-53.

102. ROBERTSON, A.M. The interpretation of geological factors for use in slope theory. *Proc. Symposium on Planning Open Pit Mines*. Johannesburg. Published by A.A. Balkema, Amsterdam, 1971, pages 55-70.

103. SERAFIM, J.L. Rock Mechanics considerations in the design of concrete dams. *Proc. Intnl. Conference on State of Stress in the Earth's Crust*. Santa Monica, California. Published by Elsevier, New York, 1964, pages 611-645.

104. HAMROL, A. A quantitative classification of the weathering and weatherability of rocks. *Proc. 5th Intnl. Conference on Soil Mechanics and Foundation Engineering*, Paris, Vol.2, 1961, pages 771-774.

105. ROCHA, M. Mechanical behaviour of rock foundations in concrete dams. *Trans. 8th International Congress on Large Dams*. Edinburgh. Vol.1, 1964, pages 785-831.

106. FOOKES, P.G., DEARMAN, W.R. and FRANKLIN, J.A. Some engineering aspects of rock weathering with field examples from Dartmoor and elsewhere. *Quarterly J. Engineering Geology*. Vol.4, 1971, pages 139-185.

107. FRANKLIN, J.A and CHANDRA, A. The Slake-durability test. *Intnl. J. Rock Mechanics and Mining Sciences*. Vol 9, 1972, pages 325-341.

108. BARTON, N.R. A review of the shear strength of filled discontinuities in rock. *Norwegian Geotechnical Institute Publication* No. 105, 1974, 38 pages.

109. RUIZ, M.D., CAMARGO, F.P., MIDEA,N.F. and NIEBLE, C.M. Some considerations regarding the shear strength of rock masses. *Proc. Intnl. Rock Mechanics Symposium*, Madrid, 1968, pages 159-169.

110. LINK, H. The sliding stability of dams. *Water Power*. 1969. Vol.21, No.3, pages 99-103, No.4, pages 135-139, No.5, pages 172-179.

111. SINCLAIR, S.R. and BROOKER,E.W. The shear strength of Edmonton shale. *Proc. Geotechnical Conference on shear properties of natural soils and rocks*. Oslo, Vol.1, 1967, pages 295-299.

112. SKEMPTON, A.W. and PETLEY, D.J. The strength along discontinuities in stiff clays. *Proc. Geotechnical Conference on shear strength properties of natural soils and rocks*. Oslo, Vol.2, 1968, pages 29-46.

113. LEUSSINK, H. and MULLER-KIRCHENBAUER, H. Determination of the shear strength behaviour of sliding planes caused by geological features. *Proc. Geotechnical Conference on shear strength properties of natural soils and rocks*. Oslo, Vol.1, 1967, pages 131-137.

114. STIMPSON, B. and WALTON, G. Clay mylonites in English coal measures. *Proc. 1st Congress Intnl. Assoc. Engineering Geology*. Paris, Vol.2, 1970, pages 1388-1393.

115. PIGOT,C.H. and MACKENZIE, I.D. A method used for an in situ bedrock shear test. *Trans. Intnl. Congress on Large Dams*. Edinburgh, Vol.1, 1964, pages 495-512.

116. BRAWNER, C.O. Case studies of stability on mining projects. *Proc. 1st Intnl. Conference on Stability in Open Pit Mining*. Vancouver, 1970. Published by AIME, New York, 1971, pages 205-226.

117. NOSE, M. Rock test in situ, conventional tests on rock properties and design of Kurobegwa No.4 Dam thereon. *Trans. Intnl. Congress on Large Dams*. Edinburgh, Vol.1, 1964, pages 219-252.

118. EVDOKIMOV, P.D. and SAPEGIN,D.D. A large-scale field shear test on rock. *Proc. 2nd Congress Intnl. Society for Rock Mechanics*. Belgrade, Vol.2, 1970, Paper 3.17.

119 DROZD, K. Variations in the shear strength of a rock mass depending upon the displacements of the test blocks. *Proc. Geotechnical Conference on shear strength properties of natural soils and rocks*. Oslo, Vol.1. 1967, pages 265-269.

120. KRSMANOVIC, D., TUFO, M. and LANGOF, Z. Shear strength of rock masses and possibilities of its reproduction on models. *Proc. 1st. Congress Intnl. Society for Rock Mechanics*. Lisbon, Vol.1,1966, pages 537-542.

121. KRSMANOVIC, D. and POPOVIC, M. Large scale field tests of the shear strength of limestone. *Proc. 1st.Congress Intnl. Society for Rock Mechanics*. Lisbon, Vol.1, 1966, pages 773-779.

122. SALAS, J.A.J. and URIEL, S. Some recent rock mechanics testing in Spain. *Trans. Intnl. Congress on Large Dams*. Edinburgh, Vol.1, 1964, pages 995-1021.

123.   BERNAIX, J.   New laboratory methods of studying the
       mechanical properties of rocks. *Intnl. J. Rock
       Mechanics and Mining Sciences*. Vol.6, 1969,
       pages 43-90.

124.   SCHULTZE, E. Large scale shear tests. *Proc. 4th Intnl.
       Conference on Soil Mechanics and Foundation Engineering.*
       London, Vol.1, 1957, pages 193-199.

125.   EURENIUS, J. and FAGERSTROM, H.   Sampling and testing
       of soft rocks with weak layers. *Geotechnique*. Vol.19,
       No.1, 1969, pages 133-139.

126.   UNDERWOOD, B.L.   Chalk foundations at four major dams
       in the Missouri River Basin. *Trans. 8th Intnl.
       Congress on Large Dams*. Edinburgh, Vol.1, 1964,
       pages 23-47.

127.   SERAFIM, J.L. and GUERREIRO, M.   Shear strength of
       rock masses at three Spanish dam sites. *Proc. Intnl.
       Rock Mechanics Symposium*. Madrid, 1968, pages 147-157.

128.   COATES, D.L., McRORIE,K.L. and STUBBINS, J.B.   Analysis
       of pit slides in some incompetent rocks. *Trans.
       Society of Mining Engineers. AIME*. 1963, pages 94-101.

129.   MULLER, L. and PACHER, P. Modellversuche zur Klarung
       der Bruchgefahr geklufteter Medien. *Rock Mechanics
       and Engineering Geology*. Supplement 2, 1965,
       pages 7-24.

130.   JOHN, K.W. Civil engineering approach to evaluate
       strength and deformability of regularly jointed rock.
       Proc. 11th Symposium on Rock Mechanics. Berkeley, 1969,
       pages 69-80.

131.   BRAY, J.W.   A study of jointed and fractured rock.
       Rock Mechanics and Engineering Geology. Vol.5, 1967,
       pages 119-136 and 197-216.

132.   BROWN, E.T.   Strength of models of rock with intermit-
       tent joints. *J. Soil Mechanics and Foundation Division.
       ASCE*. Vol.96, No.SM6, 1970, pages 1917-1934.

133.   ROSENGREN, K.J. and JAEGER, J.C.   The mechanical
       properties of an interlocked low-porosity aggregate.
       Geotechnique. Vol.18, No.3, 1968, pages 317-328.

134.   HAYASHI, M. Strength and dilatancy of brittle jointed
       mass - the extreme value stochastics and anisotropic
       failure mechanism. *Proc. 1st Congress Intnl. Society
       for Rock Mechanics*. Lisbon, Vol.1, 1966, pages 295-301.

135.   EINSTEIN, H.H., NELSON, R.A., BRUHN,R.W. and
       HIRSCHFELD, R.C. Model studies of jointed rock mass
       behaviour. *Proc. 11th Symposium on Rock Mechanics.*
       Berkeley,1970, pages 88-103.

136.   EINSTEIN, H.H. and HIRSCHFELD, R.C. Model studies on
       mechanics of jointed rock. *J. Soil Mechanics and
       Foundation Division. ASCE*. Vol.99, No. SM3, 1973,
       pages 229-248.

137. MARSAL, R.J. Mechanical properties of rockfill. In *Embankment Dam Engineering - Casagrande Volume*. Edited by R.C.Hirschfeld and S.J.Poulos, Published by J.Wiley & Sons, New York, 1973, pages 109-200

138. MARACHI, N.D., CHAN, C.K and SEED, H.B. Evaluation of properties of rockfill materials. *J. Soil Mechanics and Foundation Division*, ASCE. Vol. 98, No. SM1, 1972, pages 95-114.

139. WILKINS, J.K. A theory for the shear strength of rockfill. *Rock Mechanics*. Vol. 2, 1970, pages 205-222.

140. ASHBY, J. Sliding and toppling modes of failure in models and jointed rock slopes. *M.Sc. Thesis, London University, Imperial College*. 1971, 150 pages.

141. NASCIMENTO, U. and TEIXEIRA, H. Mechanics of internal friction in soils and rocks. *Proc. Intnl. Symposium on Rock Fracture*. Nancy, France. 1971, Paper 2-3.

142. JAEGER, J.C. The behaviour of closely jointed rock. *Proc. 11th Symposium on Rock Mechanics*. Berkeley, 1970, pages 57-68.

143. BISHOP, A.W and HENKEL, D.J. *The measurement of soil properties in the triaxial test*. Published by Edward Arnold, London. 2nd. Edition, 1962.

144. COATES, D.F., GYENGE, M and STUBBINS, J.B. Slope stability studies at Knob Lake. *Proc. Rock Mechanics Symposium*. Toronto. 1965, pages 35-46.

145. WHITMAN, R.V. and BAILEY, W.A. Use of computers for slope stability analysis. *J. Soil Mechanics and Foundation Division*, ASCE. Vol. 93, 1967, pages 475-498.

146. HOEK, E. Estimating the stability of excavated slopes in opencast mines. *Trans. Institution of Mining and Metallurgy*. London. Vol. 79, 1970, pages A109-A132.

147. HOEK, E. Progressive caving induced by mining an inclined ore body. *Trans. Institution of Mining and Metallurgy*. London. Vol. 83, 1974, pages A133-A139.

148. ROBERTS,D. and HOEK, E. A study of the stability of a disused limestone quarry face in the Mendip Hills, England. *Proc. 1st. Intnl. Conference on Stability in Open Pit Mining*. Vancouver. Published by AIME, New York, 1972, pages 239-256.

149. SKEMPTON, A.W. and HUTCHINSON, J.N. Stability of natural slopes and embankment foundations. State of the art report. *Proc. 7th Intnl. Conference on Soil Mechanics*. Mexico. Vol. 1, 1969, pages 291-340.

150. HAMEL, J.V. The Pima mine slide, Pima County, Arizona. *Geological Society of America Abstracts with Programs*. Vol. 2, No. 5, 1970, page 335.

151. HAMEL, J.V. Kimberley pit slope failure. *Proc. 4th Panamerican Conference on Soil Mechanics and Foundation Engineering*. Puerto Rico. Vol. 2, 1971, pages 117-127.

152. HAMEL, J.V. The slide at Brilliant cut. *Proc. 13th Symposium on Rock Mechanics*. Urbana, Illinois. 1971, pages 487-510.

153. LEY, G.M.M. The properties of hydrothermally altered granite and their application to slope stability in opencast mining. *M.Sc Thesis, London University, Imperial College*. 1972.

154. MIDDLEBROOK, T.A. Fort Peck slide. *Proc. American Society of Civil Engineers*. Vol. 107, paper 2144, 1942, page 723.

155. FLEMING, R.W., SPENCER, G.S and BANKS, D.C. Empirical study of the behaviour of clay shale slopes. *US Army Nuclear Cratering Group Technical Report No. 15*, 1970

156. HUTCHINSON, J.N. Field and laboratory studies of a fall in upper chalk cliffs at Joss Bay, Isle of Thanet. *Proc. Roscoe Memorial Symposium*. Cambridge, 1970.

157. HOEK, E. and RICHARDS, L.R. Rock slope design review. *Golder Associates Report to the Principal Government Highway Engineer, Hong Kong*. 1974, 150 pages.

158. STIMPSON, B., METCALFE, R.G and WALTON, G. A new technique for sealing and packing rock and soil samples. *Quarterly Journal of Engineering Geology*. Vol. 3, No. 2, 1970, pages 127-133.

Chapter 6 : Groundwater flow; permeability and pressure

Introduction

The presence of groundwater in the rock mass surrounding an open pit has a detrimental effect upon the mining programme for the following reasons:

a) *Water pressure* reduces the stability of the slopes by reducing the shear strength of potential failure surfaces as described on pages 24 and 25. Water pressure in tension cracks or similar near vertical fissures reduces stability by increasing the forces tending to induce sliding (page 26).

b) High *moisture content* results in an increased unit weight of the rock and hence gives rise to increased transport costs. Changes in moisture content of some rocks, particularly shales, can cause accelerated weathering with a resulting decrease in stability.

c) *Freezing* of groundwater during winter can cause wedging in water-filled fissures due to temperature dependent volume changes in the ice. Freezing of surface water on slopes can block drainage paths resulting in a build-up of water pressure in the slope with a consequent decrease in stability.

d) *Erosion* of both surface soils and fissure infilling occurs as a result of the velocity of flow of ground-water. This erosion can give rise to a reduction in stability and also to silting up of drainage systems.

e) *Discharge* of groundwater into an open pit gives rise to increased operating costs because of the requirement to pump this water out and also because of the difficulties of operating heavy equipment on very wet ground. Blasting problems and blasting costs are increased by wet blast holes.

f) *Liquefaction* of overburden soils or waste tips can occur when water pressure within the material rises to the point where the uplift forces exceed the weight of the soil. This can occur if drainage channels are blocked or if the soil structure undergoes a sudden volume change as can happen under earthquake conditions.

Liquefaction is critically important in the design of tailings dams and waste dumps and it is dealt with in the references numbered [159-162] listed at the end of this chapter. It will not be considered further in this book since it does not play a significant part in controlling the stability of rock slopes.

By far the most important effect of the presence of ground-water in a rock mass is the reduction in stability resulting from water pressures within the discontinuities in the rock. Methods for including these water pressures into stability calculations are dealt with in later chapters of this book. This chapter is concerned with methods for estimating or measuring these water pressures.

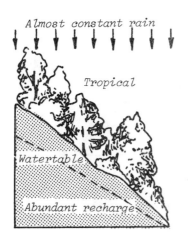

Almost constant rain

Tropical

Watertable

Abundant recharge

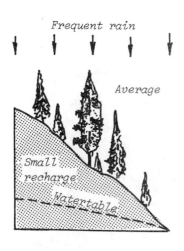

Frequent rain

Average

Small recharge

Watertable

Infrequent rain

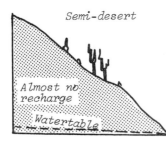

Semi-desert

Almost no recharge

Watertable

After Davis and De Wiest[163].

Groundwater flow in rock masses

There are two possible approaches to obtaining data on water pressure distributions within a rock mass:

a) Deduction of the overall groundwater flow pattern from consideration of the permeability of the rock mass and sources of groundwater.

b) Direct measurement of water levels in boreholes or wells or of water pressure by means of piezometers installed in boreholes.

As will be shown in this chapter, both methods abound with practical difficulties but, because of the very important influence of water pressure on slope stability, it is essential that the best possible estimates of these pressures should be available before a detailed stability analysis is attempted. Because of the large number of factors which control the groundwater flow pattern in a particular rock mass, it is only possible to highlight the general principles which may apply and to leave the reader to decide what combinations of these principles is relevant to his specific problem.

The hydrologic cycle

A simplified hydrologic cycle is illustrated in Figure 47 to show some typical sources of groundwater in a rock mass. This figure is included to emphasise the fact that groundwater can and does travel considerable distances through a rock mass. Hence, just as it is important to consider the regional geology of an area when starting the design on an open pit mine, so it is important to consider the regional groundwater pattern when estimating probable groundwater distributions at a particular site.

Clearly, precipitation in the catchment area of the pit is an important source of groundwater, as suggested in the sketch opposite, but other sources cannot be ignored. Groundwater movement from adjacent river systems, reservoirs or lakes can be significant, particularly if the permeability of the rock mass is highly anisotropic as suggested in Figure 48. In extreme cases, the movement of groundwater may be concentrated in open fissures or channels in the rock mass and there may be no clearly identifiable water table. The photograph reproduced in Figure 49 shows a solution channel of about 1 inch in diameter in limestone. Obviously, the hydraulic conductivity of such a channel would be so high as compared with other parts of the rock mass that the conventional picture of a groundwater flow pattern would probably be incorrect in the case of a slope in which such features occur.

These examples emphasise the extreme importance of considering the geology of the site when estimating water table levels or when interpreting water pressure measurements.

Definition of permeability

Consider a cylindrical sample of soil or rock beneath the watertable in a slope as illustrated in Figure 50. The sample has a cross-sectional area of A and a length l. Water levels in boreholes at either end of this sample are at

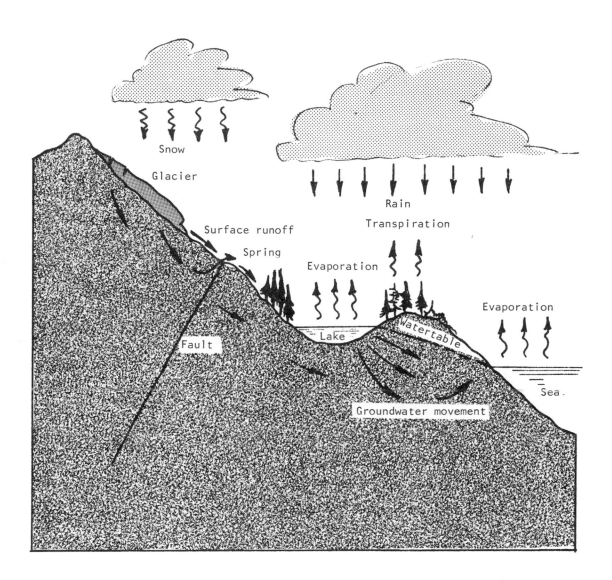

Figure 47 : Simplified representation of a hydrologic cycle showing some
typical sources of groundwater.

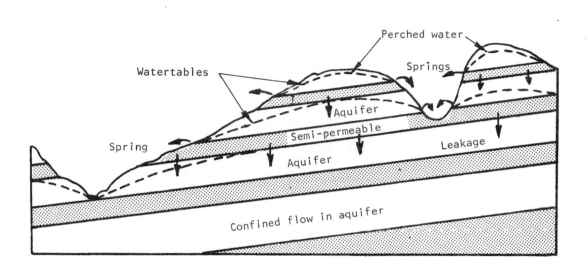

Figure 48 : Confined, unconfined and perched water in a simple stratigraphic sequence of sandstone and shale. After Davis and de Wiest[163].

Figure 49 :
Solution channel in limestone. The hydraulic conductivity of such a channel would be very high as compared with the permeability of the intact rock or of other discontinuities and it would have a major influence on the groundwater flow pattern in a rock mass.

heights h_1 and h_2 above a reference datum and the quantity of water flowing through the sample in a unit of time is Q. According to Darcy's law, the coefficient of permeability of this sample is defined as [165,166,167]:

$$k = \frac{Q \cdot l}{A(h_1 - h_2)} = \frac{V \cdot l}{(h_1 - h_2)} \tag{31}$$

where V is the discharge velocity. Substitution of dimensions for the terms in equation (31) shows that the permeability coefficient k has the same dimensions as the discharge velocity V, i.e. length per unit time. The dimension most commonly used in groundwater studies is centimetres per second and typical ranges of permeability coefficients for rock and soil are given in Table V. Figure 51 shows that the total *head* h can be expressed in terms of the pressure p at the end of the sample and the height z above a reference datum. Hence

$$h = \frac{p}{\gamma_w} + z \tag{32}$$

where γ_w is the density of water. As shown in Figure 50, h is the height to which the water level rises in a borehole standpipe.

Permeability conversion table	
To convert cm/sec to:	Multiply by
Meters/min	0.600
μ/sec	10^4
ft/sec	0.0328
ft/min	1.968
ft/year	1.034×10^6

Permeability of jointed rock

Table V shows that the permeability of intact rock is very low and hence poor drainage and low discharge would normally be expected in such material. On the other hand, if the rock is discontinuous as a result of the presence of joints, fissures or other discontinuities, the permeability can be considerably higher because these discontinuities act as channels for the water flow.

The flow of water through fissures in rock has been studied in great detail by Huitt[168], Snow[169], Louis[170], Sharp[171], Maini[172] and others and the reader who wishes to pursue this complex subject is assured of many happy hours of reading. For the purposes of this discussion, the problem is simplified to that of the determination of the equivalent permeability of a planar array of parallel smooth cracks.[170] The permeability parallel to this array is given by:

$$k = \frac{ge^3}{12v \cdot b} \tag{33}$$

where g = gravitational acceleration (981 cm/sec²)

e = opening of cracks or fissures

b = spacing between cracks and

ν is the coefficient of kinematic viscosity (0.0101 cm²/sec for pure water at 20°C)

The equivalent permeability k of a parallel array of cracks with different openings is plotted in Figure 51 which shows that the permeability of a rock mass is very sensitive to the opening of discontinuities. Since this opening changes with stress, the permeability of a rock mass will therefore be sensitive to stress.

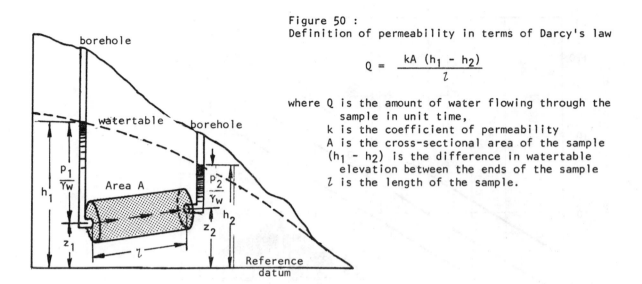

Figure 50 :
Definition of permeability in terms of Darcy's law

$$Q = \frac{kA (h_1 - h_2)}{l}$$

where Q is the amount of water flowing through the
 sample in unit time,
 k is the coefficient of permeability
 A is the cross-sectional area of the sample
 $(h_1 - h_2)$ is the difference in watertable
 elevation between the ends of the sample
 l is the length of the sample.

TABLE V – PERMEABILITY COEFFICIENTS FOR TYPICAL ROCKS AND SOILS				
	k - cm/sec	*Intact rock*	*Fractured rock*	*Soil*
Practically impermeable	10^{-10}	Slate		Homogeneous clay below zone of weathering
	10^{-9}	Dolomite		
	10^{-8}	Granite		
	10^{-7}			
Low discharge poor drainage	10^{-6}	Limestone — Sandstone		Very fine sands, organic and inorganic silts, mixtures of sand and clay, glacial till, stratified clay deposits
	10^{-5}		Clay-filled joints	
	10^{-4}			
	10^{-3}			
High discharge free drainage	10^{-2}		Jointed rock	
	10^{-1}			Clean sand, clean sand and gravel mixtures
	1.0		Open-jointed rock	
	10^{1}			
	10^{2}		Heavily fractured rock	Clean gravel

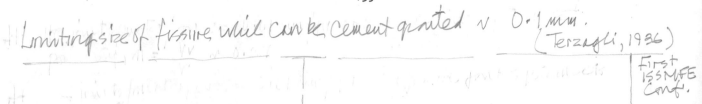

Limiting size of fissure which can be cement grouted ∨ 0·1 mm.
(Terzaghi, 1936)

First ISSMFE Conf.

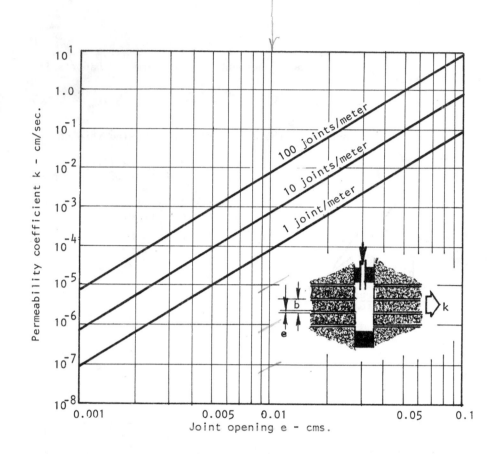

Figure 51 : Influence of joint opening e and joint spacing b on the
permeability coefficient k in the direction of a set of
smooth parallel joints in a rock mass.

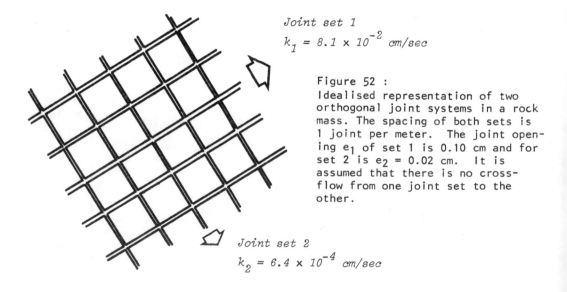

Joint set 1
$k_1 = 8.1 \times 10^{-2}$ cm/sec

Figure 52 :
Idealised representation of two
orthogonal joint systems in a rock
mass. The spacing of both sets is
1 joint per meter. The joint opening e_1 of set 1 is 0.10 cm and for
set 2 is $e_2 = 0.02$ cm. It is
assumed that there is no cross-
flow from one joint set to the
other.

Joint set 2
$k_2 = 6.4 \times 10^{-4}$ cm/sec

Let $e \sim 0.01$ cm
$b \sim 100^m$.

$\frac{e}{b} \sim 0.0001$
$= 10^{-4}$.

Let $k_f \sim 10^{-5}$ cm/sec. then $\frac{e}{b} \sim (10^{-4})(10^{-5}) \sim 10^{-9}$ cm/s.

Louis [170] points out that equation (33) only applies to laminar flow through planar parallel fissures and that it gives rise to significant errors if the flow velocity is high enough for turbulent flow to occur, if the fissure surfaces are rough or if the fissures are infilled. Louis lists no fewer than 8 equations to describe flow under various conditions. Equation (33) gives the highest equivalent permeability coefficient. The lowest equivalent permeability coefficient, for an infilled fissure system, is given by

$$k = \frac{e}{b} \cdot k_f + k_r \qquad (34)$$

where k_f is the permeability coefficient of the infilling material and

k_r is the permeability coefficient of the intact rock.

(Note that k_r has been ignored in equation (33) since it will be very small as compared with permeability of open joints).

An example of the application of equation (33) to a rock mass with two orthogonal joint systems is given in Figure 52. This shows a major joint set in which the joint opening e_1 is 0.10 cm and the spacing between joints is $b_1 = 1$ meter. The equivalent permeability k_1 parallel to these joints is $k_1 = 8 \cdot 1 \times 10^{-2}$ cm/sec. The minor joint set has a spacing $b_2 = 1$ joint per meter and an opening $e_2 = 0.02$ cm. The equivalent permeability of this set is $k_2 = 6 \cdot 5 \times 10^{-4}$ cm/sec, i.e. more than two orders of magnitude smaller than the equivalent permeability of the major joint set.

Clearly the groundwater flow pattern and the drainage characteristics of a rock mass in which these two joint sets occur would be significantly influenced by the orientation of the joint sets.

Flow nets

The graphical representation of groundwater flow in a rock or soil mass is known as a flow net and a typical example is illustrated in Figure 53. Several features of this flow net are worthy of consideration.

Flow lines are paths followed by the water in flowing through the saturated rock or soil.

Equipotential lines are lines joining points at which the total head h is the same. As shown in Figure 53, the water level is the same in boreholes or standpipes which terminate at points A and B on the same equipotential line.

Water pressures at points A and B are not the same since, according to equation (32), the total head h is given by the sum of the pressure head P/γ_w and the elevation z of the measuring point above the reference datum. The *water pressure* increases with depth along an equipotential line as shown in Figure 53.

A complete discussion on the construction or computation of flow nets exceeds the scope of this book and the interested reader is referred to the comprehensive texts by Cedergren[166]

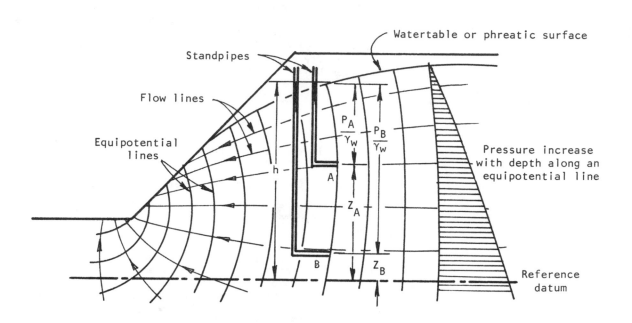

Figure 53 : Two-dimensional flow net in a slope.

Figure 54 : Electrical analogue for the study of anisotropic groundwater flow and drainage problems [171].

A simple conducting paper electrical analogue model for the study of groundwater flow. Photograph courtesy Bougainville Copper Limited.

and Haar[173] for further details. Traditional graphical methods for constructing flow nets[174] have now largely been superseded by analogue[175,176] and numerical methods[177].

An example of an electrical resistance analogue for the study of anisotropic seepage and drainage problems is illustrated in Figure 54. Some typical examples of equipotential distributions, determined with the aid of this analogue, are reproduced in Figure 55[178].

Field measurement of permeability

Determination of the permeability of a rock mass is necessary if estimates are required of groundwater discharge into an open pit or if an attempt is to be made to design a drainage system for the pit.

For evaluation of the stability of the pit slopes it is the *water pressure* rather than the volume of groundwater flow in the rock mass which is important. The water pressure at any point is independent of the permeability of the rock mass at that point but it does depend upon the path followed by the groundwater in arriving at the point (Figures 48 and 55). Hence, the anisotropy and the *distribution* of permeability in a rock mass is of interest in estimating the water pressure distribution in a slope.

In order to measure the permeability at a "point" in a rock mass, it is necessary to change the groundwater conditions at that point and to measure the time taken for the original conditions to be re-established or the quantity of water necessary to maintain the new conditions. These tests are most conveniently carried out in a borehole in which a section is isolated between the end of the casing and the bottom of the hole or between packers within the hole. The tests can be classified as follows:

a) Falling head tests in which water is poured into a vertical or near vertical borehole and the time taken for the water level to fall to its original level is determined.

b) Constant head tests in which the quantity of water which has to be poured into the borehole in order to maintain a specific water level is measured.

c) Pumping tests or Lugeon tests in which water is pumped into or out of a borehole section between two packers and the changes induced by this pumping is measured.

The first two types of test are suitable for measurement of the permeability of reasonably uniform soils or rock. Anisotropic permeability coefficients cannot be measured directly in these tests but, as shown in the example given below, allowance can be made for this anisotropy in the calculation of permeability. Pumping tests, although more expensive, are more suitable for permeability testing in jointed rock.

Falling head and constant head tests

A very comprehensive discussion on falling head and constant head permeability testing is given by Horsley[179] and a few of the points which are directly relevant to the present discussion are summarised on page 138.

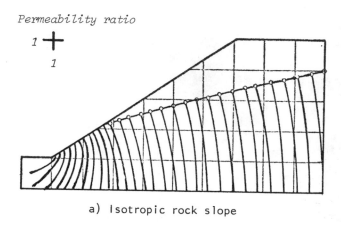

a) Isotropic rock slope

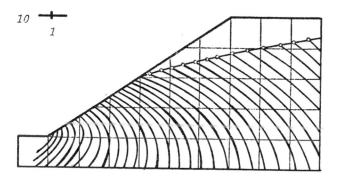

b) Anisotropic rock mass - horizontally
bedded strata.

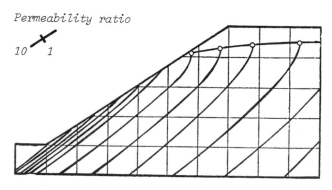

c) Anisotropic rock mass - strata
dipping parallel to slope.

Figure 55 : Equipotential distributions in slopes with
various permeability configurations.

The coefficient of permeability k is calculated from falling head and constant head tests in saturated ground (test section below water table) as follows.

Falling head: $\qquad k = \dfrac{A}{F(t_2 - t_1)} \cdot \log_e \dfrac{H_1}{H_2}$ (35)

Constant head: $\qquad k = \dfrac{q}{F\,H_c}$ (36)

where A is the cross section area of the water column. $A = \frac{1}{4}\pi d^2$ where d is the inside diameter of the casing in a *vertical* borehole. For an inclined hole, A must be corrected to account for the elliptical shape of the horizontal water surface in the casing.

F is a shape factor which depends upon the conditions at the bottom of the hole. Shape factors for typical situations are given in Figure 56.

H_1 and H_2 are water levels in the borehole measured from the rest water level, at times t_1 and t_2 respectively.

q is the flow rate and

H_c is the water level, measured from the rest water level, maintained during a constant head test.

(Note that Naperian logarithms are used in these equations and that $\log_e = 2.3026 \log_{10}$)

Consider an example of a falling head test carried out in a borehole of 7.6 cm diameter with a casing of 6.0 cm diameter. The borehole is extended a distance of 100 cms beyond the end of the casing and the material in which the test is carried out is assumed to have a ratio of horizontal to vertical permeability $k_h/k_v = 5$.

The first step in this analysis is to calculate the shape factor F from the equation given for the 4th case in Figure 56. The value of $m = \sqrt{5} = 2.24$ and substituting D = 7.6 cm and L = 100 cm,

$$F = \frac{2\pi L}{\log_e (2m\,L/D)} = \frac{628}{\log_e 58.19} = 154$$

Measurement of water levels at different times for the falling head test give the following values:

$H_1 = $ 10 metres at $t_1 = $ 30 seconds

$H_2 = $ 5 metres at $t_2 = $ 150 seconds

The cross-sectional area A of the water column is

$$A = \tfrac{1}{4}\pi(6)^2 = 28.3 \text{ cm}^2.$$

Substituting in equation (35), the horizontal permeability k_h is given by

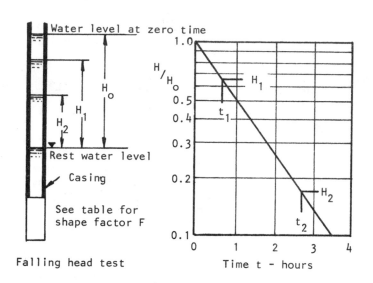

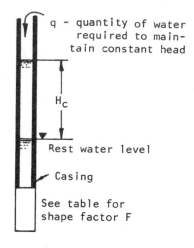

Falling head test

Time t - hours

Constant head test

End conditions		Shape factor F
	Casing flush with end of borehole in soil or rock of uniform permeability. Inside diameter of casing is d cms.	F = 2.75d
	Casing flush with boundary between impermeable and permeable strata. Inside diameter of casing is d cms.	F = 2.0d
	Borehole extended a distance L beyond the end of the casing. Borehole diameter is D.	$F = \dfrac{2\pi L}{\log_e(2L/D)}$ for $L > 4D$
	Borehole extended a distance L beyond the end of the casing in a stratified soil or rock mass with different horizontal and vertical permeabilities.	For determination of k_h: $F = \dfrac{2\pi L}{\log_e(2mL/D)}$ where $m = (k_h/k_v)^{\frac{1}{2}}$, $L > 4D$
	Borehole extended a distance L beyond the end of the casing which is flush with an impermeable boundary.	$F = \dfrac{2\pi L}{\log_e(4L/D)}$ for $L > 4D$

Figure 56 : Details of falling head and constant head tests for permeability measurement in soil and rock masses with shape factors for borehole end conditions.

$$k_h = \frac{28.3 \log_e 2}{154 (150-30)} = 1.06 \times 10^{-3} \text{ cm/sec.}$$

Since the ratio of horizontal to vertical permeability has been estimated, from examination of the core, as $k_h/k_v = 5$, $k_v = 2.12 \times 10^{-4}$ cm/sec.

Laboratory tests on core samples are useful in checking this *ratio* of horizontal to vertical permeability but, because of the disturbance to the sample, it is unlikely that the *absolute values* of permeability measured in the laboratory will be as reliable as those determined by the borehole tests described. Laboratory methods for permeability testing are described in standard texts such as that by Lambe[180].

Pumping tests in boreholes

In a rock mass in which the groundwater flow is concentrated within regular joint sets, the permeability will be highly directional. If the joint opening e could be measured in situ, the permeability in the direction of each joint set could be calculated directly from equation (33). Unfortunately such measurements are not possible under field conditions and the permeability must therefore be determined by pumping tests.

A pumping test for the measurement of the permeability in the direction of a particular set of discontinuities such as joints involves drilling a borehole perpendicular to these discontinuities as shown in Figure 57. It is assumed that most of the flow is concentrated within this one joint set and that cross-flow through other joint sets, past the packers and through the intact rock surrounding the hole is negligible. A section of the borehole is isolated between packers or a single packer is used to isolate a length at the end of the hole and water is pumped into or out of this cavity.

A variety of borehole packers are available commercially[181] but the authors consider that many of these packers are too short to eliminate leakage. Leakage past packers is one of the most serious sources of error in pumping tests and every effort should be made to ensure that an effective seal has been achieved before measurements are commenced. A simple, inexpensive and highly effective packer has been described by Harper and Ross-Brown[182] and the principal features are illustrated in Figure 58. This packer is manufactured from rubber hosing which is normally used in the building industry for forming voids in concrete*. It consists of inner and outer rubber tubes enclosing a diagonally braided cotton core and this arrangement allows an increase in diameter of approximately 20% when the hose is inflated. Because of its low cost and simplicity, long packers can be used and packer lengths of 10 feet (3 metres) have proved extremely effective in pumping tests in 3 inch (7.6 cm) diameter boreholes.

The permeability of the discontinuities perpendicular to the borehole is calculated as follows:

---

* Available in a wide range of diameters from Ductube Company Limited, Daneshill Road, Lound, Near Retford, Nottinghamshire, England and from Petrometallic, Cambrai, France.

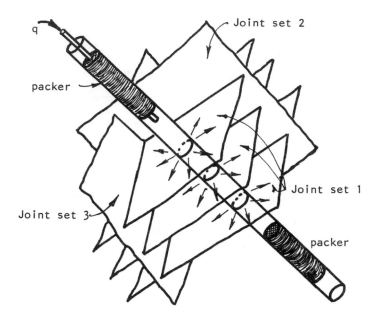

Figure 57 : Pumping test in regularly jointed rock.
The borehole is drilled at right angles
to the joint set in which the permeability
is to be measured.

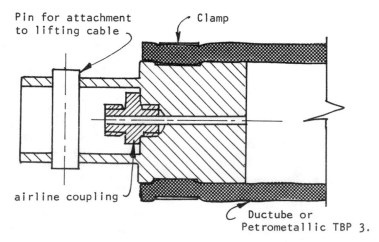

Figure 58 : Section through the end of a packer for
sealing the bottom end of a pumping test
cavity. The upper packer end has additional fittings for pressure inlet and piezometer cables.

$$k = \frac{q \ \text{Log}_e \ (2^{R/D})}{2\pi L \ (H_1 - H_2)} \qquad (37)$$

where q is the pumping rate required to maintain a constant pressure in the test cavity

L is the length of the test cavity

H_1 is the total head in the test cavity

D is the borehole diameter

H_2 is the total head measured at a distance R from the borehole.

The most satisfactory means of obtaining the value of H_2 is to measure it in a borehole parallel to and at a distance R from the test hole. Where a pattern of boreholes is available, as is the case on many opencast mine sites, this does not present serious problems. Techniques for water pressure measurement are dealt with in the following section of this chapter.

When only one borehole is available, an approximate solution to equation (37) can be obtained by using the shape factor F for a stratified system (Figure 56). Substituting this value into equation (36) gives

$$k = \frac{q. \ \text{Log}_e \ (2m^{L/D})}{2\pi L \ H_c} \qquad (38)$$

where, in this case, $m = (k/k_p)^{\frac{1}{2}}$

k is the permeability at right angles to the borehole (quantity required)

k_p is the permeability parallel to the borehole which, if cross flow is neglected, is equal to the permeability of the intact rock

H_c is the constant head above the original groundwater level in the borehole.

The value of the term Log_e (2m $^{L/D}$) in this equation does not have a major influence upon the value of k and hence a crude estimate of m is adequate. Consider the example where L = 4D; the values of Log_e (2m $^{L/D}$) are as follows:

k/k_p	1.0	10^2	10^4	10^6	10^8	10^{10}	10^{12}
m	1.0	10^1	10^2	10^3	10^4	10^5	10^6
Log_e (2m $^{L/D}$)	2.1	4.4	6.7	9.0	11.3	13.6	15.9

A reasonable value of k for most practical applications is given by assuming $k/k_p = 10^6$, m = 10^3 which gives

$$k = \frac{1.4q}{L \ H_c} \qquad (39)$$

In deriving equation (39), it has been assumed that the test
cavity of length L intersects a large number of discontinuities (say 100) and that the value k represents a reasonable
average permeability for the rock mass (in the direction at
right angles to the borehole). When the discontinuity
spacing varies along the length of the hole, water flow will
be concentrated in zones of closely spaced discontinuities
and the use of an average permeability value can give
misleading results. Under these circumstances, it is
preferable to express the permeability in terms of the
permeability k_j of individual discontinuities where

$$k_j = \frac{k}{n} \qquad (40)$$

   n is the number of discontinuities which intersect
   the test cavity of length L.

The value of n can be estimated from the borehole core log
and, assuming that the discontinuity opening (e in equation
33) remains constant, the variation in permeability along
the borehole can then be estimated.

Before leaving this question of permeability testing it must
be pointed out that the discussion which has been presented
has been grossly simplified. This has been done deliberately
since the literature dealing with this subject is copious,
complex and confusing. A number of techniques, more
sophisticated than those which have been described here,
are available for the evaluation of permeability but the
authors believe that these are best left in the hands of
experienced specialist consultants. The simple tests which
have been described are generally adequate for open pit
stability and drainage studies.

Measurement of water pressure

The importance of water pressure in relation to the stability
of slopes has been emphasised in several of the previous
chapters. If a reliable estimate of stability is to be
obtained or if the stability of a slope is to be controlled
by drainage, it is essential that water pressures within
the slope should be measured. Such measurements are most
conveniently carried out by *piezometers* installed in boreholes.

A variety of piezometer types is available and the choice
of the type to be used for a particular installation depends
upon a number of practical considerations. A detailed
discussion on this matter has been given by Terzaghi and
Peck[183] and only the most important considerations will be
summarised here.

The most important factor to be considered in choosing a
piezometer is the time lag of the complete installation.
This is the time taken for the pressure in the system to
reach equilibrium after a pressure change and it depends
upon the permeability of the ground and the volume change
associated with the pressure change. Open holes can be used
for pressure measurement when the permeability is greater
than 10^{-4}cm/sec but, for less permeable ground, the time lag
is too long. In order to overcome this problem, a pressure
measuring device or piezometer is installed in a sealed
section of the borehole. The volume change within this
sealed section, caused by the operation of the piezometer

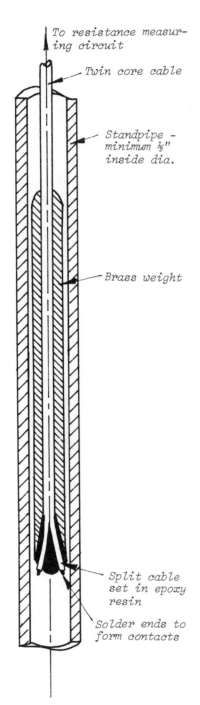

To resistance measuring circuit

Twin core cable

Standpipe - minimum ½" inside dia.

Brass weight

Split cable set in epoxy resin

Solder ends to form contacts

A simple probe for water level detection.

should be very small in order that the response of the complete installation to pressure changes in the surrounding rock should be rapid. If a device which requires a large volume change for its operation is used, the change in pressure induced by this change in volume may give rise to significant errors in measurement.

Some of the common types of piezometer are briefly discussed below.

a)   Open piezometers or observation wells

As discussed above, open ended cased holes can be used to measure water pressure in rock or soil in which the permeability is greater than about 10^{-4} cm/sec. All that is required for these measurements is a device for measuring water level in the borehole. A very simple probe consisting of a pair of electrical contacts housed in a brass weight is illustrated in the margin sketch. When the contacts touch the water, the resistance of the electrical circuit drops and this can be measured on a standard "Avometer" or similar instrument. The depth of water below the collar of the hole is measured by the length of cable and it is convenient to mark the cable in feet or metres for this purpose. Portable water level indicators, consisting of a probe, a marked cable and a small resistance measuring instrument, are available from Soil Instruments Ltd., Townsend Lane, London N.W.9 or from Soiltest Inc., 2205 Lee Street, Evanston, Illinois 60202, U.S.A.

b)   Standpipe piezometers

When the permeability of the ground in which water pressure is to be measured is less than 10^{-4} cm/sec, the time lag involved in using an open hole will be unacceptable and a standpipe piezometer such as that illustrated in Figure 59 should be used. This device consists of a perforated tip which is sealed into a section of borehole as shown. A small diameter standpipe passing through the seals allows the water level to be measured by means of the same type of water level indicator as described above under open hole piezometers. Because the volume of water within the standpipe is small, the response time of this piezometer installation will be adequate for most applications likely to be encountered on an open pit mine site.

An advantage of the standpipe piezometer is that, because of the small diameter of the standpipe, a number can be installed in the same hole. Hence different sections can be sealed off along the length of the borehole and the water pressure within each section monitored. This type of installation is important when it is suspected that water flow is confined to certain layers within a rock mass.

c)   Closed hydraulic piezometers

When the permeability of the ground falls below about 10^{-6} cm/sec, the time lag of open ended boreholes or standpipe piezometers becomes unacceptable. For example, approximately 5 days would be required for a typical standpipe piezometer to reach an acceptable state of equilibrium after a change of water pressure in a rock or soil mass having a permeability of 10^{-7} cm/sec.

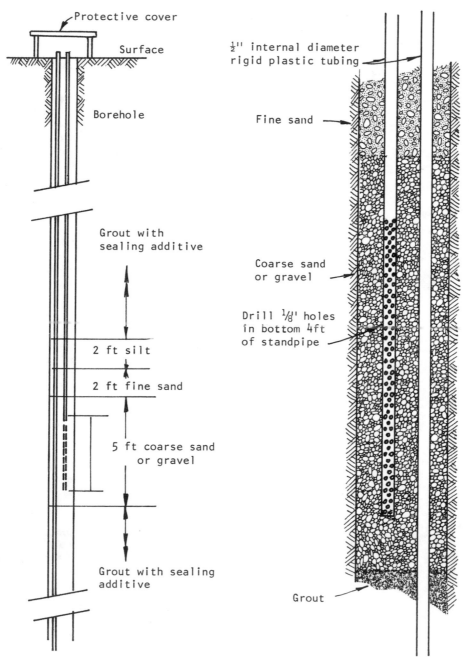

Protective cover

Surface

Borehole

Grout with
sealing additive

2 ft silt

2 ft fine sand

5 ft coarse sand
or gravel

Grout with sealing
additive

½'' internal diameter
rigid plastic tubing

Fine sand

Coarse sand
or gravel

Drill ⅛' holes
in bottom 4ft
of standpipe

Grout

Note : The two sealing layers of fine sand and silt above the piezometer
section can be replaced by bentonite pellets which form a gel
in contact with water and form an effective seal . Pelletized
bentonite is available commercially as "Peltonite" from Rocktest
ltd., Lambert, Quebec, Canada.

Figure 59 : Typical standpipe piezometer installation details .

An improved time lag can be obtained by using a closed hydraulic piezometer such as that described by Bishop et al[184]. This type of piezometer is completely filled with de-aired water and is suitable for measurement of small water pressures. Such piezometers are generally used for pore pressure measurement during construction of embankments or dams where they can be installed during construction and left in place.

d)   Air actuated piezometers

A very rapid response time can be achieved by use of air actuated piezometers in which the water pressure is measured by a balancing air pressure acting against a diaphragm. As shown in Figure 60, an air valve allows air to escape when the air and water pressures on either side of the diaphragm are equal[185]. A commercially available air piezometer is illustrated in Figure 61. Similar types of instrument are available from other suppliers and these devices are playing an increasingly important role in slope stability studies.

e)   Electrically indicating piezometer

An almost instantaneous reponse time is obtained from piezometers in which the deflection of a diaphragm as a result of water pressure is measured electrically by means of some form of strain gauge attached to the diaphragm. A wide variety of such devices is available commercially and they are ideal for measuring the water pressure within the test cavity during a pumping test[172]. Because of their relatively high cost and because of the possibility of electrical faults, these piezometers are less satisfactory for permanent installation in boreholes.

General comments

A frequent mistake made by engineers or geologists in examining rock or soil slopes is to assume that groundwater is not present if no seepage appears on the slope face. In many cases, the seepage rate may be lower than the evaporation rate and hence the slope surface may appear completely dry and yet there may be water at significant pressure within the rock mass. Remember that it is water pressure and not rate of flow which is responsible for instability in slopes and it is essential that measurement or calculation of this water pressure should form part of site investigation for stability studies. Drainage, which is discussed in a later chapter, is one of the most effective and most economical means available for improving the stability of open pit mine slopes. Rational design of drainage systems is only possible if the water flow pattern within the rock mass is understood and measurement of permeability and water pressure provides the key to this understanding.

Installation of a plastic tube standpipe piezometer in a drill hole.

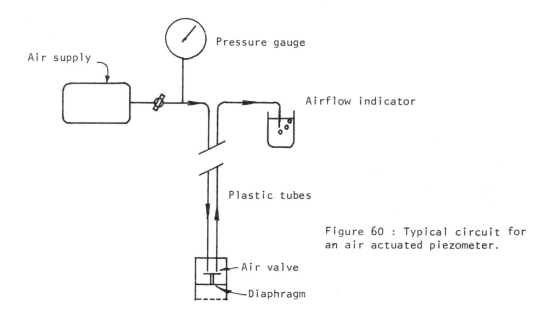

Air supply

Pressure gauge

Airflow indicator

Plastic tubes

Air valve

Diaphragm

Figure 60 : Typical circuit for an air actuated piezometer.

Figure 61 : Model P-100 Pneumatic Piezometer and Model C-102 read-out unit manufactured by Thor Instrument Company Inc., Seattle.

Piezometer measures $\frac{1}{2}$'' x $2\frac{1}{2}$'' (12mm x 62mm).

Chapter 6 references

Selected references on permeability and water pressure measurement.

159. BRAWNER, C.O. The influence and control of groundwater in open pit mining. *Proc. 5th Canadian Symposium on Rock Mechanics,* Toronto, 1968.

160. DOBRY, R. and ALVAREZ, L. Seismic failures of Chilean tailings dams. *Proc. American Society of Civil Engineers.* Vol. 93, No. SM6, 1967.

161. CASAGRANDE, L. and MACIVER, B.N. Design and construction of tailings dams. *Proc. 1st Symposium on Stability for Open Pit Mining,* Vancouver, 1970.

162. SEED, H.B. and LEE, K.L. Liquefaction of saturated sands during cyclic loading. *Journal Soil Mechanics and Foundation Div. Proc. ASCE.* Vol. 92, No. SM6, 1966, pages 105-134.

163. DAVIS, S.N. and DE WIEST, R.J.M. *Hydrogeology.* John Wiley & Sons, New York & London, 1966, 463 pages.

164. MORGENSTERN, N.R. The influence of groundwater on stability. *Proc. 1st Symposium on Stability of Open Pit Mining,* Vancouver, 1970. Published by AIME, New York, 1971.

165. SCHEIDEGGER, A.E. *The physics of flow through porous media.* MacMillan, New York, 1960.

166. CEDERGREN, H.R. *Seepage, drainage and flow nets.* John Wiley and Sons, New York, 1967.

167. DAVIS, S.N. Porosity and permeability of natural materials, in *Flow through porous media,* edited by R. de Wiest, Academic Press, London, 1969, pages 54-89.

168. HUITT, J.L. Fluid flow in simulated fracture. *Journal American Inst. Chemical Eng.* Vol. 2, 1956, pages 259-264.

169. SNOW, D.T. Rock fracture spacings, openings and porosities. *Journal Soil. Mech. Foundation Div. Proc. ASCE.* Vol. 94, 1968, pages 73-91.

170. LOUIS, C. A study of groundwater flow in jointed rock and its influence on the stability of rock masses. Doctorate thesis, University of Karlsruhe, 1967, (in German). English translation *Imperial College Rock Mechanics Research Report No, 10,* September, 1969. 90 pages.

171. SHARP, J.C. Fluid flow through fissured media. *Ph.D. Thesis, University of London* (Imperial College), 1970.

172. MAINI, Y.N. In situ parameters in jointed rock - their measurement and interpretation. *Ph.D. Thesis, University of London* (Imperial College), 1971.

173. HAAR, M.E. *Groundwater and Seepage,* McGraw Hill Co., New York, 1962.

174. TAYLOR, D.W. *Fundamentals of Soil Mechanics*. John Wiley & Sons, New York, 1948.

175. KARPLUS, W.J. *Analog Simulation. Solution of Field Problems*. McGraw Hill Co., New York, 1968.

176. MEEHAN, R.L. and MORGENSTERN, N.R. The approximate solution of seepage problems by a simple electrical analogue method. *Civil Engineering and Public Works Review*. Vol. 63, 1968, pages 65-70.

177. ZIENKIEWICZ, O.C., MAYER, P. and CHEUNG, Y.K. Solution of anisotropic seepage by finite elements. *Journal Eng. Mech. Div., ASCE*, Vol. 92, EM1, 1966, pages 111-120.

178. SHARP, J.C., MAINI, Y.N. and HARPER, T.R. Influence of groundwater on the stability of rock masses. *Trans. Institute Mining and Metallurgy*, London. Vol. 81, Bulletin No. 782, 1972, pages A13 - 20.

179. HORSLEV, M.S. Time lag and soil permeability in groundwater measurements. *U.S. Corps of Engineers Waterways Experiment Station, Bulletin No. 36*, 1951, 50 pages.

180. LAMBE, T.W. *Soil testing for engineers*. John Wiley & Sons, New York, 1951.

181. MUIR WOOD, A.M. In-situ testing for the Channel Tunnel. *Proc. Conference on in-situ investigations in soils and rocks*, London, 1969. Published by the Institution of Civil Engineers, London, 1969. pages 79-86.

182. HARPER, T.R. and ROSS-BROWN, D.M. An inexpensive durable borehole packer. *Imperial College Rock Mechanics Research Report No. D24*, 1972, 5 pages.

183. TERZAGHI, K. and PECK, R. *Soil Mechanics in Engineering Practice*. John Wiley & Sons Inc., New York, 1967, 729 pages.

184. BISHOP, A.W., KENNARD, M.F. and PENMAN, A.D.M. Pore-pressure observations at Selset Dam. *Proc. Conference on Pore Pressure and Suction in Soils*. Butterworth, London, 1960, pages 91-102.

185. WARLAM, A.A. and THOMAS, E.W. Measurement of hydrostatic uplift pressure on spillway weir with air piezometers. *Instruments and apparatus for soil and rock mechanics*. American Society for Testing and Materials Special Technical Publication No. 392, 1965, pages 143-151.

Chapter 7 : Plane failure

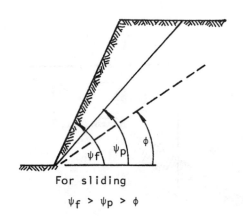

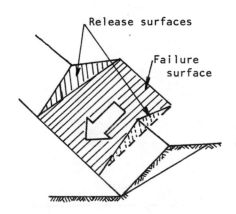

For sliding

$\psi_f > \psi_p > \phi$

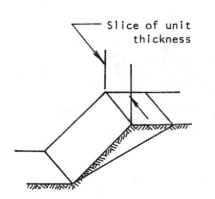

Introduction

A plane failure is a comparatively rare sight in rock slopes because it is only occasionally that all the geometrical conditions required to produce such a failure occur in an actual slope. The wedge type of failure, considered in Chapter 8, is a much more general case and many rock slope engineers treat the plane failure as a special case of the more general wedge failure analysis.

While this is probably the correct approach for the experienced slope designer who has a wide range of design tools at his disposal, it would not be right to ignore the two-dimensional case in this general discussion on slope failure. There are many valuable lessons to be learned from a consideration of the mechanics of this simple failure mode and it is particularly useful for demonstrating the sensitivity of the slope to changes in shear strength and groundwater conditions - changes which are less obvious when dealing with the more complex mechanics of a three-dimensional slope failure.

General conditions for plane failure

In order that sliding should occur on a *single plane*, the following geometrical conditions must be satisfied:

a. The plane on which sliding occurs must strike parallel or nearly parallel (within approximately ± 20°) to the slope face.

b. The failure plane must "daylight" in the slope face. This means that its dip must be *smaller* than the dip of the slope face, i.e. $\psi_f > \psi_p$.

c. The dip of the failure plane must be *greater* than the angle of friction of this plane, i.e. $\psi_p > \phi$.

d. Release surfaces which provide negligible resistance to sliding must be present in the rock mass to define the lateral boundaries of the slide. Alternatively, failure can occur on a failure plane passing through the convex "nose" of a slope.

In analysing two-dimensional slope problems, it is usual to consider a slice of unit thickness taken at right angles to the slope face. This means that the area of the sliding surface can be represented by the length of the surface visible on a vertical section through the slope and the volume of the sliding block is represented by the area of the figure representing this block on the vertical section.

Plane failure analysis

The geometry of the slope considered in this analysis is defined in Figure 62. Note that two cases must be considered.

a. A slope having a tension crack in its upper surface.

b. A slope with a tension crack in its face.

The transition from one case to another occurs when the

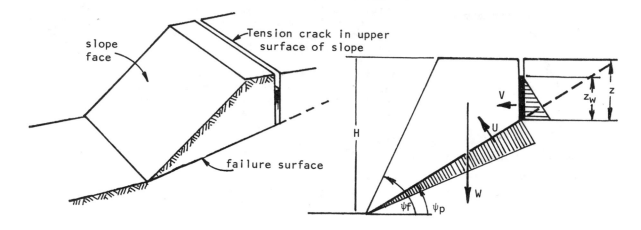

Figure 62a : Geometry of slope with tension crack in upper slope surface.

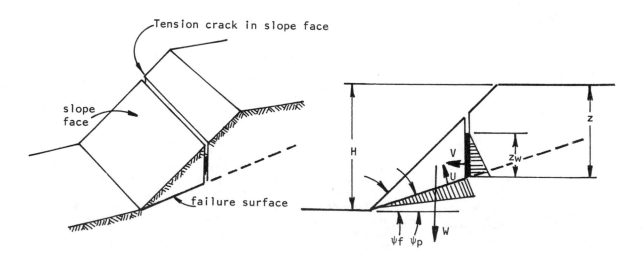

Figure 62b : Geometry of slope with tension crack in slope face.

tension crack coincides with the slope crest, i.e. when

$$z/H = (1 - Cot\psi_f.Tan\psi_p) \qquad (41)$$

The following assumptions are made in this analysis:

a. Both sliding surface and tension crack strike parallel to the slope surface.

b. The tension crack is vertical and is filled with water to a depth z_w.

c. Water enters the sliding surface along the base of the tension crack and seeps along the sliding surface, escaping at atmospheric pressure where the sliding surface daylights in the slope face. The pressure distribution induced by the presence of water in the tension crack and along the sliding surface is illustrated in Figure 62.

d. The forces W (the weight of the sliding block), U (uplift force due to water pressure on the sliding surface) and V (force due to water pressure in the tension crack) all act through the centroid of the sliding mass. In other words, it is assumed that there are no moments which would tend to cause rotation of the block and hence failure is by sliding only. While this assumption may not be strictly true for actual slopes, the errors intro-duced by ignoring moments are small enough to neglect. However, in steep slopes with steeply dipping discontinuities, the possibility that toppling failure may occur should be kept in mind.

e. The shear strength of the sliding surface is defined by cohesion c and a friction angle ϕ which are related by the equation $\tau = c + \sigma Tan\phi$ as discussed on page 22. In the case of a rough surface having a curvilinear shear strength curve, the apparent cohesion and apparent friction angle, defined by a tangent to the curve are used. This tangent should touch the curve at a normal stress value which corresponds to the normal stress acting on the failure plane. In this case, the analysis is only valid for the slope height used to determine the normal stress level. The normal stress acting on a failure surface can be determined from the graph given in Figure 63.

f. A slice of unit thickness is considered and it is assumed that release surfaces are present so that there is no resistance to sliding at the lateral boundaries of the failure.

The factor of safety of this slope is calculated in the same way as that for the block on an inclined plane considered on page 27. In this case the factor of safety, given by the total force resisting sliding to the total force tending to induce sliding, is

$$F = \frac{cA + (W.Cos\psi_p - U - V.Sin\psi_p)Tan\phi}{W.Sin\psi_p + V.Cos\psi_p} \qquad (42)$$

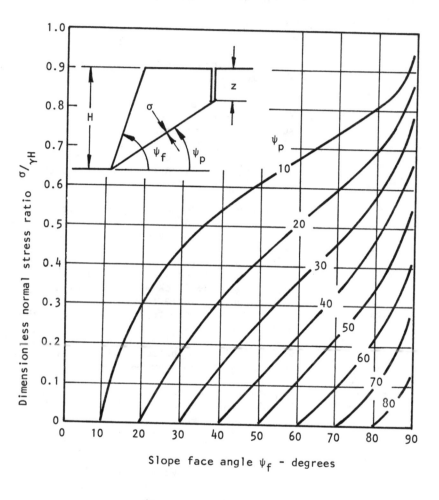

$$\frac{\sigma}{\gamma H} = \frac{\left((1 - (z/H)^2)Cot\psi_p - Cot\psi_f\right)Sin\psi_p}{2(1 - z/H)}$$

where $z/H = 1 - \sqrt{Cot\psi f \cdot Tan\psi p}$ (see page 161)

Figure 63:   Normal stress acting on the failure plane in
a rock slope.

where, from Figure 62 :

$$A = (H - z).\text{Cosec}\psi_p \qquad (43)$$

$$U = \tfrac{1}{2}\gamma_w.\ z_w(H - z).\text{Cosec}\psi_p \qquad (44)$$

$$V = \tfrac{1}{2}\gamma_w.z^2_w \qquad (45)$$

For the tension crack in the upper slope surface (Figure 62a)

$$W = \tfrac{1}{2}\gamma H^2\left((1 - (z/H)^2)\text{Cot}\psi_p - \text{Cot}\psi_f\right) \qquad (46)$$

and, for the tension crack in the slope face (Figure 62b)

$$W = \tfrac{1}{2}\gamma H^2\left((1 - z/H)^2\text{Cot}\psi_p(\text{Cot}\psi_p.\text{Tan}\psi_f - 1)\right) \qquad (47)$$

When the geometry of the slope and the depth of water in the tension crack are known, the calculation of a factor of safety is a simple enough matter. However, it is sometimes necessary to compare a range of slope geometries, water depths and the influence of different shear strengths. In such cases, the solution of equations 42 to 47 can become rather tedious. In order to simplify the calculations, equation 42 can be rearranged in the following dimensionless form :

$$F = \frac{(2c/\gamma H).P + \left(Q.\text{Cot}\psi_p - R(P + S)\right)\text{Tan}\phi}{Q + R.S\text{Cot}\psi_p} \qquad (48)$$

where

$$P = (1 - z/H).\text{Cosec}\psi_p \qquad (49)$$

When the tension crack is in the upper slope surface :

$$Q = \left((1 - (z/H)^2)\text{Cot}\psi_p - \text{Cot}\psi_f\right)\text{Sin}\psi_p \qquad (50)$$

When the tension crack is in the slope face :

$$Q = \left((1 - z/H)^2\text{Cos}\psi_p(\text{Cot}\psi_p.\text{Tan}\psi_f - 1)\right) \qquad (51)$$

$$R = \frac{\gamma_w}{\gamma}.\frac{z_w}{z}.\frac{z}{H} \qquad (52)$$

$$S = \frac{z_w}{z}.\frac{z}{H}\ \text{Sin}\psi_p \qquad (53)$$

The ratios P,Q,R and S are all dimensionless which means that they depend upon the geometry but not upon the size of the slope. Hence, in cases where the cohesion c = 0, the factor of safety is independent of the size of the slope. The important principle of dimensionless grouping, illustrated in these equations, is a useful tool in rock engineering and extensive use will be made of this principle in the study of wedge and circular failures.

In order to facilitate the application of these equations to practical problems, values for the ratios P, Q and S, for a range of slope geometries, are presented in graphical form in Figure 64. Note that both tension crack positions are included in the graphs for the ratio Q and hence the values of Q may be determined for any slope configuration without having first to check on the tension crack position.

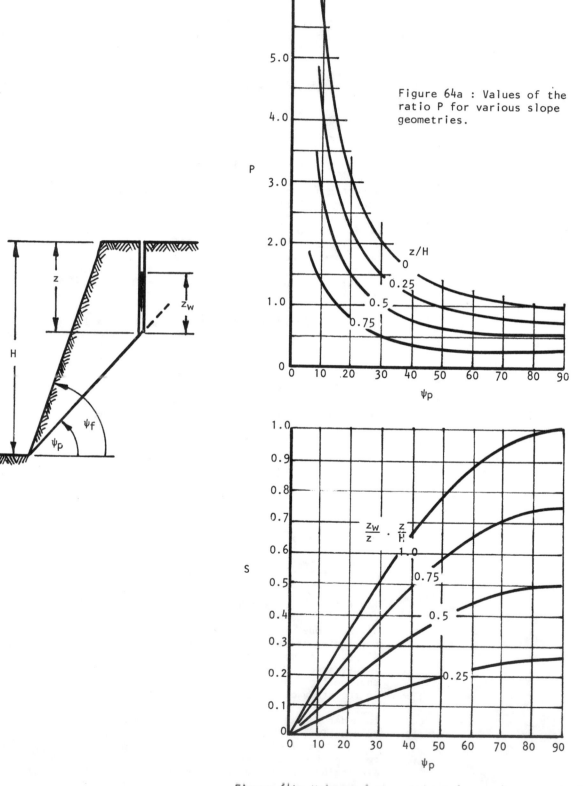

Figure 64a : Values of the
ratio P for various slope
geometries.

Figure 64b: Values of the ratio S for various geometries

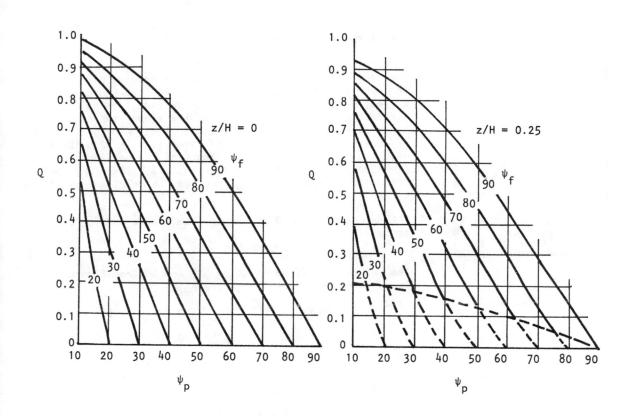

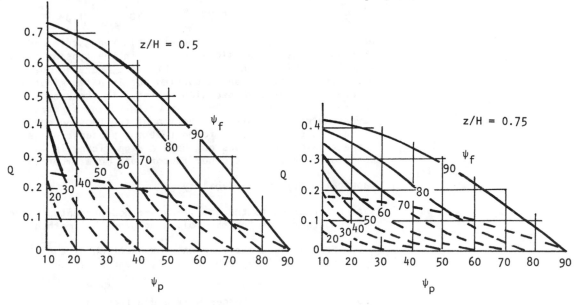

Figure 64c : Value of the ratio Q for various slope geometries.

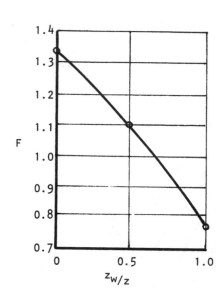

One point to keep in mind when using these graphs is that the depth of the tension crack is always measured from the top of the slope as illustrated in Figure 62b-

Consider the example in which a 100 foot high slope with a face angle $\psi_f = 60^\circ$ is found to have a bedding plane running through it at a dip $\psi_p = 30^\circ$. A tension crack occurs 29 feet behind the crest of the slope and, from an accurately drawn cross-section of the slope, the tension crack is found to have a depth of 50 feet. The unit weight of rock $\gamma = 160$ lb/ft^3, that of water is $\gamma_w = 62.5$ lb/ft^3. Assuming that the cohesive strength of the bedding plane $c = 1000$ lb/ft^2 and the friction angle $\phi = 30^\circ$, find the influence of water depth z_w upon the factor of safety of the slope.

The values of P and Q are found from Figure 64, for z/H = 0.5 to be :

$$P = 1.0 \text{ and } Q = 0.36$$

The values of R (from equation 52) and S (from Figure 64b), for a range of values of z_w/z, are :

z_w/z	1.0	0.5	0
R	0.195	0.098	0
S	0.26	0.13	0

The value of $2c/\gamma H = 0.125$

Hence, the factor of safety for different depths of water in the tension crack, from equation 48, varies as follows :

z_w/z	1.0	0.5	0
F	0.77	1.10	1.34

These values are plotted in the graph in the margin and the sensitivity of the slope to water in the tension crack is obvious. Simple analyses of this sort, varying one parameter at a time, can be carried out in a few minutes and are useful aids to decision making. In the example considered, it would be obviously worth taking steps to prevent water from entering the top of the tension crack. In other cases, it may be found that the presence of water in the tension crack does not have a significant influence upon stability and that other factors are more important.

Graphical analysis of stability

As an alternative to the analytical method presented above, some readers may prefer the following graphical method :

    a.   From an accurately drawn cross-section of the slope, scale the lengths H, X, D, A, z and z_w shown in Figure 65a.

    b.   Calculate the forces W, V and U from these dimensions by means of the equations given in Figure 65a. Also calculate the magnitude of the cohesive force A.c.

    c.   Construct the force diagram illustrated in Figure 65b as follows :

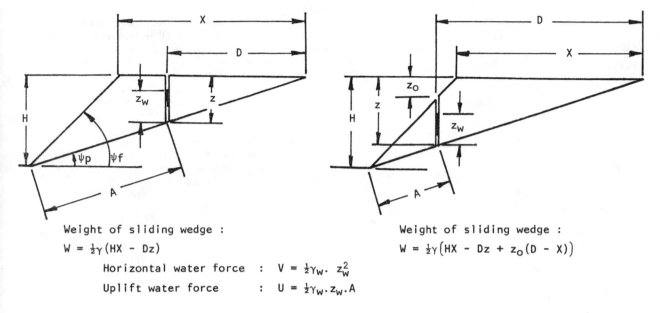

Weight of sliding wedge :

$W = \frac{1}{2}\gamma(HX - Dz)$

Weight of sliding wedge :

$W = \frac{1}{2}\gamma\big(HX - Dz + z_0(D - X)\big)$

Horizontal water force : $V = \frac{1}{2}\gamma_w . z_w^2$

Uplift water force : $U = \frac{1}{2}\gamma_w . z_w . A$

Figure 65a : Slope geometry and equations for calculating forces acting on slope.

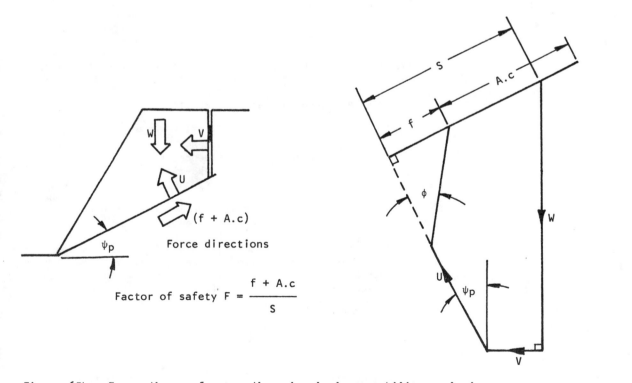

Force directions

Factor of safety $F = \dfrac{f + A.c}{S}$

Figure 65b : Force diagram for two-dimensional slope stability analysis.

i) Draw a vertical line to represent the weight W of the sliding wedge. The scale should be chosen to suit the size of the drawing board used.

ii) At right angles to the line representing W, draw a line to represent the force V due to water pressure in the tension crack.

iii) Measure the angle ψ_p as shown in Figure 65b and draw a line to represent the uplift force U due to water pressure on the sliding surface.

iv) Project the line representing U (shown dashed in Figure 65b) and, from the upper extremity of the line representing W, construct a perpendic- lar to the projection of the U line.

v) From the upper extremity of the U line, draw a line at an angle ϕ to intersect the line from W to the projection fo the U line.

vi) The length f in Figure 65b represents the frictional force which resists sliding along the failure plane.

vii) The cohesive resisting force A.c can be drawn parallel to f. Although this step is not essential, drawing A.c on the force diagrams ensures that there is no error in converting to and from the various scales which may have been used in this analysis since it provides a visual check of the magnitude of A.c.

viii) The length of the line marked S on the force diagram represents the total force tending to induce sliding down the plane.

ix) The factor of safety F of the slope is given by the ratio of the lengths (f + A.c) to S.

An example of the application of this graphical technique will be given later in this chapter.

Influence of groundwater on stability

In the preceding discussion it has been assumed that it is only the water present in the tension crack and that along the failure surface which influences the stability of the slope. This is equivalent to assuming that the rest of the rock mass is impermeable, an assumption which is certainly not always justified. Consideration must, therefore, be given to water pressure distribution other than that upon which the analysis so far presented is based.

The current state of knowledge in rock engineering does not permit a precise definition of the groundwater flow patterns in a rock mass. Consequently, the only possibility open to the slope designer is to consider a number of realistic extremes in an attempt to bracket the range of possible factors of safety and to assess the sensitivity of the slope to variations in groundwater conditions.

a. *Dry slopes*

The simplest case which can be considered is that in which the slope is assumed to be completely drained. In practical terms, this means that there is no water pressure in the tension crack or along the sliding surface. Note that there may be moisture in the slope but, as long as no *pressure* is generated, it will not influence the stability of the slope.

Under these conditions, the forces V and U are both zero and equation 42 reduces to :

$$F = \frac{c.A}{W.Sin\psi_p} + Cot\psi_p.Tan\phi \qquad (54)$$

Alternatively, equation 48 reduces to :

$$F = \frac{2c}{\gamma H} . \frac{P}{Q} + Cot\psi_p.Tan\phi \qquad (55)$$

b.   Water in tension crack only

A heavy rain storm after a long dry spell can result in the rapid build-up of water pressure in the tension crack which will offer little resistance to the entry of surface flood water unless effective surface drainage has been provided. Assuming that the remainder of the rock mass is relatively impermeable, the only water pressure which will be generated during and immediately after the rain will be that due to water in the tension crack. In other words, the uplift force U = 0.

The uplift force U could also be reduced to zero or nearly zero if the failure surface was impermeable as a result of clay filling. In either case, the factor of safety of the slope is given by

$$F = \frac{c.A + (W.Cos\psi_p - V.Sin\psi_p)Tan\phi}{W.Sin\psi_p + V.Cos\psi_p} \qquad (56)$$

or, alternatively

$$F = \frac{2c/\gamma H.P + (Q.Cot\psi_p - RS)Tan\phi}{Q + RS.Cot\psi_p} \qquad (57)$$

c.   Water in tension crack and on sliding surface

These are the conditions which were assumed in deriving the general solution presented on the preceding pages. The pressure distribution along the sliding surface has been assumed to decrease linearly from the base of the tension crack to the intersection of the failure surface and the slope face. This water pressure distribution is probably very much simpler than that which occurs in an actual slope but, since the actual pressure distribution is unknown, this assumed distribution is as reasonable as any other which could be made.

It is possible that a more dangerous water pressure distribution could exist if the face of the slope became frozen in winter so that, instead of the zero pressure condition which has been assumed at the face, the water pressure at the face would be that due to the full head of water in the slope. Such extreme water pressure conditions may occur from time to time and the slope designer should keep this possibility in mind. However, for general slope design, the use of this water pressure distribution would result in an excessively conservative slope and hence the triangular pressure distribution used in the general analysis is presented as the basis for normal slope design.

d. *Saturated slope with heavy recharge*

If the rock mass is heavily fractured so that it becomes relatively permeable, a groundwater flow pattern similar to that which would develop in a porous system could occur (see Figure 55 on page 137). The most dangerous conditions which would develop in this case would be those given by prolonged heavy rain.

Flow nets for saturated slopes with heavy surface recharge have been constructed and the water pressure distributions obtained from these flow nets have been used to calculate the factors of safety of a variety of slopes. The process involved is too tedious to include in this chapter but the results can be summarised in a general form. It has been found that the factor of safety for a permeable slope, saturated by heavy rain and subjected to surface recharge by continued rain, can be approximated by equation 42 (or 48), assuming that the tension crack is water-filled, i.e. $z_w = z$.

In view of the uncertainties associated with the actual water pressure distributions which could occur in rock slopes subjected to these conditions, there seems little point in attempting to refine this analysis any further.

Critical tension crack depth

In the analysis which has been presented, it has been assumed that the position of the tension crack is known from its visible trace on the upper surface or on the face of the slope and that its depth can be established by constructing an accurate cross-section of the slope. When the tension crack position is unknown, due for example, to the presence of a waste dump on the top of the slope, it becomes necessary to consider the most probable position of a tension crack.

The influence of tension crack depth and of the depth of water in the tension crack upon the factor of safety of a typical slope is illustrated in Figure 66 (based on the example considered on page 157).

When the slope is dry or nearly dry, the factor of safety reaches a minimum value which, in the case of the example considered, corresponds to a tension crack depth of 0.42H. This critical tension crack depth for a dry slope can be found by minimising the right hand side of equation 54 with respect to z/H. This gives the critical tension crack depth as :

A mountain top tension crack above a large landslide.

$$z_c/H = 1 - \sqrt{\mathrm{Cot}\psi_f . \mathrm{Tan}\psi_p} \qquad (58)$$

From the geometry of the slope, the corresponding position of the tension crack is :

$$b_c/H = \sqrt{\mathrm{Cot}\psi_f . \mathrm{Cot}\psi_p} - \mathrm{Cot}\psi_f \qquad (59)$$

Critical tension crack depths and locations for a range of dry slopes are plotted in Figure 67.

162

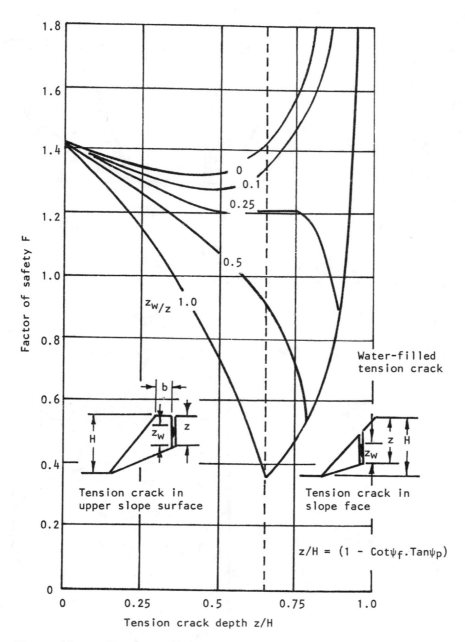

Figure 66 : Influence of tension crack depth and of depth of water
in the tension crack upon the factor of safety of a
slope. (Slope geometry and material properties as for
example on page 154).

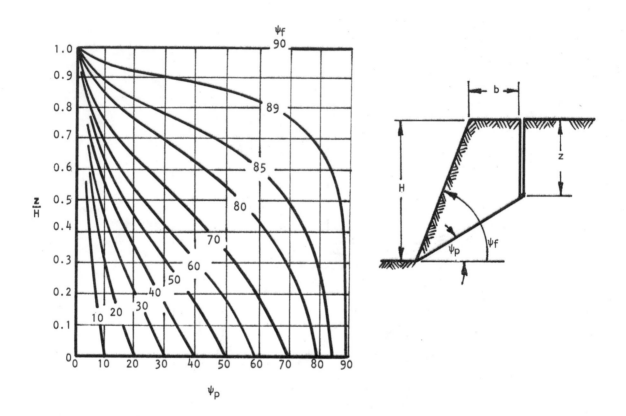

Figure 67a: Critical tension crack depth for a dry slope.

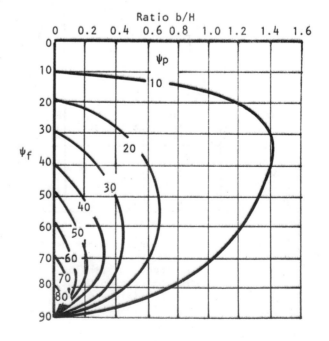

Figure 67b; Critical tension
crack location for a dry slope.

A small tension crack on the bench of a slate quarry indicating the onset of instability. Photograph by Dr. R.E.Goodman.

A large tension crack on an open pit mine bench in which considerable hroizontal and vertical movement has occurred.

Figure 66 shows that, once the water level z_w exceeds about one quarter of the tension crack depth, the factor of safety of the slope does not reach a minimum until the tension crack is water-filled. In this case, the minimum factor of safety is given by a water-filled tension crack which is coincident with the crest of the slope (b = 0).

It is most important, when considering the influence of water in a tension crack, to consider the sequence of tension crack formation and water filling. Field observations suggest that tension cracks usually occur behind the crest of a slope and, from Figure 66, it must be concluded that these tension cracks occur as a result of movement in a dry or nearly dry slope. If this tension crack becomes water-filled as a result of a subsequent rain storm, the influence of the water pressure will be in accordance with the rules laid down earlier in this chapter. The depth and location of the tension crack are, however, independent of the groundwater conditions and are defined by equations 58 and 59.

If the tension crack forms *during* heavy rain or if it is located on a pre-existing geological feature such as a vertical joint, equations 58 and 59 no longer apply. In these circumstances, when the tension crack position and depth are unknown, the only reasonable procedure is to assume that the tension crack is coincident with the slope crest and that it is water-filled.

The tension crack as an indicator of instability

Anyone who has examined excavated rock slopes cannot have failed to notice the frequent occurrences of tension cracks in the upper surfaces of these slopes. Some of these cracks have been visible for tens of years and, in many cases, do not appear to have had any adverse influence upon the stability of the slope. It is, therefore, interesting to consider how such cracks are formed and whether they can give any indication of slope instability.

In a series of very detailed model studies on the failure of slopes in jointed rocks, Barton[91] found that the tension crack was generated as a result of small shear movements within the rock mass. Although these individual movements were very small, their cumulative effect was a significant displacement of the slope surfaces - sufficient to cause separation of vertical joints behind the slope crest and to form "tension" cracks. The fact that the tension crack is caused by shear movements in the slope is important because it suggests that, when a tension crack becomes visible in the surface of a slope, it must be assumed that shear failure has initiated within the rock mass.

It is impossible to quantify the seriousness of this failure since it is only the start of a very complex progressive failure process about which very little is known. It is quite probable that, in some cases, the improved drainage resulting from the opening up of the rock structure and the interlocking of individual blocks within the rock mass could give rise to an *increase* in stability. In other cases, the initiation of failure could be followed by a very rapid decrease in stability with a consequent failure of the slope.

In summary, the authors recommend that the presence of a

tension crack should be taken as an indication of potential instability and that, in the case of an important slope, this should signal the need for detailed investigation into the stability of that particular slope.

Critical failure plane inclination

When a through-going discontinuity such as a bedding plane exists in a slope and the inclination of this discontinuity is such that it satisfies the conditions for plane failure defined on page 150, the failure of the slope will be controlled by this feature. However, when no such feature exists and when a failure surface, if it were to occur, would follow minor geological features and, in some places, would pass through intact material, how could the inclination of such a failure path be determined?

The first assumption which must be made concerns the shape of the failure surface. In a soft rock slope or a soil slope with a relatively flat slope face ($\psi_f < 45°$), the failure surface would have a circular shape. The analysis of such failure surfaces will be dealt with in Chapter 9.

In steep rock slopes, the failure surface is almost planar and the inclination of such a plane can be found by partial differentiation of equation 42 with respect to ψ_p and by equating the resulting differential to zero. For dry slopes this gives the critical failure plane inclination ψ_{pc} as

$$\psi_{pc} = \tfrac{1}{2}(\psi_f + \phi) \qquad (60)$$

The presence of water in the tension crack will cause the failure plane inclination to be reduced by up to 10% and, in view of the uncertainties associated with this failure surface, the added complication of including the influence of groundwater is not considered justified. Consequently, equation 60 can be used to obtain an estimate of the critical failure plane inclination in steep slopes which do not contain through-going discontinuity surfaces. An example of the application of this equation in the case of chalk cliff failure will be given later in this chapter.

Figure 68: Two-dimensional model used by Barton[91] for the study of slope failure in jointed rock masses.

Influence of under-cutting the toe of a slope

It is not unusual for the toe of a slope to be under-cut, either intentionally by mining or by natural agencies such as the weathering of underlying strata or, in the case of sea cliffs, by the action of waves. The influence of such undercutting on the stability of a slope is important in many practical situations and an analysis of this stability is presented here.

In order to provide as general a solution as possible, it is assumed that the geometry of the slope is that illustrated in the margin sketch. A previous failure is assumed to have left a face inclined at ψ_f and a vertical tension crack depth z_1. As a result of an under-cut of ΔM, inclined at an angle ψ_0, a new failure occurs on a plane inclined at ψ_p and involves the formation of a new tension crack of depth z_2.

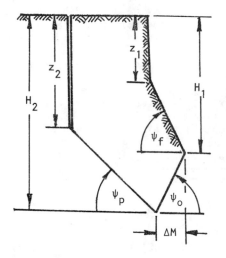

Geometry of under-cut slope

The factor of safety of this slope is given by equation 42 but it is necessary to modify the expression for the weight term as follows ;

$$W = \tfrac{1}{2}\gamma\left((H_2^2 - z_2^2)\text{Cot}\psi_p - (H_1^2 - z_1^2)\text{Cot}\psi_f + (H_1 + H_2)\Delta M\right) \quad (61)$$

Note that, for $\psi_0 > 0$,

$$\Delta M = (H_2 - H_1)\text{Cot}\psi_0 \quad (62)$$

The critical tension crack depth, for a dry under-cut slope, is given by

$$z_2 = \frac{c \cdot \text{Cos}\phi}{\gamma\text{Cos}\psi_p \cdot \text{Sin}(\psi_p - \phi)} \quad (63)$$

The critical failure plane inclination is

$$\psi_p = \tfrac{1}{2}\left(\phi + \text{Arctan}\frac{H_2^2 - z_2^2}{(H_1^2 - z_1^2)\text{Cot}\psi_f - (H_1 + H_2)\Delta M}\right) \quad (64)$$

The application of this analysis to an actual slope problem is presented at the end of this chapter.

Reinforcement of a slope

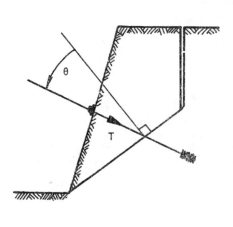

Reinforcement of a slope

When it has been established that a particular slope is unstable, it becomes necessary to consider whether it is possible to stabilise the slope by drainage or by the application of external loads. Such external loads may be applied by the installation of rock bolts or cables anchored into the rock mass behind the failure surface or by the construction of a waste rock berm to support the toe of the slope.

The factor of safety of a slope with external loading of magnitude T, inclined at an angle θ to the failure plane as shown in the sketch opposite, is approximated by :

$$F = \frac{cA + (W.\text{Cos}\psi_p - U - V.\text{Sin}\psi_p + T.\text{Cos}\theta)\text{Tan}\phi}{W.\text{Sin}\psi_p + V.\text{Cos}\psi_p - T.\text{Sin}\theta} \quad (65)$$

This equation is correct for the condition of limiting

equilibrium (F = 1) but there are certain theoretical problems in using it for other values of F. These problems are
discussed fully in Appendix 3 at the end of this book.

Analysis of failure on a rough plane

As discussed in Chapter 5, most rock surfaces exhibit a
non-linear relationship between shear strength and effective
normal stress. This relationship may be defined by Ladanyi
and Auchambault's equation (21 on page 87) or by Barton's
equation (26 on page 89). In order to apply either of these
equations to the analysis of failure on a rough surface
plane it is necessary to know the effective normal stress σ
acting on this plane.

Consider the slope geometry illustrated in Figure 62. The
effective normal stress acting on the failure surface can
be determined from equations 43 to 47 and is given by :

$$\sigma = \frac{W.\cos\psi_p - U - V.\sin\psi_p}{A} \qquad (66)$$

Alternatively, from equations 49 to 53

$$\sigma = \frac{\gamma H}{2P} \left(Q \cot\psi_p - R(P + S) \right) \qquad (67)$$

Having determined the value of σ, the shear strength τ of
the failure surface is calculated from equation 21 or 26.
The factor of safety of the slope is given by modifying
equations 42 and 48 as follows ;

$$F = \frac{\tau.A}{W.\sin\psi_p + V.\cos\psi_p} \qquad (68)$$

or

$$F = \frac{2P.\tau}{\gamma H(Q + R.S.\cot\psi_p)} \qquad (69)$$

The application of these equations is best illustrated by
means of a practical example. Consider a slope defined by
H = 100 ft., z = 50 ft., ψ_f = 60° and ψ_p = 30°. The unit
weight of the rock γ = 160 lb/ft^3 and that of water γ_w = 62.5
lb/ft^3. Two cases will be considered :

Case 1 : A drained slope in which z_w = 0

Case 2 : A slope with a water filled tension crack defined
by z_w = z.

The values given by substitution in equations 43 to 46 and
49, 50, 52 and 53 are as follows :

Case 1 : A = 100 ft^2/ft, U = 0, V = 0, W = 577350 lb/ft.
P = 1.00, Q = 0.36, R = 0 and S = 0.

Case 2 : A = 100 ft^2/ft, U = 156250 lb/ft, V = 78125 lb/ft,
W = 577350 lb/ft, P = 1.00, Q = 0.36, R = 0.195
and S = 0.25

Substitution in equations 66 and 67 gives the effective normal stress on the faiure plane as σ = 5000 lb/ft^2 for Case 1 and σ = 3049 lb/ft^2 for Case 2.

Assume that the shear strength of the surface is defined by Barton's equation (26 on page 89) with ϕ = 30°, JRC = 10 and σ_J = 720,000 lb/ft^2. Substitution of these values gives τ = 6305 lb/ft^2 for case 1 and τ = 4155 lb/ft^2 for case 2. Substituting these values of τ into equations 68 or 69 gives:

Case 1 : F = 2.18

Case 2 : F = 1.17

The application of this analysis to a practical example will be discussed later in this chapter.

Practical example number 1

Stability of porphyry slopes in a Spanish open pit mine

In order to assist the mine planning engineers in designing an extension to the Atalaya open pit operated by Rio Tinto Española in southern Spain, an analysis was carried out on the stability of porphyry slopes forming the nothern side of the pit (left hand side of the pit in the photograph reproduced in Figure 69). A summary of this analysis is presented in this example.

At the time of this design study (1969), the Atalaya pit was 260 meters deep and the porphyry slopes, inclined at an overall angle of approximately 45° as shown in Figure 70, appeared to be stable. The proposed mine plan called for deepening the pit to in excess of 300 meters and required that, if at all possible, the porphyry slopes should be left untouched. The problem, therefore, was to decide whether these slopes would remain stable at the proposed mining depth.

Since no slope failure had taken place in the porphyry slopes of the Atalaya pit, deciding upon the factor of safety of the existing slopes posed a difficult problem. Geological mapping and shear testing of discontinuities in the porphyry provided a useful guide to the possible failure modes and the range was too wide to permit the factor of safety to be determined with a reasonable degree of confidence.

Consequently, it was decided to use a technique similar to that employed by Salamon and Munro[186] for the analysis of coal pillar failures in South Africa. This method involved collecting data on slope heights and slope angles for both stable and unstable slopes in porphyry in order to establish a pattern of slope behaviour based upon full scale slopes. The data on unstable slopes had to be collected from other open pit mines in the Rio Tinto area in which failures had occurred in porphyrys judged to be similar to those in the Atalaya pit. The collected slope height versus slope angle data are plotted in Figure 71.

In order to establish the theoretical relationship between slope height and slope angle, the following assumptions are made :

    a. Because the geological mapping had failed to reveal any dominant through-going structures which could control the stability of the slopes in question but

Figure 69: Rio Tinto Espanola's Atalaya open pit mine.

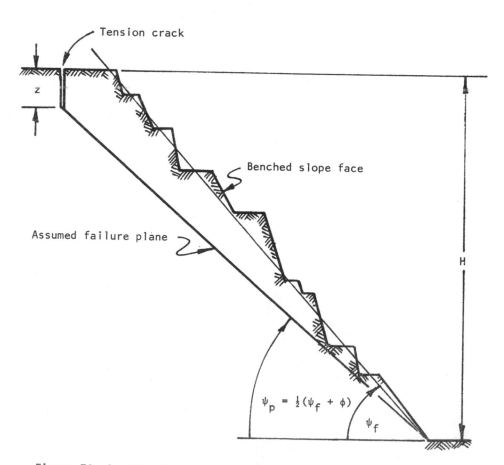

Figure 70: Section through a typical porphyry slope in the Atalaya open pit at Rio Tinto in Spain.

had revealed the presence of a number of intersecting joint sets, it was assumed that failure, if it
were to occur, would be on a composite planar
surface inclined at $\psi_p = \frac{1}{2}(\psi_f + \phi)$ as defined by
equation 60 on page 165.

b. From the shear strength data a friction angle
$\phi = 35^\circ$ was chosen as the starting point for this
analysis.

c. Because of the presence of underground workings, the
porphyry slopes were assumed to be fully drained and
it was assumed that tension cracks would occur in
accordance with the critical conditions defined in
equation 58 on page 161. It was assumed that these
tension cracks would occur in all slopes, including
those with factors of safety in excess of unity, and
the typical failure geometry is illustrated in
Figure 70.

The factor of safety for a dry slope is defined by
equation 55 on page 159 which, for the purposes of this
analysis, can be rearranged in the following form :

$$H = \frac{2c.P}{\gamma Q(F - Cot\psi_p.Tan\phi)} \qquad (70)$$

Solving equations 60,58,49 and 50 for a range of slope
angles, assuming $\gamma = 2.95$ tonnes/m^3, gives :

ψ_f	ψ_p	z/H	P	Q	H
85	60.0	0.610	0.450	0.238	1.28c/(F - 0.404)
80	57.5	0.474	0.624	0.268	1.58c/(F - 0.446)
70	52.5	0.311	0.868	0.261	2.25c/(F - 0.537)
60	47.5	0.206	1.077	0.221	3.30c/(F - 0.641)
50	42.5	0.123	1.300	0.159	5.54c/(F - 0.764)
40	37.5	0.044	1.572	0.007	152c/(F - 0.913)

The problem now is to find a value for the cohesion c which
gives the best fit for a limiting curve (F = 1) passing
through the slope height/slope angle points for unstable
slopes. The two points at $\psi_f = 61^\circ$ and 66° and H = 40m
and 35m respectively are ignored in this curve fitting since
they were identified as individual bench failures on through
going discontinuities and they would not, therefore, belong
to the same family as the other slopes.

A number of trial calculations showed that the best fit for
the F = 1 curve to the seven failure points shown in
Figure 71 is given by a cohesive strength c = 14 tonnes/m^2.

The shear strength relationship defined by c = 14 tonnes/m^2
and $\phi = 35^\circ$ has been plotted in Figure 72 which also shows
peak and residual strength values determined by shear
testing at Imperial College. Note that the shear strength
relationship determined by back analysis appears to fall
between the peak and residual shear strength values
determined in the laboratory. Care should, however, be
exercised not to draw too many conclusions from this Figure
since the range of normal stresses in the slope failures
which were back analysed was approximately 20 to 50 tonnes/m^2.
The scatter of test results in this stress range is too
large to allow a more detailed analysis of the results to be

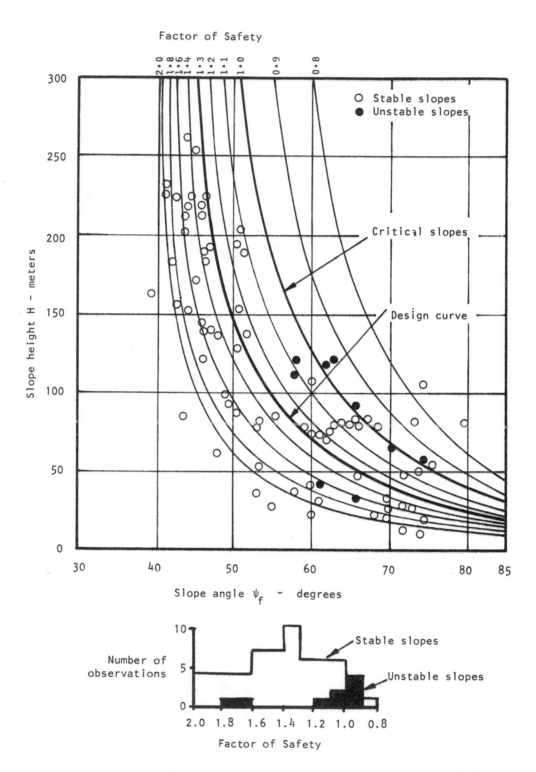

Figure 71 : Relationship between slope heights and slope angles
for porphyry slopes in the Rio Tinto area, Spain.

carried out.

Figure 72 illustrates an important historical point in slope
stability analysis since it reflects the shear testing
philosophy of the late 1960s. The importance of testing
at very low normal stress levels and of the non-linearity of
the shear strength curve had not been recognised at that
time and shear tests were frequently carried out at stress
levels which were several times higher than the normal
stresses acting in actual slopes. This philosophy was
carried over from underground rock mechanics and from
studies of intact rock fracture in which testing was usually
carried out at high normal stresses. As discussed in
Chapter 5, studies by Patton, Barton, Ladanyi and Archambault
and others have contributed greatly to our understanding of
shear strength behaviour at low normal stresses and there is
no doubt that the Rio Tinto analysis, presented on the
preceding pages, would follow slightly different lines if it
were to be repeated today. An example of an analysis using
a curvilinear shear strength relationship is given later in
this chapter. Incidentally, the authors feel that, in spite
pf the crude analysis carried out on the Rio Tinto slopes,
the engineering decisions summarised in Figure 71 are still
sound and the overall approach is still valid.

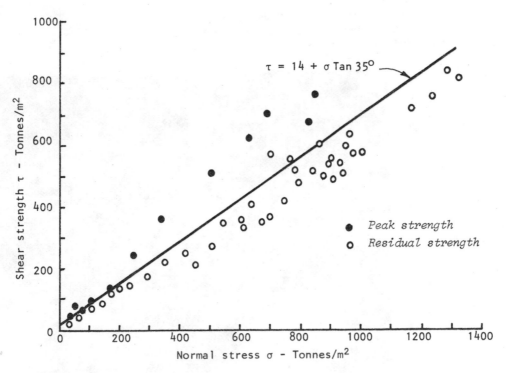

Figure 72 : Shear strength characteristics of porphyry from Rio Tinto.

Figure 73: Small scale failures of individual benches are not usually
significant in open pit mining unless they cause disruption
of haul roads.

Figure 74: The open pit designer is concerned primarily with
minimising the risk of overall slope failure.
(Kennecott Copper photograph published by Broadbent & Armstrong[187]).

Substitution of the value of c = 14 tonnes/m^2 into the
relationships for the slope height H, listed after
equation 70, gives the curves for different factors of
safety which have been plotted in Figure 71. By counting
the number of points falling between factor of safety
increments, it is possible to construct the histogram
reproduced in the lower part of Figure 71. This histogram
confirms that the seven unstable slopes are clustered around
a factor of safety F = 1 while the stable slopes show a peak
between 1.3 and 1.4.

From a general consideration of the anticipated working life
of the slope and of the possible consequences of slope
failure during the mining operations, it was concluded that
a factor of safety of 1.3 would be acceptable for the
porphyry slopes in the Atalaya pit and, hence, the design
curve presented to the mine planning engineers is that shown
as a heavy line in Figure 71. This curve shows that, for the
slope heights in excess of 250m under consideration, the
factor of safety changes very little for a change in slope
angle. It was therefore concluded that the proposed deepening of the pit would not decrease the overall stability of
the porphyry slopes, provided that no major changes in rock
mass properties or drainage conditions were encountered in
this deepening process.

Before leaving this example it is important to point out
that this analysis deals with the stability of the *overall*
pit slope and not with possible failures of *individual*
benches. In a large pit such as the Atalaya pit, it would
be totally uneconomic to attempt to analyse the stability
of each bench and, in any case, small bench failures are not
particularly important in large pits provided that they do
not influence haul roads. On the other hand, a failure of
of the wedge illustrated in Figure 70, involving
approximately 20,000 tonnes/meter of face (from equation 46
assuming γ = 2.95 tonnes/m^3) would obviously represent a
very serious problem which has to be avoided.

Practical example number 2

Investigation of the stability of a limestone quarry face

Figure 75 shows a hillside limestone quarry in the Mendip
Hills in England, owned and operated by the Amalgamated
Roadstone Corporation*. This photograph was taken in 1968
after a slope failure had occurred during a period of
exceptionally heavy rain.

In 1970, it was decided to expand the quarry facilities and
this involved the installation of new plant on the floor of
the quarry. In view of the large horizontal movement of
material which had occurred in the 1968 slope failure (as
shown in Figure 75), it was considered that an investigation
of the stability of the remainder of the slope was necessary.
This example gives a summary of the most important aspects
of this stability study, full details of which have been
published by Roberts and Hoek[148].

The 1968 failure occurred after a week or more of steady
soaking rain had saturated the area. This was followed by

---

* Now Amey Roadstone Corporation

an exceptionally heavy downpour which flooded the upper
quarry floor, filling an existing tension crack in the
slope crest. The geometry of the failure is illustrated in
Figure 76. As seen in Figure 75, the failure is basically
two-dimensional, the sliding surface being a bedding plane
striking parallel to the slope crest amd dipping into the
excavation at 20°. A vertical tension crack existed 41 feet
behind the slope crest at the time of the failure.

In order to provide shear strength data for the analysis of
the stability of the slope under which the new plant was to
be installed, it was decided to analyse the 1968 failure
by means of the graphical method described in Figure 65
on age 158. Because the dimensions of the proposed slope
were reasonably similar to those of the 1968 failure, it
was assumed that a linear shear strength relationship,
defined by a cohesive strength and angle of friction, would
be sufficiently accurate for this analysis.

Assuming a rock density of 0.08 tons/ft^3 (160 lb/ft^3)
and a water density of 0.031 tons/ft^3 (62.4 lb/ft^3) ;

Weight of sliding mass $W = \frac{1}{2}\gamma(XH - Dz) = 404.8$ tons/ft.

Horizontal water force $V = \frac{1}{2}\gamma_w.z_w^2 \quad = 65.5$ tons/ft.

Uplift of water force $U = \frac{1}{2}\gamma_w.z_w.A \quad = 110.8$ tons/ft.

From the force diagram, Figure 77a, the shear strength
mobilised in the 1968 failure can be determined and this is
plotted in Figure 77b.

From an examination of the surface upon which failure had
taken place in 1968, it was concluded that the friction
angle was probably 20° ± 5°. This range of friction angles
and the associated cohesive strengths, shown in Figure 77b,
are used to determine the stability of the overall slopes in
this illustrative example.

Having established the range of shear strengths mobilised
in the 1968 failure, these values were now used to check the
stability of the 210 ft. high slopes under which the new
plant was to be installed. The geometry of the slope
analysed is illustrated in Figure 78 which shows that, in
order to provide for the worst possible combination of
circumstances, it was assumed that the bedding plane on
which the 1968 slide had occurred daylights in the toe of
the slope.

Figure 79 shows typical force diagrams for dry and saturated
slopes, assuming a slope face angle $\psi_f = 50°$ and a friction
angle $\phi = 25°$. A range of such force diagrams was
constructed and the factors of safety determined from these
constructions are plotted in Figure 80. In this figure the
full lines are for a friction angle $\phi = 20°$, considered the
most probable value, while the dashed lines define the
influence of a 5° variation on either side of this angle.

It is clear from Figure 80 that 58° slopes are unstable
under the heavy rainfall conditions which caused the slopes
to become saturated in 1968. Drainage of the slope, particularly the control of surface water which could enter the
top of an open tension crack, is very beneficial but, since
it cannot be guaranteed that such drainage could be fully
effective, it was recommended that the slope should also be
benched back to an overall angle of 45°.

Figure 75 : Air photograph of Amalgamated Roadstone Corporation's
Batts Combe limestone quarry in Somerset, England,
showing details of the 1968 slope failure (Roberts
and Hoek[148]).

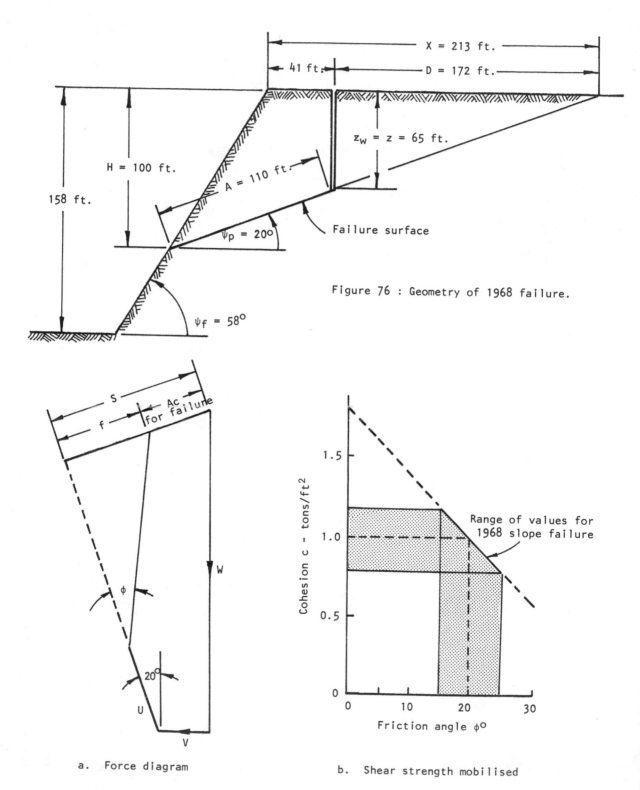

Figure 76 : Geometry of 1968 failure.

a.   Force diagram

b.   Shear strength mobilised

Figure 77 : Determination of shear strength mobilised in 1968 failure.

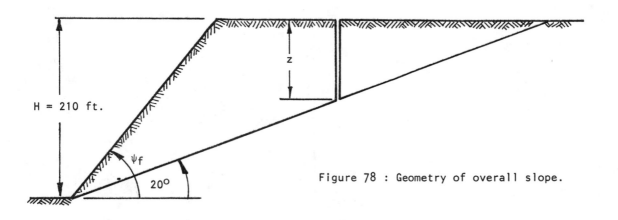

Figure 78 : Geometry of overall slope.

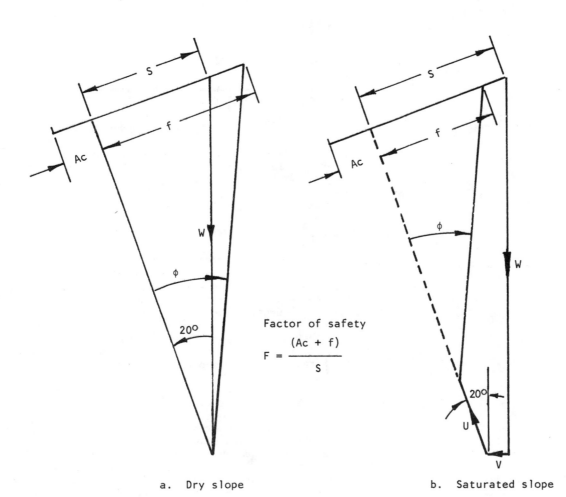

Factor of safety

$$F = \frac{(Ac + f)}{S}$$

a. Dry slope

b. Saturated slope

Figure 79 : Force diagrams for design of overall quarry slopes.

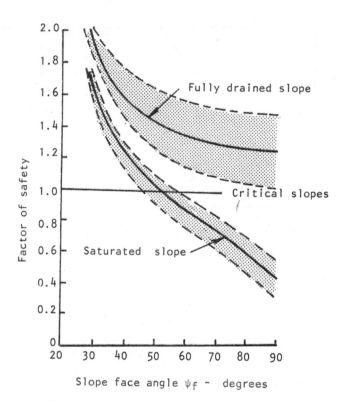

Figure 80 : Factor of safety for dry and saturated
slopes with different face angles.

Practical example number 3

Choice of remedial measures for critical slopes

When a slope above an important highway or a haul road in an
open pit mine or above a civil engineering structure is
found to be potentially unstable, an urgent decision on the
effective and economical remedial measures which can be
employed is frequently required. The following example
illustrates one of the methods which may be adopted in
arriving at such a decision. Although this example is
hypothetical, it is based upon a number of actual problems
with which the authors have been concerned.

The first stage in the analysis is obviously to check that
the slope is actually unstable and whether any remedial
measures are required. Sometimes it is obvious that a
potential failure problem exists because failures of limited
extent have already taken place in part of the slope - this
was the case in the quarry stability problem discussed in
practical example 2. In other cases, a suspicion may have
been created by failures of adjacent slopes or even by the
fact that the engineer in charge of the slope has recently
attended a conference on slope stability and has become
alarmed about the stability of the slopes in his charge.
Whatever the cause, once a doubt has been cast upon the
stability of an important slope, it is essential that its

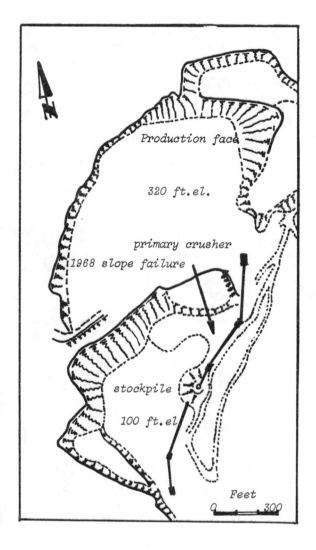

Figure 81a : Batts Combe quarry plan in 1970 showing the location of the 1968 slope failure which destroyed part of the conveyor system (see Figure 75).

Figure 81b : Plan of proposed benching of lower slopes in Batts Combe quarry. Slopes are to be benched back to an overall slope of 45° with pre-splitting of final faces. Surface drainage on upper quarry floor and provision of horizontal drain holes in bench faces if piezometers indicate high sub-surface water levels.

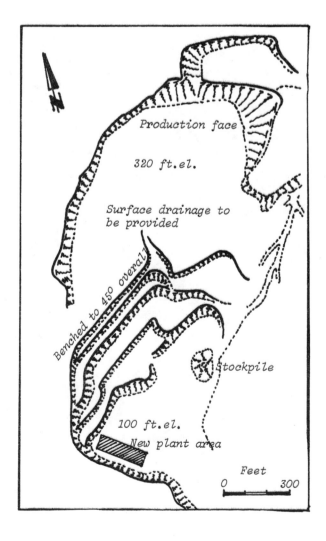

overall stability should be investigated and that
appropriate remedial measures should be implemented if these
are found to be necessary.

Consider the following examples :

A 60 meter high slope has an overall face angle of 50°, made
up from three 20 meter benches with 70° faces. The slope
is in reasonably fresh granite but several sets of steeply
dipping joints are visible and sheet jointing similar to
that described by Terzaghi[17] is evident. The slope is in an
area of high rainfall intensity and low seismicity. An
acceleration of 0.08g has been suggested as the maximum to
which this slope is likely to be subjected. A small slide
in a nearby slope has caused attention to be focussed onto
this particular slope and concern has been expressed in case
a major slide could occur and could result in serious damage
to an important civil engineering structure at the foot of
the slope. The rock slope engineer called in to examine the
problem is required to assess both the short and the long
term stability of the slope and to recommend appropriate
remedial measures, should these prove necessary. No previous
geological or engineering studies have been carried out on
this slope and no boreholes are known to exist in the area.

Faced with this problem and having no geological or engineering data from which to work, the first task of the rock
slope engineer is to obtain a representative sample of
structural geology data in order that the most likely failure
mode can be established. Time would not usually allow a
drilling programme to be mounted, even if drilling equipment
and operators of the required standard were readily available
in the area. Consequently, the collection of structural data
would have to be based upon surface mapping as described in
Chapter 4, page 65. In some circumstances, this mapping can
be carried out using the photogammetric techniques described
on page 69.

It is assumed that structural mapping is carried out and
that the following geometrical and structural features have
been identified :

Feature	dip °	dip direction °
Overall slope face	50	200
Individual benches	70	200
Sheet joint	35	190
Joint set J1	80	233
Joint set J2	80	040
Joint set J3	70	325

The stereoplot of this data is given in Figure 82 and
a friction circle of 30° is included on this plot. Note that,
although the three joint sets provide a number of steep
release surfaces which would allow blocks to separate from
the rock mass, none of their lines of intersection, ringed
in Figure 82, fall within the zone designated as potentially
unstable. On the other hand, the sheet joint great circle
passes through the zone of potential instability and, since
its dip direction is close to that of the slope face, it can
be concluded that the most likely failure mode is that
involving a planar slide on the sheet joint surface in the
direction indicated in Figure 82.

The stability check carried out in Figure 82 suggests that

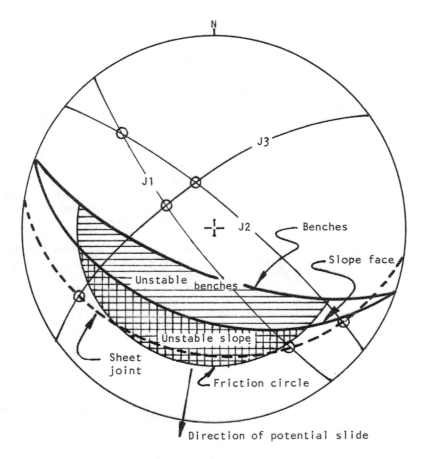

Figure 82 : Stereoplot of geometrical and geological data
for example number 3.

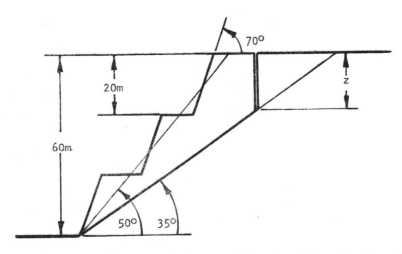

Figure 83 : Geometry assumed for two-dimensional analysis
of the slope defined in Figure 82.

both the overall slope and the individual benches are potentially unstable and it is therefore clearly necessary to carry out further checks on both.

Because of the presence of the three steeply dipping joint sets, the possibility of a tension crack forming in the upper surface of the slope must be regarded as high. One possible failure mode is that illustrated as Model I in Figure 84. This theoretical model assumes that a tension crack occurs in the dry state in the most critical position and that this crack is filled to depth z_w with water during a period of exceptionally heavy rain. A simultaneous earthquake subjects the slope to an acceleration of 0.08g. The factor of safety of this slope is given by equation 71 in Figure 84, derived from equation 42 on page 152 with provision for the earthquake loading.

In deriving equation 71, it has been assumed that the acceleration induced by an earthquake can be replaced by an equivalent static force of αW. This is almost certainly a gross over-simplification of the actual loading to which the slope is subjected during an earthquake[188,189] but it is probable that it tends to over-estimate the loading and hence it errs on the side of safety. In view of the poor quality of the other input data in this problem, there is no justification for attempting to carry out a more detailed analysis of earthquake loading.

Since no boreholes exist on this hypothetical site, the sub-surface groundwater conditions are totally unknown. To allow for the possibility that substantial sub-surface water may be present, an alternative theoretical model is proposed. This is illustrated as Model II in Figure 84 and, again this model includes the effect of earthquake loading.

Having decided upon the most likely failure mode and having proposed one or more theoretical models to represent this failure mode, the rock engineer is now in a position to substitute a range of possible values into the factor of safety equations in order to determine the sensitivity of the slope to the different conditions to which it is likely to be subjected.

Summarising the available input data :

Slope height	H	= 60m
Overall slope angle	ψ_f	= 50°
Bench face angle	ψ_f	= 70°
Bench height	H	= 20m
Failure plane angle	ψ_p	= 35°
Rock density	γ	= 2.6 tonnes/m³
Water density	γ_w	= 1.0 tonnes/m³
Earthquake acceleration	α	= 0.08g

Substituting in equations 71 and 72 :

Overall slopes Model I

$$F = \frac{80.2c + (1850 - 40.1z_w - 0.287z_w{}^2)\operatorname{Tan}\phi}{1529 + 0.410z_w^2} \qquad (74)$$

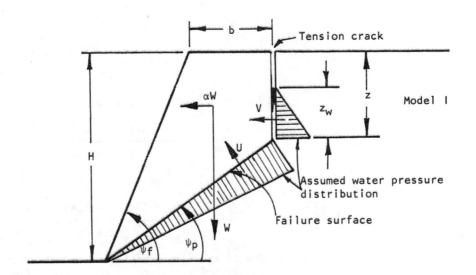

$$F = \frac{cA + \left(W(Cos\psi_p - \alpha Sin\psi_p) - U - V\,Sin\psi_p\right)Tan\,\phi}{W(Sin\psi_p + \alpha Cos\psi_p) + V\,Cos\psi_p} \qquad (71)$$

Where

$$z = H(1 - \sqrt{Cot\psi_f.Tan\psi_p}) \qquad (58)$$

$$A = (H - z)\,Cosec\psi_p \qquad (43)$$

$$W = \tfrac{1}{2}\gamma H^2\left((1 - (z/H)^2)Cot\psi_p - Cot\psi_f\right) \qquad (46)$$

$$U = \tfrac{1}{2}\gamma_w.z_w.A \qquad (44)$$

$$V = \tfrac{1}{2}\gamma_w.z_w^2 \qquad (45)$$

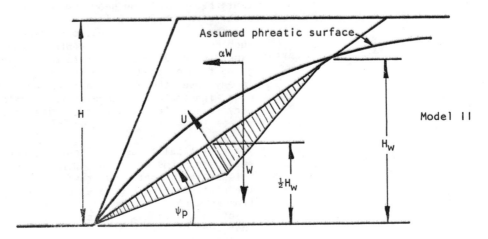

$$F = \frac{cA + \left(W(Cos\psi_p - \alpha Sin\psi_p) - U\right)Tan\,\phi}{W(Sin\psi_p + \alpha Cos\psi_p)} \qquad (72)$$

Where

$$U = \tfrac{1}{4}\gamma_w.H_w^2 Cosec\,\psi_p \qquad (73)$$

Figure 84 : Theoretical models for example number 3.

Overall slopes Model II

$$F = \frac{104.6c + (2132 - 0.436H_w^2)\tan\phi}{1762} \qquad (75)$$

Individual benches Model I

$$F = \frac{17.6c + (287.1 - 8.8z_w - 0.287z_w^2)\tan\phi}{237.3 + 0.410z_w^2} \qquad (76)$$

Individual benches Model II

$$F = \frac{34.9c + (428.0 - 0.436H_w^2)\tan\phi}{353.7} \qquad (77)$$

One of the most useful studies which can be carried out with the aid of equations 74 to 77 is to find the shear strength which would have to be mobilised for failure of the overall slope or for the individual benches. Figure 85 gives the results of such a study and the numbered lines on this plot represent the following conditions :

1 - Overall slope, Model I, dry, $z_w = 0$.
2 - Overall slope, Model I, saturated, $z_w = z = 14m$.
3 - Overall slope, Model II, dry, $H_w = 0$.
4 - Overall slope, Model II, saturated, $H_w = H$ 60m.
5 - Individual bench, Model I, dry, $z_w = 0$.
6 - Individual bench, Model I, saturated, $z_w = z = 9.9m$.
7 - Individual bench, Model II, dry, $H_w = 0$.
8 - Individual bench, Model II, saturated, $H_w = H = 20m$.

The reader may feel that a consideration of all these possibilities is unnecessary but it is only coincidental that, because of the geometry of this particular slope, the shear strength values found happen to fall reasonably close together. In other cases, one of the conditions may be very much more critical than the others and it would take a very experienced slope engineer to detect this condition without going through the calculations required to produce Figure 85. In any case, these calculations should only take about one hour with the aid of an electronic calculator and this is a very reasonable investment of time when lives and property may be in danger.

The elliptical figure in Figure 85 surrounds the range of shear strengths which the authors consider to be reasonable for partially weathered granite. These values are based on the plot given in Figure 43 on page 115 and on experience from working with granites. Note that a high range of friction angles has been chosen because experience suggests that even heavily kaolinised granites (point 11 in Figure 43) exhibit high friction values because of the angular nature of the mineral grains.

It is clear from Figure 85 that simultaneous heavy rain and earthquake loading could cause the shear strength required to maintain stability to rise to a dangerous level. Considering the rapidity with which granite weathers, particularly in tropical environments, with a consequent reduction in available cohesive strength, these results suggest that the slope is unsafe and that steps should be taken to increase

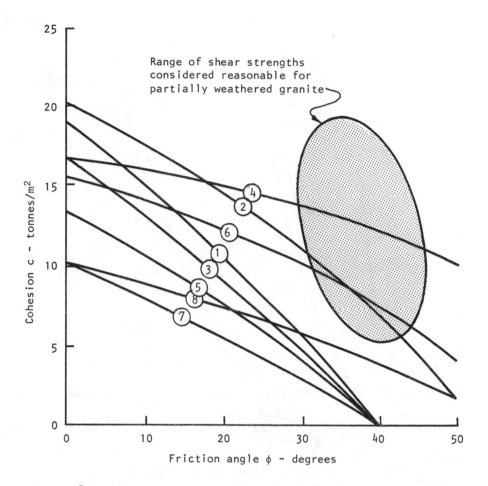

Figure 85 : Shear strength mobilised for failure of slope considered
in practical example number 3.

its stability.

Four basic methods for improving the stability of the slope
can be considered. These methods are the following :

    a.   Reduction of slope height.
    b.   Reduction of slope face inclination.
    c.   Drainage of slope.
    d.   Reinforcement of slope with bolts and cables.

In order to compare the effectiveness of these different
methods, it is assumed that the sheet joint surface has a
cohesive strength of 10 tonnes/m^2 and a friction angle of
$35°$. The increase in factor of safety for a reduction in
slope height, slope angle and water level can be found by
altering one of these variables at a time in equations 71
and 72. The influence of reinforcing the slope is obtained
by modifying these equations as in equations 78 and 79 on
page 188.

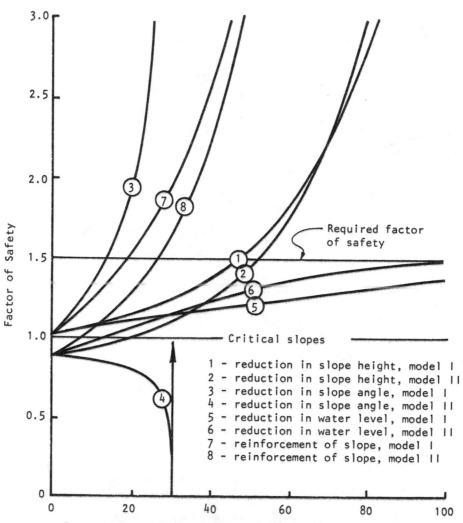

Figure 86 : Comparison between alternative methods of increasing stability of overall slope considered in example 3.

Model 1

$$F = \frac{cA + \left(W(\cos\psi_p - \alpha\sin\psi_p) - U - V\sin\psi_p + T\cos\theta\right)\tan\phi}{W(\sin\psi_p + \alpha\cos\psi_p) + V\cos\psi_p - T\sin\theta} \qquad (78)$$

Model II

$$F = \frac{cA + \left(W(\cos\psi_p - \alpha\sin\psi_p) - U + T\cos\theta\right)\tan\phi}{W(\sin\psi_p + \alpha\cos\psi_p) - T\sin\theta} \qquad (79)$$

Where T is the total reinforcing force applied by bolts or cables and θ is the inclination of this force to the normal to the failure surface, as illustrated in the sketch opposite.

Figure 86 gives the results of the comparison between the different methods which could be considered for increasing the stability of the overall slope. In each case, the change is expressed as a percentage of the total range of the variable (H = 60m, ψ_f = 50°, z_w/z = 1, H_w = 60m) except for the reinforcing load. This is expressed as a percentage of the weight of the wedge of rock being supported. In calculating the effect of the reinforcement, it has been assumed that the cables or bolts are installed horizontally, i.e. θ = 55°. The influence of the inclination θ upon the reinforcing load required to produce a factor of safety of 1.5 is shown in the graph given in the margin.

Figure 86 shows that reduction in slope height (lines 1 and 2) only begins to show significant benefits once the height reduction exceeds about 40%. In many practical situations, a height reduction of this magnitude may be totally impossible, particularly when the slope has been cut into a mountainside. In any case, once one has reduced the slope height by 40%, more than 60% of the mass of the material forming the unstable wedge will have been removed and it would then be worth removing the rest of the wedge and the remains of the problem. Obviously, this solution would be very expensive but it does have the merit of providing a permanent solution to the problem.

Reducing the angle of the slope face can be very effective, as shown by line 3, but it can also be very dangerous as shown by line 4. This wide variation in response to what is normally regarded as a standard method for improving the stability of a slope raises a very interesting problem which deserves more detailed examination.

Equations 58 and 46 (Figure 84) both contain the term $\cot\psi_f$ and hence both z and W are decreased as the slope face angle ψ_f is reduced. A reduction in tension crack depth reduces both water forces U and V and the final result is a dramatic increase in factor of safety for a decrease in slope face inclination. Note that, if the tension crack occurs *before* the slope is flattened, the tension crack z will remain unaltered at 14m and the water forces U and V will remain at their maximum values. Under these conditions the factor of safety will still be increased for a reduction in slope face inclination but not to the same extent as shown by line 3 in Figure 86.

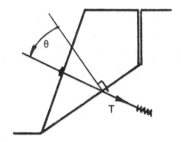

Slope reinforcement

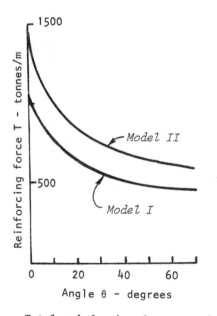

Total reinforcing force required for a factor of safety of 1.5.

In the case of Model II in Figure 84, it is only the weight term which is altered by the reduction in slope angle and, because the uplift force term U Tan ϕ is greater than the cohesive force cA, the factor of safety actually *reduces* as the slope face is flattened. As the slope face angle approaches the failure plane angle, the thin sliver of material resting on the failure plane will be floated off by the excess water force U. Although many practical arguments could be put forward to show that this extreme behaviour would be very unlikely, the example does illustrate the danger of indiscriminate alteration of the slope geometry without having first considered the possible consequences. The practical conclusion to be drawn from this discussion is that, if Model II in Figure 84 is representative of the conditions which exist in an actual slope, partial flattening of the slope would achieve no useful purpose. The wedge of rock resting on the failure plane would have to be *removed entirely* if it was decided that flattening the slope was the only means to be used for increasing the stability.

Drainage of the slope is probably the cheapest remedial measure which can be employed and, as shown in Figure 86, complete drainage, if this could be achieved, would increase the factor of safety to very nearly the required value. Unfortunately, complete drainage can never be achieved and hence, in this particular slope, drainage would have to be supplemented by some other remedial measure such as bolting in order to produce an acceptable level of safety. In any event, nothing would be lost by the provision of some drainage and the authors would recommend careful consideration of surface water control and also the drilling of horizontal drain holes to intersect the potential failure surface.

Reinforcing the slope by means of bolts or cables may create a useful illusion of safety but unless the job is done properly, the result could be little more than an illusion. In order to achieve a factor of safety of 1.5, assuming the bolts or cables to be installed in a horizontal plane, the total force required amounts to about 500 tonnes per meter of slope length. In other words, the complete reinforcement of a 100 meter face would require the installation of 500 one tonne capacity cables. Simultaneous drainage of the slope, even if only partially successful, would reduce this number by about half but reinforcing a slope of this size would obviously be a very costly process.

Considering all the facts now available, the authors would offer the following suggestions to the engineer responsible for the hypothetical slope which has been under discussion in this example :

a. Immediate steps should be taken to have a series of standpipe piezometers installed in vertical drillholes from the upper slope surface or from one of the benches. The importance of groundwater has been clearly demonstrated in the calculations which have been presented and it is essential that further information on possible groundwater flow patterns should be obtained.

b. If diamond drilling equipment of reasonable quality is readily available, the vertical holes for the piezometers should be cored. A geologist should be present during this drilling programme and should log the core

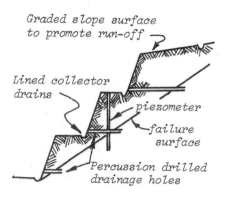

Lined surface water diversion drain

Graded slope surface to promote run-off

Lined collector drains

piezometer

failure surface

Percussion drilled drainage holes

Possible drainage measures

immediately upon removal from the core barrel. Particular attention should be given to establishing the exact position of the sheet joint or joints so that an accurate cross-section of the slope can be constructed. If adequate diamond drilling equipment is not available, the piezometer holes may be percussion drilled.

c. As soon as the piezometers are in position and it has been demonstrated that groundwater is present in the slope, horizontal drainholes should be percussion drilled into the bench faces to intersect the sheet joints. These holes can be drilled at an initial spacing of about 10 meters and their effectiveness checked by means of the piezometers. The hole spacing can be increased or decreased according to the water level changes observed in the piezometers.

d. During this groundwater control programme, a careful examination of the upper surface of the slope should be carried out to determine whether open tension cracks are present and whether any recent movements have taken place in the slope. Such movements would be detected by cracks in concrete or plaster or by displacements of vertical markers such as telephone poles. If the upper surface of the slope is covered by overburden soil, it may be very difficult to detect cracks and it may be necessary to rely upon the reports of persons resident on or close to the top of the slope.

e. Depending upon the findings of this examination of the upper slope surface, a decision could then be made on what surface drainage measures should be taken. If open tension cracks are found, these should be filled with gravel and capped with an impermeable material such as clay. The existence of such cracks should be taken as evidence of severe danger and serious consideration should be given to remedial measures in addition to drainage.

f. Further geological mapping to confirm the geological structure of the slope, together with evidence on groundwater and tension cracks, would provide information for a review of the situation to decide upon the best means of permanent stabilisation, in addition to the drainage measures which have already been implemented.

Practical example number 4

Chalk cliff failure induced by undercutting

Hutchinson[156] has described the details of a chalk cliff failure at Joss Bay on the Isle of Thanet in England. This failure, induced by the undercutting action fo the sea, provides an interesting illustration of the analysis of undercutting on page 166 and Hutchinson's data is reanalysed on the following pages.

The failure is illustrated in the photograph reproduced in Figure 87 and a cross-section, reconstructed from the paper by Hutchinson, is given in Figure 88. Apart from a thin capping of overburden and the presence of a few flint bands

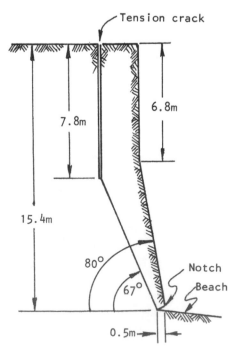

Figure 87: Chalk cliff failure at Joss Bay, Isle of Thanet, England. (Photograph reproduced with permission of Dr. J.N.Hutchinson, Imperial College, London.)

Figure 88: Cross section of chalk cliff failure at Joss Bay.

the chalk is reasonably uniform. Bedding is within one degree of horizontal and two major joint sets, both almost vertical, are present. The cliff face is parallel to one of these joint sets.

Measurement of water levels in wells near the coast together with the lack of face seepage caused Hutchinson to conclude that the chalk mass in which the failure occurred could be taken as fully drained. Since the failure does not appear to have been associated with a period of exceptionally heavy rain, as was the case of the quarry failure discussed in example number 2, the possibility of a water-filled tension crack is considered to be remote and will not be included in this analysis. The interested reader is left to check the influence of various water pressure distributions upon the behaviour of this slope.

Laboratory tests on samples taken from the cliff face gave a density of 1.9 tonnes/m^3 and a friction angle of about 42° for the peak strength and 30° for the residual strength. The cohesive strength ranged from 13.3 tonnes/m^2 for the peak strength to zero for the residual strength. Since this failure can be classed as a fall in which relatively little movement may have taken place before failure, as opposed to a slide in which the shear strength on the failure plane is reduced to its residual value by movements before the actual failure, there is considerable justification for regarding the peak strength of the chalk as relevant for this analysis. The purpose of this analysis is to determine the shear strength mobilised in the actual failure and to compare this with the laboratory values.

Summarising the available input data :

H	- slope height ($H_1 = H_2$)	15.4m
z_1	- original tension crack depth	6.8m
z_2	- new tension crack depth	7.8m
ΔM	- depth of undercut	0.5m
ψo	- inclination of undercut	0°
ψf	- slope face angle	80°
ψ_p	- failure plane angle	67°

The effective friction angle of the chalk mass can be determined by rearranging equation 64 on page 166.

$$\phi = 2\psi_p - Arctan \frac{H_2^2 - z_2^2}{(H_1^2 - z_1^2)Cot\psi_f - (H_1 + H_2)\Delta M} \qquad (80)$$

Substitution gives $\phi = 49.9^o$.

This value is significantly higher than the friction angle of 42° measured on laboratory specimens but the influence of the roughness of the actual failure surface must be taken into account in comparing the results. The photograph reproduced in Figure 87 shows this surface is very rough indeed and the difference between the laboratory value and the friction angle mobilised in the failure is not surprising.

The cohesion mobilised at failure can be estimated by rearranging equation 63 on page 166 :

$$c = \frac{\gamma z_2 Cos\psi_p.Sin(\psi_p - \phi)}{Cos \phi} \qquad (81)$$

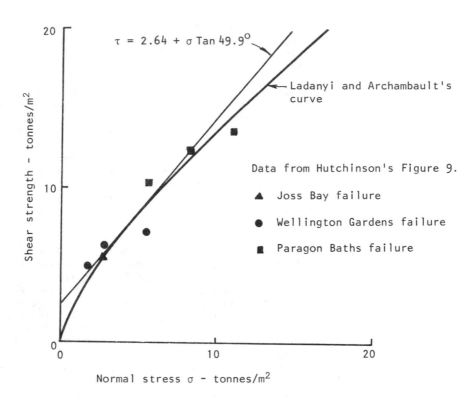

$\tau = 2.64 + \sigma \,\mathrm{Tan}\,49.9^{\circ}$

Ladanyi and Archambault's curve

Data from Hutchinson's Figure 9.

▲ Joss Bay failure

● Wellington Gardens failure

■ Paragon Baths failure

Figure 89 : Relationship between shear strength and normal stress for chalk cliff failures analysed by Hutchinson[156].

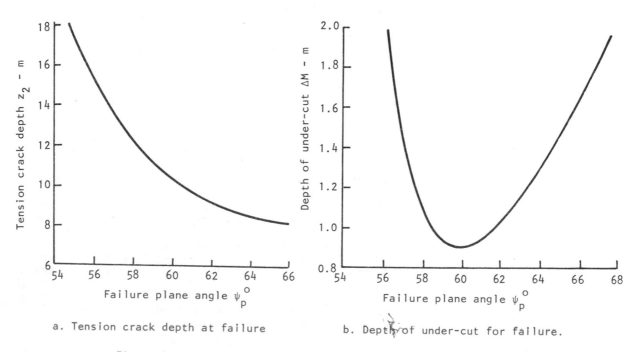

a. Tension crack depth at failure

b. Depth of under-cut for failure.

Figure 90 : Tension crack depths and under-cut depths required for failure for different failure plane inclinations.

Substituting $z_2 = 7.8m$, $\psi_p = 67^o$ and $\phi = 49.9^o$ gives $c = 2.64$ tonnes/m^2. As would be expected, this value is considerably lower than the value of $c = 13.3$ tonnes/m^2 determined by laboratory shear tests on intact chalk.

Hutchinson's paper (Figure 9) contains further data on the shear strength mobilised in chalk cliff failures at Wellington Gardens and Paragon Baths and this data is reproduced in Figure 89. The dashed curve, calculated from Ladanyi and Archambault's equation (28 on page 104) is a good fit to this rock mass strength data and it will be seen that the line defined by $\tau = 2.64 + \sigma \text{Tan } 49.9^o$ is a tangent to the dashed curve. The evidence presented in Figure 89 suggests that the values of cohesion and friction angle determined from the failure geometry illustrated in Figure 88 are reasonable.

Before leaving this example, it is instructive to consider what will happen to the Joss Bay cliff as the sea continues to undercut its toe. The input data for the next step in the failure process is now as follows :

H = slope height ($H_1 = H_2$)   15.4m
z_1 = original tension crack depth   7.8m
ψ_f = slope face angle   67°
c = cohesive strength of chalk mass   2.65 tonnes/m
ϕ = friction angle of chalk mass   49.9^o

The unknowns in this analysis are

z_2 = new tension crack depth
ψ_p = failure plane angle
ΔM = depth of undercut

Since there are three unknowns and only two equations (63 and 64) the solution to this problem is obtained in the following manner :

a. From equation 63, the depth of the tension crack z_2 is calculated for a range of possible failure plane angles (ψ_p). The results of this calculation are plotted in Figure 90. Since z_2 must lie between z_1 and H, Figure 90 shows that the angle of the failure plane ψ_p must lie between 67° and 56°.

b. Rearranging equation 64 gives :

$$\Delta M = \frac{(H^2 - z_1^2)\text{Cot}\psi_f}{2H} - \frac{H^2 - z_2^2}{2H \text{ Tan}(2\psi_p - \phi)} \qquad (82)$$

Solving for a range of corresponding values of ψ_p and z_2 gives the depth of the undercut shown in Figure 90b.

It is clear, from this figure, that a further cliff failure will occur when the undercut reaches a depth of approximately 0.9m and that the corresponding failure plane angle will be $\psi_p = 60^o$ and the new tension crack depth will be $z_2 = 10.2m$. This new failure geometry is illustrated in Figure 91.

The consequence of the cliff failure illustrated in Figure 91 is serious for property owners on the cliff-top and, hence, the problem of stabilisation of the cliff face must be considered.

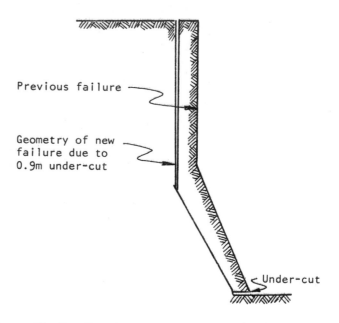

Figure 91: Predicted geometry of next cliff
failure due to under-cutting.

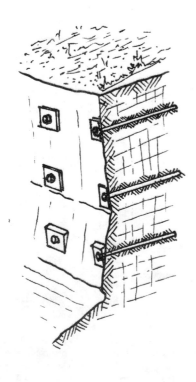

*Suggested reinforcement
of cliff face using 5m ×
5m pattern of 5m long bolts.*

In the analysis which has been presented on the previous
pages, it has been assumed that the chalk mass is dry. The
presence of groundwater in the cliff, and particularly in
the tension crack, would result in a serious reduction of
face stability. Consequently, the first step in stabilising
the slope is to ensure that it remains completely drained.
Attention to surface water to ensure that pools cannot
collect near the slope crest is important and, if possible,
horizontal drains should be drilled into the face to allow
free drainage of any water which does find its way into the
rock mass which would be involved in a further failure.

In view of the fact that the stability of this slope is so
sensitive to undercutting, it is tempting to suggest that
this undercutting should be prevented by the provision of a
concrete wall along the toe of the cliff. In some cases
this may be a practical solution but, in others, it may be
impossible to provide a secure foundation for such a wall.

Assuming that the protection of the toe of the cliff
illustrated in Figure 91 is not possible, the only remaining
alternative is to stabilise the cliff face by reinforcement.
Since the mass of material involved in any further failure
will be relatively small - say 50 tonnes per meter of slope -
the stabilising force need not be very large.

Because of the dilatant nature of the failure process, it is
suggested that the most effective reinforcement would be
provided by fully grouted bolts or cables lightly tensioned
to ensure that all contacts were closed. The onset of
failure would induce tension in this reinforcement which
would inhibit further failure. It is suggested that the
load capacity of these bolts or cables should be approximately

25% of the mass of material which could fail and hence a pattern of 20 tonne bolts or cables, each 5 meters long, installed on a 5 meter grid, should provide adequate reinforcement for this slope. If the chalk mass is closely jointed so that there is a danger of ravelling between washers, the pattern can be changed to a closer grid of lower capacity bolts or, alternatively, corrosion resistant wire mesh can be clamped beneath the washers to reinforce the chalk surface.

Practical example number 5

Block sliding on clay layers

In mining a horizontal bedded coal deposit, sliding of blocks of material on clay seams occurs during periods of high rainfall. The clay seams have a high montmorillonite content and have been slickensided by previous shear displacements, consequently very low residual shear strength values of $c = 0$ and $\phi = 10^{\circ}$ are considered appropriate for the analysis of failures[114].

The geometry of the block is illustrated in the margin sketch and it is assumed that the clay seam is horizontal.

- H is the height of the block
- ψ_f is the angle of the face of the block
- B is the distance of a vertical crack behind the crest of the slope
- z_w is the depth of water in the tension crack
- W is the weight of the block
- V is the horizontal force due to water in the tension crack
- U is the uplift force due to water pressure on the base

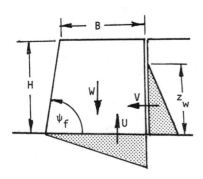

Geometry of failure and water pressure distribution

The factor of safety of the block is given by :

$$F = \frac{(W - U) \, \text{Tan} \, \phi}{V} \tag{83}$$

Where

$W = \gamma BH - \frac{1}{2}\gamma H^2 \text{Cot} \, \psi_f$

$U = \frac{1}{2}\gamma_w z_w (B + H \text{Cot} \, \psi_f)$

$V = \frac{1}{2}\gamma_w z_w^2$

γ is the density of the rock

γ_w is the density of water

Hence

$$F = \frac{\left((2B/H - \text{Cot}\psi_f) - \gamma_w/\gamma \cdot z_w/H \cdot B/H(1 + \text{Cot}\psi_f)\right) \text{Tan} \, \phi}{\gamma_w/\gamma \cdot (z_w/H)^2} \tag{84}$$

This equation has been solved for a range of values of B/H and z_w/H, assuming $\psi_f = 80^{\circ}$, $\gamma_w/\gamma = 0.4$ and $\phi = 10^{\circ}$, and the results are plotted in Figure 92.

The extreme sensitivity of the factor of safety to changes in water level depth in the tension crack is evident in this figure. This means that drainage, even if it is not very efficient, should do a great deal to improve the stability of the slope. Horizontal holes through the base of the block may be the most economical drainage system and the effectiveness of such drain-holes can be checked by monitoring the movement across a tension crack before and after drilling of the hole.

It must be emphasised that it is not the *quantity* of water which is important in this case but the *pressure* of water in the tension crack. Hence, in a low permeability rock mass, the drain may only produce a trickle of water but, if it has reduced the water pressure in the tension crack, it will stabilise the slope.

As the ratio B/H decreases, the weight of the block W decreases and hence the factor of safety of the slope decreases, as shown in Figure 92. This is a factor over which the mine operator has no control but it would be interesting to relate the ratio B/H to the frequency of observed block failures.

An increase in the block face angle ψf results in an increase in the weight of the block and improvement in slope stability. This suggests that, for this type of failure, the face angle should be kept as steep as possible.

In some cases of block failure on clay seams, the blocks have been observed to move *up-hill* on seams dipping into the rock mass. This requires a combination of a very low friction angle in the clay and a relatively high water level in the tension crack. Since the shear strength of the clay cannot be altered, drainage is the obvious remedial measure in such cases.

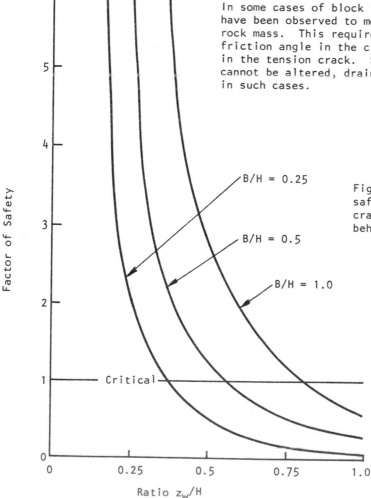

Figure 92: Sensitivity of factor of safety to depth of water in tension crack and distance of tension crack behind slope crest.

Chapter 7 references

186. SALAMON, M.D.G. and MUNRO, A.H. A study of the strength of coal pillars. *J. South African Inst. Min. Metall.* Vol.68, 1967, pages 55-67.

187. BROADBENT, C.D. and ARMSTRONG, C.W. Design and application of microseismic devices. *Proc. 5th Canadian Symposium on Rock Mechanics.* 1968.

188. IDRISS, I.M. and SEED, H.B. The response of earth banks during earthquakes. *Report Soil Mech. and Bituminous Materials Lab. University of California. Berkley.* April 1966.

189. FINN, W.D.L. Static and dynamic stresses in slopes *Proc. 1st Congress, Intnl. Soc. Rock Mechanics, Lisbon.* 1966. Vol. 2, page 167.

Chapter 8 : Wedge failure

Introduction

The previous chapter was concerned with slope failure resulting from sliding on a single planar surface dipping into the excavation and striking parallel or nearly parallel to the slope face. It was stated that the plane failure analysis is valid provided that the strike of the failure plane is within ± 20° of the strike of the slope face. This chapter is concerned with the failure of slopes in which structural features upon which sliding can occur strike across the slope crest and where sliding takes place along the line of intersection of two such planes.

This problem has been extensively discussed in geotechnical literature and the authors have drawn heavily upon the work of Londe, John, Wittke, Goodman and others listed in references 190-200 at the end of this chapter. The reader who has examined this literature may have been confused by some of the mathematics which have been presented. It must, however, be appreciated that our understanding of the subject has grown rapidly over the past decade and that many of the simplifications which are now clear were not at all obvious when some of these papers were written. The basic mechanics of failure are very simple but, because of the large number of variables involved, the mathematical treatment of the mechanics can become very complex unless a very strict sequence is adhered to in the development of the equations.

In this chapter, the basic mechanics of failure involving the sliding of a wedge along the line of intersection of two planar discontinuities are presented in a form which the non-specialist reader should find easy to follow. Unfortunately the very simple equations which are presented to illustrate the mechanics are of limited practical value because the variables used to define the wedge geometry cannot easily be measured in the field. Consequently, the second part of the chapter deals with the stability analysis in terms of the dips and dip directions of the planes and the slope face. In the transformation of the equations which is necessary in order to accommodate this information the basic mechanics becomes obscure but it is hoped that the reader should be able to follow the logic involved in the development of these equations.

In the chapter itself, the discussion is limited to the case of the sliding of a simple wedge such as that illustrated in Figure 93, acted upon by friction, cohesion and water pressure. The influence of a tension crack and of external forces due to bolts, cables or seismic accelerations results in a significant increase in the complexity of the equations and, since it would only be necessary to consider these influences on the fairly rare occasions when the critical slopes are being examined, the complete solution to the problem has been presented in Appendix I at the end of the book. The analytical treatment of the problem presented in part III of the Appendix is particularly suitable for processing by computer and, once the reader has understood the basic mechanics of the problem, he should have no difficulty in having this general solution programmed for almost any type of computer, including the desk top machines which are now available.

Figure 93 : A typical wedge failure involving
sliding along the line of intersection of two planar discontinuities.

Figure 94 : Sets of intersecting discontinuities can
sometimes give rise to the formation of
families of wedge failures.

(Photograph reproduced with permission of Mr. K.M. Pare)

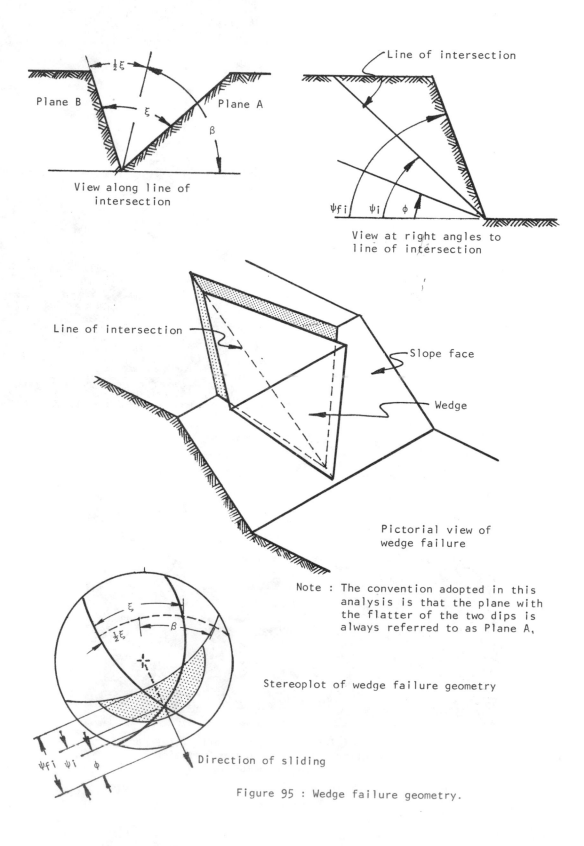

View along line of intersection

Line of intersection

View at right angles to line of intersection

Line of intersection

Slope face

Wedge

Pictorial view of wedge failure

Note : The convention adopted in this analysis is that the plane with the flatter of the two dips is always referred to as Plane A.

Stereoplot of wedge failure geometry

Direction of sliding

Figure 95 : Wedge failure geometry.

Definition of wedge geometry

Typical wedge failures are illustrated in Figures 93 and 94 which show, in the one case, the through-going planar discontinuities which are normally assumed for the analytical treatment of this problem and, in the other case, the wedge formed by sets of closely spaced structural features. In the latter case, the analytical treatment would still be based upon the assumption of through-going planar features although it would have to be realised that the definition of the dips and dip directions and the locations of these planes may present practical difficulties. The failure illustrated in Figure 94 would probably have involved the fairly gradual ravelling of small loose blocks of rock and it is unlikely that this failure was associated with any violence. On the other hand, the failure illustrated in Figure 93 probably involved a fairly sudden fall of a single wedge which would only have broken up on impact and which would, therefore, constitute a threat to anyone working at the toe of the slope.

The geometry of the wedge, for the purpose of analysing the basic mechanics of sliding, is defined in Figure 95. Note that, throughout this book, the *flatter* of the two planes is called Plane A while the *steeper* plane is called Plane B.

As in the case of plane failure, a condition of sliding is defined by $\psi_{fi} > \psi_i > \phi$, where ψ_{fi} is the inclination of the slope face, measured in the view at right angles to the line of intersection, and ψ_i is the dip of the line of intersection. Note the ψ_{fi} would only be the same as ψ_f, the true dip of the slope face, if the dip direction of the line of intersection was the same as the dip direction of the slope face.

Analysis of wedge failure

The factor of safety of the wedge defined in Figure 95, assuming that sliding is resisted by *friction only* and that the friction angle ϕ is the same for both planes, is given by

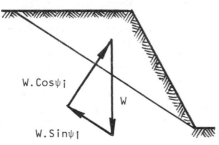

W.Cosψ_i

W.Sinψ_i

*View at right angles to
line of intersection*

$$F = \frac{(R_A + R_B)\ \mathrm{Tan}\,\phi}{W.\mathrm{Sin}\psi_i} \qquad (85)$$

where R_A and R_B are the normal reactions provided by planes A and B as illustrated in the sketch opposite.

In order to find R_A and R_B, resolve horizontally and vertically in the view along the line of intersection :

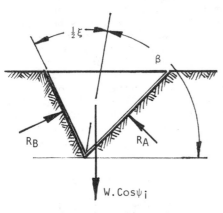

$\frac{1}{2}\xi$

β

R_B R_A

W.Cosψ_i

*View along line of
intersection*

$$R_A.\mathrm{Sin}(\beta - \tfrac{1}{2}\xi) = R_B.\mathrm{Sin}(\beta + \tfrac{1}{2}\xi) \qquad (86)$$

$$R_A.\mathrm{Cos}(\beta - \tfrac{1}{2}\xi) - R_B.\mathrm{Cos}(\beta + \tfrac{1}{2}\xi) = W.\mathrm{Cos}\psi_i \qquad (87)$$

Solving for R_A and R_B and adding :

$$R_A + R_B = \frac{W.\mathrm{Cos}\psi_i.\mathrm{Sin}\beta}{\mathrm{Sin}\tfrac{1}{2}\xi} \qquad (88)$$

Hence

$$F = \frac{\mathrm{Sin}\beta}{\mathrm{Sin}\tfrac{1}{2}\xi} \cdot \frac{\mathrm{Tan}\phi}{\mathrm{Tan}\psi_i} \qquad (89)$$

In other words :

$$F_w = K.F_p \qquad\qquad (90)$$

where F_w is the factor of safety of a wedge supported by friction only. F_p is the factor of safety of a plane failure in which the slope face is inclined at ψ_{fi} and the failure plane is inclined at ψ_i.

K is the wedge factor which, as shown by equation 89, depends upon the included angle of the wedge and upon the angle of tilt of the wedge. Values for the wedge factor K, for a range of values of β and ξ are plotted in Figure 96.

As shown in the stereoplot given in Figure 95, measurement of the angles β and ξ can be carried out on the great circle, the pole of which is the point representing the line of intersection of the two planes. Hence, a stereoplot of the features which define the slope and the wedge geometry can provide all the information required for the determination of the factor of safety. It should, however, be remembered that the case which has been dealt with is very simple and that, when different friction angles and the influence of cohesion and water pressure are allowed for, the equations become more complex. Rather than develop these equations in terms of the angles β and ξ, which cannot be measured directly in the field, the more complete analysis is presented in terms of directly measurable dips and dip directions.

Before leaving this simple analysis, the reader's attention is drawn to the important influence of the wedging action as the included angle of the wedge decreases below 90°. The increase by a factor of 2 or 3 on the factor of safety determined by plane failure analysis is of great practical importance. Some authors have suggested that a plane failure analysis is acceptable for *all* rock slopes because it provides a lower bound solution which has the merit of being conservative. Figure 96 shows that this solution is so conservative as to be totally uneconomic for most practical slope designs. It is therefore recommended that, where the structural features which are likely to control the stability of a rock slope do not strike parallel to the slope face, the stability analysis should be carried out by means of the three-dimensional methods presented in this book or published by the authors listed in references 190 to 202 at the end of this chapter.

Wedge analysis including cohesion and water pressure

Figure 97 shows the geometry of the wedge which will be considered in the following analysis. Note that the upper slope surface in this analysis can be obliquely inclined with respect to the slope face, thereby removing a restriction which has been present in all the stability analyses which have been discussed so far in this book. The total height of the slope, defined in Figure 97, is the total difference in vertical elevation between the upper and lower extremities of the line of intersection along which sliding is assumed to occur.

The water pressure distribution assumed for this analysis is based upon the hypothesis that the wedge itself is impermeable and that water enters the top of the wedge

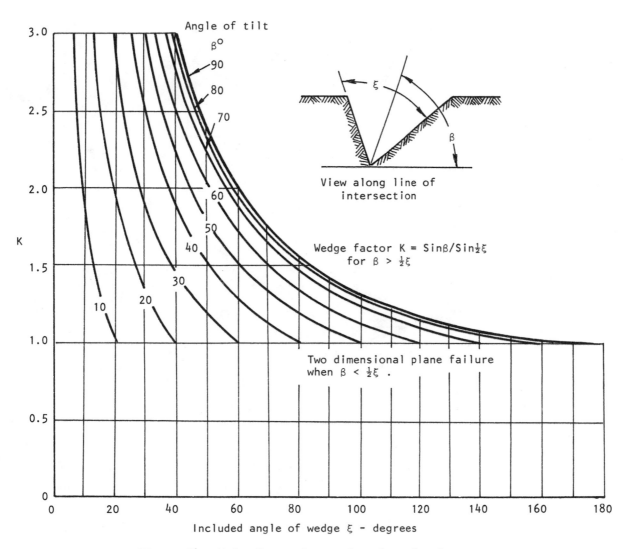

Figure 96 : Wedge factor K as a function of wedge geometry.

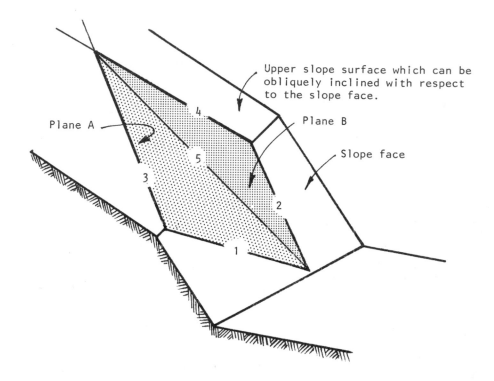

a. Pictorial view of wedge showing the numbering of intersection
   lines and planes.

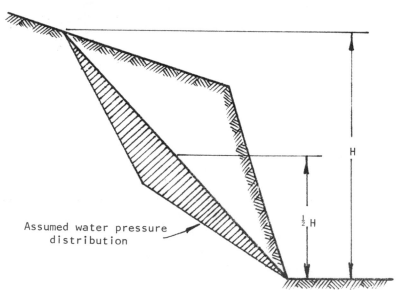

b. View normal to the line of intersection 5 showing the
    total wedge height and the water pressure distribution.

Figure 97 : Geometry of wedge used for stability analysis including the influence
            of cohesion and of water pressure on the failure surfaces.

along lines of intersection 3 and 4 and leaks from the
slope face along lines of intersection 1 and 2. The resulting pressure distribution is shown in Figure 97b - the
maximum pressure occuring along the line of intersection 5
and the pressure being zero along lines 1,2,3 and 4. This
water pressure distribution is believed to be representative
of the extreme conditions which could occur during very
heavy rain.

The numbering of the lines of intersection of the various
planes involved in this problem is of extreme importance
since total confusion can arise in the analysis if these
numbers are mixed-up. The numbering used throughout this
book is as follows :

> 1 - intersection of plane A with the slope face
> 2 - intersection of plane B with the slope face
> 3 - intersection of plane A with upper slope surface
> 4 - intersection of plane B with upper slope surface
> 5 - intersection of planes A and B

It is assumed that sliding of the wedge always takes place
along the line of intersection numbered 5.

The factor of safety of this slope is derived from the
detailed analysis presented in part III of Appendix I at
the end of this book and is :

$$F = \frac{3}{\gamma H}(c_A.X + c_B.Y) + (A - \frac{\gamma_w}{2\gamma}.X)\text{Tan}\phi A + (B - \frac{\gamma_w}{2\gamma}.Y)\text{Tan}\phi_B \tag{91}$$

where

> c_A and c_B are the cohesive strengths of planes A and B
>
> ϕ_A and ϕ_B are the angles of friction on planes A and B
>
> γ is the unit weight of the rock
>
> γ_w is the unit weight of water
>
> H is the total height of the wedge (see Figure 97)
>
> X,Y,A, and B are dimensionless factors which depend
> upon the geometry of the wedge.

$$X = \frac{\text{Sin}\theta_{24}}{\text{Sin}\theta_{45}.\text{Cos}\theta_{2na}} \tag{92}$$

$$Y = \frac{\text{Sin}\theta_{13}}{\text{Sin}\theta_{35}.\text{Cos}\theta_{1nb}} \tag{93}$$

$$A = \frac{\text{Cos}\psi_a - \text{Cos}\psi_b.\text{Cos}\theta_{na.nb}}{\text{Sin}\psi_5.\text{Sin}^2\theta_{na.nb}} \tag{94}$$

$$B = \frac{\text{Cos}\psi_b - \text{Cos}\psi_a.\text{Cos}\theta_{na.nb}}{\text{Sin}\psi_5.\text{Sin}^2\theta_{na.nb}} \tag{95}$$

where ψ_a and ψ_b are the dips of planes A and B respectively
and ψ_5 is the dip of the line of intersection 5.

The angles required for the solution of these equations
can most conveniently be measured on a stereoplot of the
data which defines the geometry of the wedge and the slope.

Consider the following example :

Plane	dip°	dip direction°	Properties
A	45	105	$\phi_A = 20°$, $c_A = 500 lb/ft^2$
B	70	235	$\phi_B = 30°$, $c_B = 1000 lb/ft^2$
Slope face	65	185	$\gamma = 160 lb/ft^3$
Upper surface	12	195	$\gamma_w = 62.5 lb/ft^3$

The total height of the wedge H = 130 feet.

The stereoplot of the great circles representing the four planes involved in this problem is presented in Figure 98 and all the angles required for the solution of equations 92 to 95 are marked in this figure.

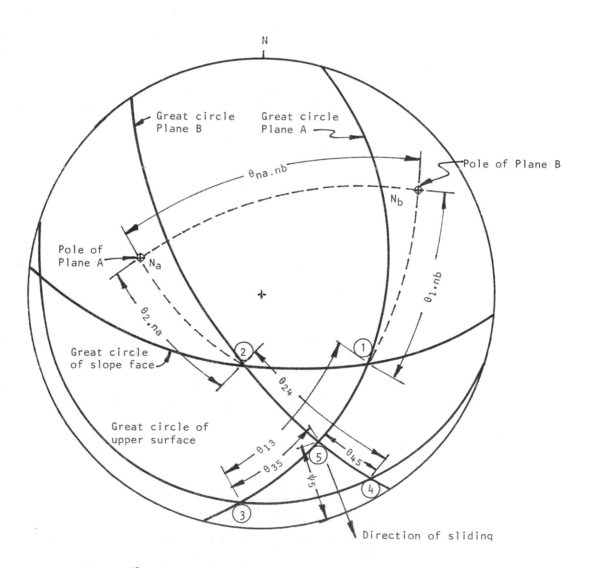

Figure 98 : Stereoplot of data required for wedge stability analysis.

WEDGE STABILITY CALCULATION SHEET

INPUT DATA	FUNCTION VALUE	CALCULATED ANSWER
$\psi_a = 45°$ $\psi_b = 70°$ $\psi_5 = 31.2°$ $\theta na.nb = 101°$	$Cos\,\psi_a = 0.7071$ $Cos\,\psi_b = 0.3420$ $Sin\,\psi_5 = 0.5180$ $Cos\,\theta na.nb = -0.191$ $Sin\,\theta na.nb = 0.982$	$A = \dfrac{Cos\psi_a - Cos\psi_b\cdot Cos\theta na.nb}{Sin\psi_5\cdot Sin^2\theta na.nb} = \dfrac{0.7071 + 0.342 \times 0.191}{0.5180 \times 0.9636} = 1.5475$ $B = \dfrac{Cos\psi_b - Cos\psi_a\cdot Cos\theta na.nb}{Sin\psi_5\cdot Sin^2\theta na.nb} = \dfrac{0.3420 + 0.7071 \times 0.191}{0.5180 \times 0.9636} = 0.9557$
$\theta_{24} = 65°$ $\theta_{45} = 25°$ $\theta_{2.na} = 50°$	$Sin\,\theta_{24} = 0.9063$ $Sin\,\theta_{45} = 0.4226$ $Cos\,\theta_{2.na} = 0.6428$	$X = \dfrac{Sin\theta_{24}}{Sin\theta_{45}\cdot Cos\theta_{2.na}} = \dfrac{0.9063}{0.4226 \times 0.6428} = 3.3363$
$\theta_{13} = 62°$ $\theta_{35} = 31°$ $\theta_{1.nb} = 60°$	$Sin\,\theta_{13} = 0.8829$ $Sin\,\theta_{35} = 0.5150$ $Cos\,\theta_{1.nb} = 0.5000$	$Y = \dfrac{Sin\theta_{13}}{Sin\theta_{35}\cdot Cos\theta_{1.nb}} = \dfrac{0.8829}{0.5150 \times 0.500} = 3.4287$
$\phi_A = 30°$ $\phi_B = 20°$ $\gamma = 160\ \text{lb/ft}^3$ $\gamma_w = 62.5\ \text{lb/ft}^3$ $c_A = 500\ \text{lb/ft}^2$ $c_B = 1000\ \text{lb/ft}^2$ $H = 130\ \text{ft}$	$Tan\,\phi_A = 0.5773$ $Tan\,\phi_B = 0.3640$ $\gamma_w/2\gamma = 0.1953$ $3c_A/\gamma H = 0.0721$ $3c_B/\gamma H = 0.1442$	$F = \dfrac{3c_A}{\gamma H}\cdot X + \dfrac{3c_B}{\gamma H}\cdot Y + \left(A - \dfrac{\gamma_w}{2\gamma}\cdot X\right)Tan\phi_A + \left(B - \dfrac{\gamma_w}{2\gamma}\cdot Y\right)Tan\phi_B$ $F = 0.2405 + 0.4944 + 0.8934 - 0.3762 + 0.3478 - 0.2437 = 1.3562$

Determination of the factor of safety is most conveniently carried out on a calculation sheet such as that presented on page 208. Setting the calculations out in this manner not only enables the user to check all the data but it also shows how each variable contributes to the overall factor of safety. Hence, if it is required to check the influence of the cohesion on both planes falling to zero, this can be done by setting the two groups containing the cohesion values c_A and c_B to zero, giving a factor of safety of 0.62. Alternatively, the effect of drainage can be checked by putting the two water pressure terms (i.e. those containing γ_w) to zero, giving F = 1.98.

As has been emphasised in previous chapters, this ability to check the sensitivity of the factor of safety to changes in material properties or in slope loading is probably as important as the ability to calculate the factor of safety itself.

Wedge stability charts for friction only

If the cohesive strength of the planes A and B is zero and the slope is fully drained, equation 77 reduces to

$$F = A.\mathrm{Tan}\phi_A + B.\mathrm{Tan}\phi_B \qquad (96)$$

The dimensionless factors A and B are found to depend upon the dips and dip directions of the two planes and values of these two factors have been computed for a range of wedge geometries and the results are presented as a series of charts on the following pages.

In order to illustrate the use of these charts, consider the following example :

	dip°	dip direction°	friction angle°
Plane A	40	165	35
Plane B	70	285	20
Differences	30	120	

Hence, turning to the charts headed "Dip difference 30°" and reading off the values of A and B for a difference in dip direction of 120°, one finds that

A = 1.5 and B = 0.7

Substitution in equation 96 gives the factor of safety as F = 1.30. The values of A and B give a direct indication of the contribution which each of the planes makes to the total factor of safety.

Note that the factor of safety calculated from equation 96 is independent of the slope height, the angle of the slope face and the inclination of the upper slope surface. This rather surprising result arises because the weight of the wedge occurs in both the numerator and denominator of the factor of safety equation and, for the friction only case, this term cancels out, leaving a dimensionless ratio which defines the factor of safety (see equation 89 on page 202). This simplification is very useful in that it enables the user of these charts to carry out a very quick check on the stability of a slope on the basis of the dips and dip

directions of the discontinuities in the rock mass into
which the slope has been cut. A example of such an analysis
is presented later in this chapter.

Many trial calculations have shown that a wedge having a
factor of safety in excess of 2.0, as obtained from the
friction only stability charts, is unlikely to fail under
even the most severe combination of conditions to which the
slope is likely to be subjected. Consider the example
discussed on pages 207 to 209 in which the factor of safety
for the worst conditions (zero cohesion and maximum water
pressure) is 0.62. This is 50% of the factor of safety of
1.24 for the friction only case. Hence, had the factor of
safety for the friction only case been 2.0, the factor of
safety for the worst conditions would have been 1.0, assuming
that the ratio of the factors of safety for the two cases
remains constant.

On the basis of such trial calculations, the authors suggest
that the friction only stability charts can be used to define
those slopes which are adequately stable and which can be
ignored in subsequent analyses. Such slopes, having a factor
of safety in excess of 2.0, pass into category 3 in the
chart presented in Figure 6 on page 14. Slopes with a
factor of safety, based upon friction only, of less than 2.0
must be regarded as potentially unstable and pass into
category 4 of Figure 6, i.e. these slopes require further
detailed examination.

In many practical problems involving the design of the
overall slopes of an open pit mine or the cuttings for a
highway, it will be found that these friction only stability
charts provide all the information which is required. It is
frequently possible, having identified a potentially
dangerous slope, to eliminate the problem by a slight
re-alignment of the pit benches or of the road cutting.
Such a solution is clearly only feasible if the potential
danger is recognised before excavation of the slope is
started and the main use of the charts is during the site
investigation and preliminary planning stage of a slope
project.

Once a slope has been excavated, these charts will be of
limited use since it will be fairly obvious if the slope
is unstable. Under these conditions, a more detailed study
of the slope will be required and use would then have to be
made of the method described on pages 203 to 208 or of one
of the methods described in Appendix I. In the authors'
experience, relatively few slopes require this detailed
analysis and the reader should beware of wasting time on
such an analysis when the simpler methods presented in this
chapter would be adequate. A full stability analysis may
look very impressive in a report but, unless it has enabled
the slope engineer to take positive remedial measures, it
may have served no useful purpose.

Practical example of wedge analysis

During the feasibility study for a proposed open pit mine,
the mine planning engineer responsible for the pit layout
has requested guidance on the maximum safe angles which may
be used for the design of the overall pit slopes. Extensive
geological mapping of outcrops on the site together with a
certain amount of core logging has established that there

WEDGE STABILITY CHARTS FOR FRICTION ONLY

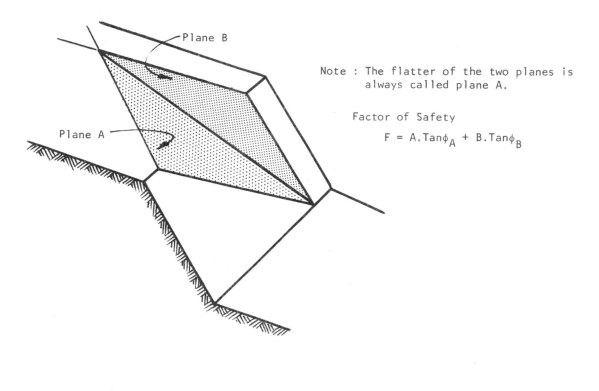

Note : The flatter of the two planes is
always called plane A.

Factor of Safety

$$F = A.Tan\phi_A + B.Tan\phi_B$$

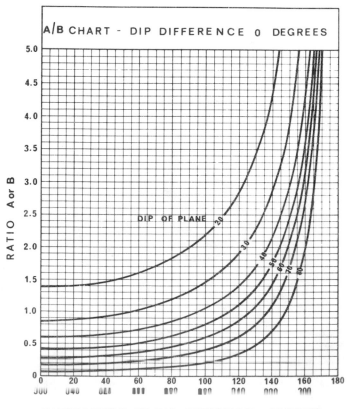

DIFFERENCE IN DIP DIRECTION - DEGREES

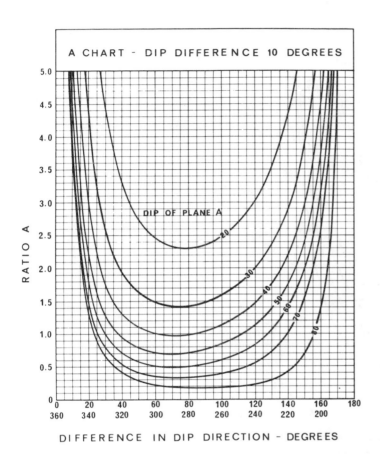

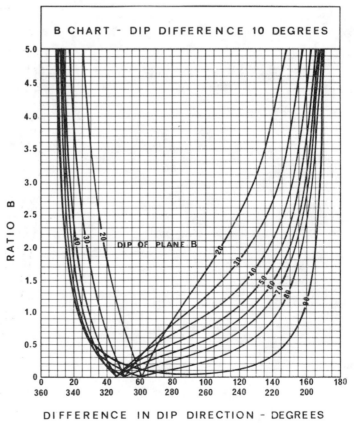

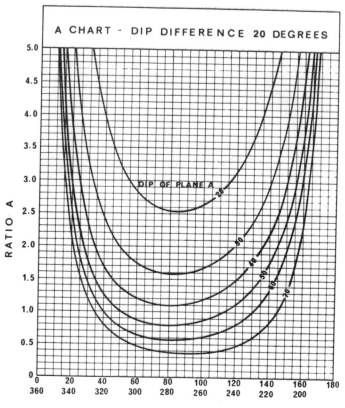

DIFFERENCE IN DIP DIRECTION - DEGREES

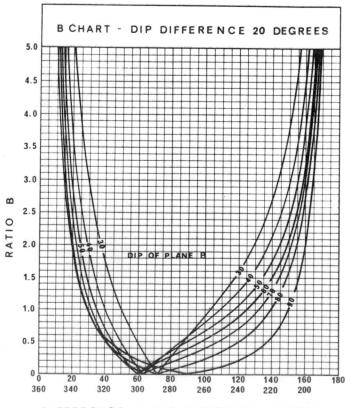

DIFFERENCE IN DIP DIRECTION - DEGREES

214

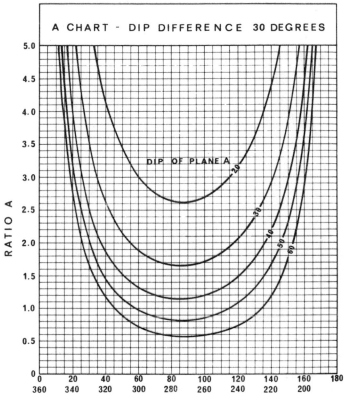

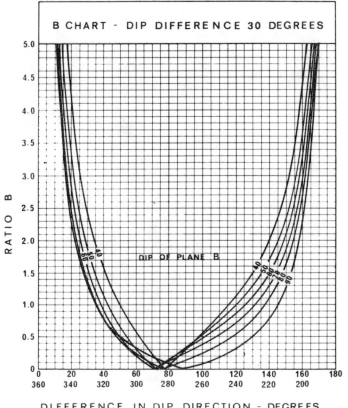

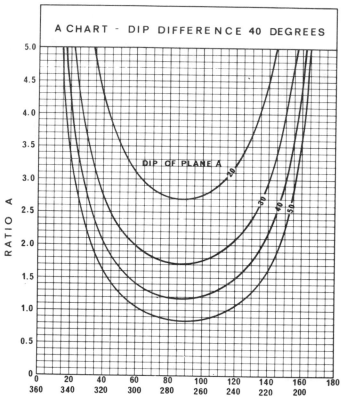

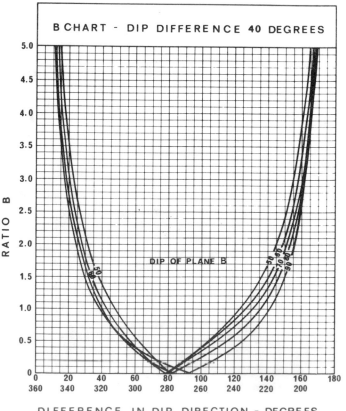

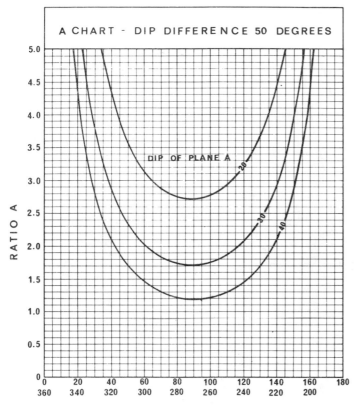

DIFFERENCE IN DIP DIRECTION - DEGREES

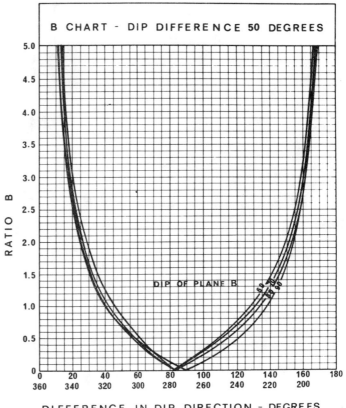

DIFFERENCE IN DIP DIRECTION - DEGREES

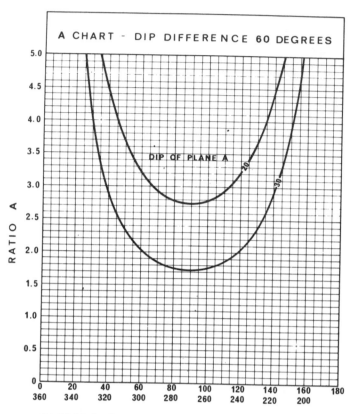

DIFFERENCE IN DIP DIRECTION - DEGREES

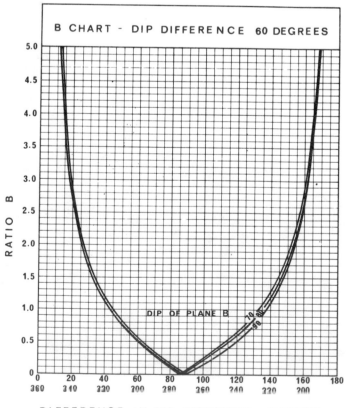

DIFFERENCE IN DIP DIRECTION - DEGREES

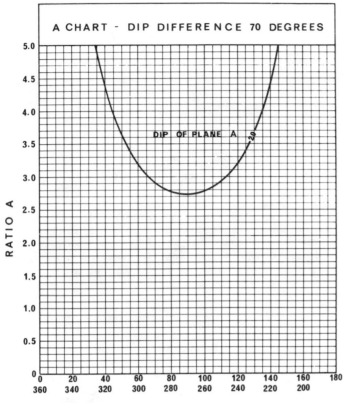

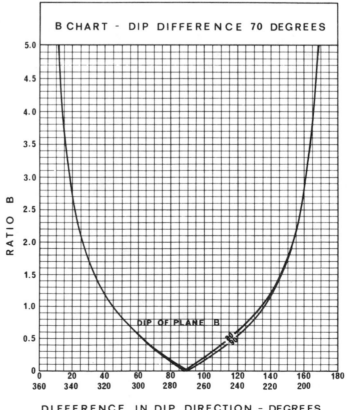

are five sets of geological discontinuities in the rock
mass surrounding the ore body. The dips and dip directions
of these discontinuities are as follows :

Discontinuity set	dip°	dip direction°
1	66 ± 2	298 ± 2
2	68 ± 6	320 ± 15
3	60 ± 16	360 ± 10
4	58 ± 6	76 ± 6
5	54 ± 4	118 ± 2

Note that, because this mapping covers the entire site which
extends over several acres, the scatter in the dip and dip
direction measurements is considerable and must be taken
into account in the analysis. This scatter can be reduced
by more detailed mapping in specific locations, for example,
Figure 21 on page 61, but this may not be possible because
of shortage of time or because suitable outcrops are not
available.

Figure 99 shows the pole locations for these five sets of
discontinuities. Also shown on this figure are the extent
of the scatter in the pole measurements and the great circles
corresponding to the most probable pole positions. The
dashed figure surrounding the great circle intersections is
obtained by rotating the stereoplot to find the extent to
which the intersection point is influenced by the scatter
around the pole points. The technique described on page 47
is used to define this dashed figure. The intersection of
great circles 2 and 5 has been excluded from the dashed
figure because it defines a line of intersection dipping at
less than 20° and this is considered to be less than the
angle of friction.

The factors of safety for each of the discontinuity intersections is determined from the wedge charts (some inter-
polation is necessary) and the values are given in the
circles over the intersection points. Because all of the
planes are relatively steep, some of the factors of safety
are dangerously low (assuming a friction angle of 30°).
Since it is unlikely that slopes with a factor of safety of
less that 0.5 could be economically stabilised, the only
practical solution is to cut the slopes in these regions to
a flat enough overall angle to eliminate the problem.

The construction given in Figure 100 is that which is used
to find the maximum safe slope angle for different parts of
the pit. This construction involves positioning the great
circle representing the slope face for a particular dip
direction in such a way that the unstable region (shaded)
is avoided. The maximum safe slope angles are marked around
the perimeter of this figure and their positions correspond
to the position on the pit perimeter.

Figure 101 shows the suggested pit layout as presented to
the mine planning engineer by the rock slope engineer. The
pit floor shape and elevation is that originally specified
by the mine planning engineer on the basis of the shape of
the ore body. This layout is for the overall slopes only,
no benches or haul road have been included. It must also
be pointed out that the slopes on the north-eastern side of

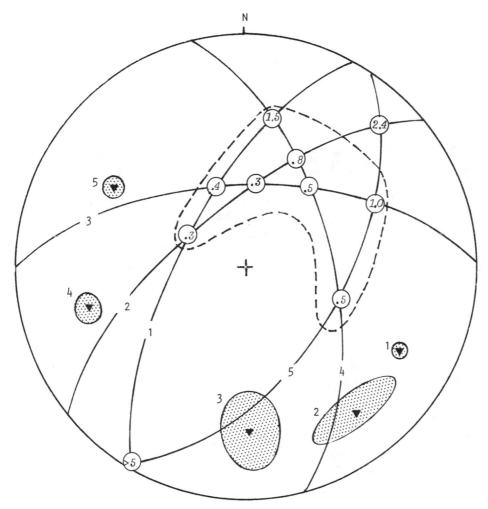

Notes :
a. Black triangles mark most likely position of poles of five
   sets of discontinuities present in rock mass.
b. Shaded area surrounding pole position defines extent of
   scatter in measurements.
c. Factors of safety for each combination of discontinuities
   given in *italics* in circle over corresponding intersection.
d. Dashed line surrounds area of potential instability.

Figure 99 : Stereoplot of geological data for the preliminary
             design of an open pit

221

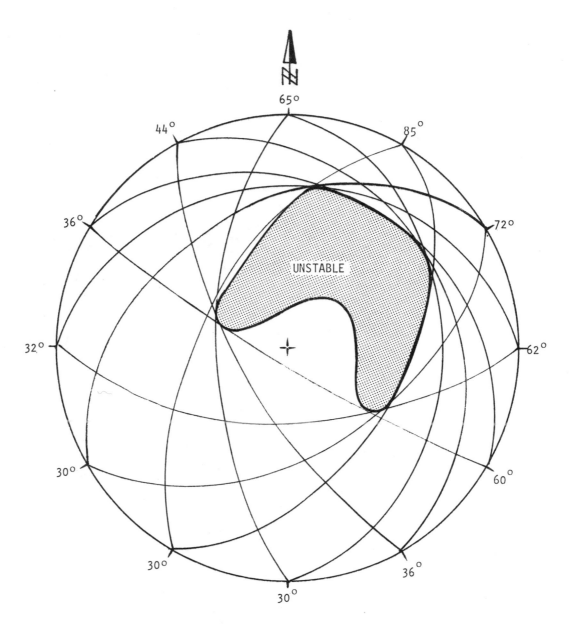

Note : Figures around the perimeter are the recommended
stable slope angles for the corresponding position
on the pit perimeter.

Figure 100 : Stereoplot of great circles representing stable slopes
around an open pit in a rock mass containing the five
sets of discontinuities defined in Figure 99.

the pit have been specified at 70° instead of the 85° suggested by Figure 100. This laying back results from a consideration of the maximum slope height-slope angle relationship presented in Figure 7 on page 20.

On no account should the layout suggested in Figure 101 be regarded as the final pit plan. The next stage in the feasibility study would obviously be to consider the implications of this suggested pit shape on the overall stripping ratio and hence the economics of the operation. This could easily result in a re-definition of the economic ore body shape and the need for a new pit layout.

Once the general pit shape has been decided upon, the next step is to consider the layout of both production and final benches and to make provision for a haul road or for an alternative transportation system.

Wedge failures in the benches forming the south-western part of this pit would be unavoidable since any faces cut steeper than 30° would allow the wedge intersections to daylight and, considering the factors of safety shown in Figure 99, stabilisation of these benches would not be economically feasible. It could, of course, happen that the assumption of friction only is grossly conservative and that the factors of safety are much too low. It may, therefore, be worth carrying out further stability studies on the south-western side of the pit to determine whether any cohesive strength could be relied upon. Back analysis of local quarry slopes, if such quarries exist in the area, would provide the most reliable source of cohesive strength data. Alternatively, shear strength testing would have to be carried out.

If further studies showed that the benches in the south-western part of the pit would be reasonably stable, this side of the pit would provide a good haul road route since this would permit the stripping ratio to be kept to a minimum by retaining the steep overall slopes on the north-eastern side of the pit.

On the other hand, many open pit operators dislike steep slopes and it may be decided, without further stability studies, to sacrifice on the stripping ratio and to place the haul road on the north-eastern side of the pit. While this would result in a considerable flattening of this side of the pit, it would ensure trouble-free benches since, with the reduced height of benches, 80° bench faces could be tolerated and, according to Figure 100, such benches would be safe. This solution would probably be the most satisfactory from an operational point of view - provided that the ore body grade was high enough to stand the high stripping ratio.

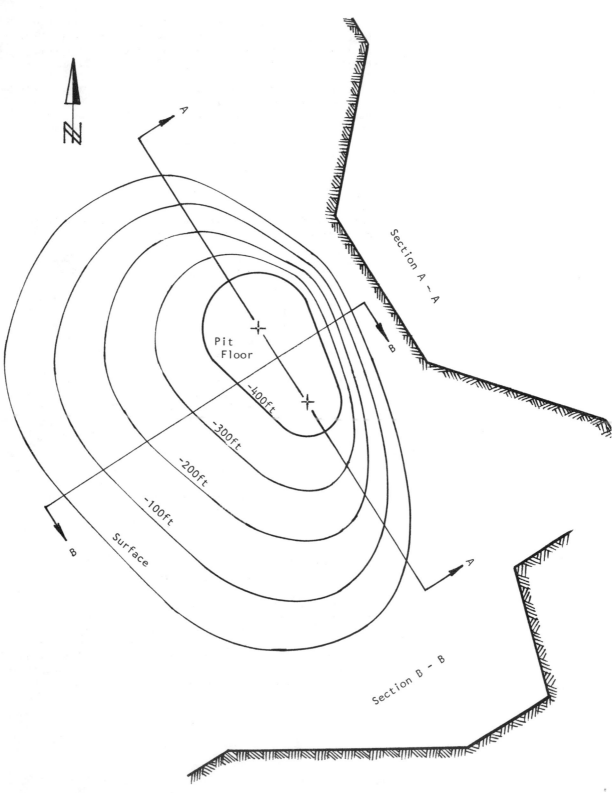

Figure 101 : Design of overall pit slopes according to safe angles
defined in Figure 100. Note that no benches or haul
roads are included in this pit layout.

Chapter 8 references

190. LONDE, P. Une méthode d'analyse à trois dimensions de la stabilité d'une rive rocheuse. *Annales des Ponts et Chaussées*. Paris. 1965, pages 37-60.

191. LONDE, P., VIGIER,G. and VORMERINGER, R. The stability of rock slopes, a three-dimensional study. *J. Soil Mech. and Foundation Div. ASCE*. Vol.95, No. SM 1, 1969, pages 235-262.

192. LONDE, P., VIGIER,G. and VORMERINGER, R. Stability of slopes - graphical methods. *J. Soil Mech. and Foundation Div. ASCE*. Vol. 96, No. SM 4, 1970, pages 1411-1434.

193. JOHN, K.W. Engineering analysis of three-dimensional stability problems utilising the reference hemisphere. *Proc. 2nd Congress. Intnl. Soc. Rock Mech.* Belgrade. 1970, Vol. 2, pages 314-321.

194. WITTKE, W.W. Method to analyse the stability of rock slopes with and without additional loading.(in German) *Felsmechanik und Ingenieurgeologie*. Supp.II, Vol. 30, 1965, pages 52-79.

    English translation in Imperial College Rock Mechanics Research Report No. 6, July 1971.

195. GOODMAN, R.E. The resolution of stresses in rock using stereographic projection. *Intnl. J. Rock Mech. Mining Sci*. Vol. 1, 1964, pages 93-103.

196. GOODMAN, R.E. and TAYLOR, R.L. Methods of analysis of rock slopes and abutments : a review of recent developments. in *Failure and Breakage of Rocks*. Edited by C.Fairhurst. AIME, 1967, pages 303-320.

197. HEUZE, F.E. and GOODMAN, R.E. Three-dimensional approach for the design of cuts in jointed rock. *Proc. 13th Sympos. Rock Mech.* Urbana, Illinois, 1971.

198. HENDRON, A.J., CORDING, E.J. and AIYER, A.K. Analytical and graphical methods for the analysis of slopes in rock masses. *U.S. Army Engineering Nuclear Cratering Group*. Tech. Rep. No.36, 1971, 168 pages.

199. SRIVASTAVA, L.S. Stability of rock slopes and excavations. *J. Eng. Geology*. Indian Soc.Engineering Geology. Vol.1/1, 1966, pages 57-72.

200. SAVKOV, L.V. Considerations of fracture in the calculation of rock slope stabilities. *Soviet Mining Science* 1967, pages 1-6.

201. HOEK, E., BRAY, J.W. and BOYD, J.M. The stability of a rock slope containing a wedge resting on two intersecting discontinuities. *Quarterly J. Engineering Geology*. Vol.6, No.1, 1973.

202. HOEK, E. Methods for the rapid assessment of the stability of three-dimensional rock slopes. *Quarterly J. Engineering Geology*. Vol.6, No.3, 1973.

203. TAYLOR, C.L. Geometric analysis of geological
     separation for slope stability investigations.
     Bull. Ass. Engineering Geologists. Vol. VII, Nos.1 & 2,
     1970, pages 76-85.

204. TAYLOR, C.E. Geometric analysis of rock slopes. *Proc.
     21st Annual Highway Geology Symposium*. University
     Kansas, April, 1970.

205. WILSON, S.D. The application of soil mechanics to
     the stability of open pit mines. *Colorado School of
     Mines Quarterly*. Vol. 54, No.3, 1959, pages 95-113.

206. MULLER, L. The European approach to slope stability
     problems in open-pit mines. *Colorado School of Mines
     Quarterly*. Vol.54, No.3, 1959, pages 117-133.

207. MULLER, L. and JOHN, K.W. Recent developments of
     stability studies of steep rock slopes in Europe.
     Trans. Soc. Min. Engineers, AIME. Vol.226, No.3,
     1963, pages 326-332.

208. MULLER, L. Application of rock mechanics in the
     design of rock slopes. *Intnl. Conf. State of Stress
     in the Earth's Crust*. Santa Monica. 1963. Elsevier,
     New York. 1964.

209. PETZNY, H. On the stability of rock slopes (in
     German) *Felsmechanik und Ingenieurgeoligie*. Suppl.III,
     1967.

Chapter 9 : Circular failure

Introduction

Although this book is concerned primarily with the stability of rock slopes, the reader will occasionally be faced with a slope problem involving soft materials such as overburden soils or crushed waste. In such materials, failure occurs along a surface which approaches a circular shape and this chapter is devoted to a brief discussion on how stability problems involving these materials are dealt with.

In a review on the historical development of slope stability theories, Golder[210] has traced the subject back almost 300 hundred years. During the past half century, a vast body of literature on this subject has accumulated and no attempt will be made to summarise this material in this chapter. Standard soil mechanics text books such as those by Taylor[174], Terzaghi[211] and Lambe and Whitman[212] all contain excellent chapters on the stability of soil slopes and it is suggested that at least one of these books should occupy a prominent place on the bookshelf of anyone who is concerned with slope stability. In addition to these books a number of important papers dealing with specific aspects of soil slope stability have been published and a selected list of these is given under references 213 to 233 at the end of this chapter.

The approach adopted in this chapter is to present a series of the slope stability charts for circular failure. These charts enable the user to carry out a very rapid check on the factor of safety of a slope or upon the sensitivity of the factor of safety to changes in groundwater conditions or slope profile. These charts should only be used for the analysis of circular failure in materials where the properties do not vary through the soil or waste rock mass and where the conditions assumed in deriving the charts, discussed in the next section, apply. A more elaborate form of analysis is presented at the end of this chapter for use in cases where the material properties vary within the slope or where part of the slide surface is at a soil/rock interface and the shape of the failure surface differs significantly from a simple circular arc.

Conditions for circular failure

In the previous chapters it has been assumed that the failure of rock slopes is controlled by geological features such as bedding planes and joints which divide the rock body up into a discontinuous mass. Under these conditions, the failure path is normally defined by one or more of the discontinuities. In the case of a soil, a strongly defined structural pattern no longer exists and the failure surface is free to find the line of least resistance through the slope. Observations of slope failures in soils suggests that this failure surface generally takes the form of a circle and most stability theories are based upon this observation.

The conditions under which circular failure will occur arise when the individual particles in a soil or rock mass are very small as compared with the size of the slope and when these particles are not interlocked as result of their shape. Hence, crushed rock in a large waste dump will tend to behave as a "soil" and large failures will occur in a circular mode. Alternatively, the finely ground waste

circular slip

Figure 102 : Shallow surface failure in large waste
dumps are generally of a circular type.

Figure 103 : Circular failure in the highly
altered and weathered rock forming
the upper benches of an open pit
mine.

material which has to be disposed of after completion of a milling and metal recovery process will exhibit circular failure surfaces, even in slopes of only a few feet in height. Highly altered and weathered rocks will also tend to fail in this manner and it is appropriate to design the overburden slopes around an open pit mine on the assumption that failure would be by a circular failure process.

Derivation of circular failure charts

The following assumptions are made in deriving the stability charts presented in this chapter :

a. The material forming the slope is assumed to be homogeneous, i.e. its mechanical properties do not vary with direction of loading

b. The shear strength of the material is characterised by a cohesion c and a friction angle ϕ which are related by the equation $\tau = c + \sigma.\mathrm{Tan}\phi$.

c. Failure is assumed to occur on a circular failure surface which passes through the toe of the slope*.

d. A vertical tension crack is assumed to occur in the upper surface or in the face of the slope.

e. The locations of the tension crack and of the failure surface are such that the factor of safety of the slope is a minimum for the slope geometry and ground-water conditions considered.

f. A range of groundwater conditions, varying from a dry slope to a fully saturated slope under heavy recharge, are considered in the analysis. These conditions are defined later in this chapter.

Defining the factor of safety of the slope as

$$F = \frac{\text{Shear strength available to resist sliding}}{\text{Shear stress mobilised along failure surface}}$$

and rearranging this equation, we get

$$\tau_{mb} = \frac{c}{F} + \frac{\sigma.\mathrm{Tan}\phi}{F} \qquad (97)$$

where τ_{mb} is the shear stress mobilised along the failure surface.

Since the shear strength available to resist sliding is dependent upon the distribution of the normal stress σ along this surface and, since this normal stress distribution is unknown, the problem is statically indeterminate. In order to obtain a solution it is necessary to assume a specific normal stress distribution and then to check whether this distribution gives meaningful practical results.

---

*Terzaghi[211], page 170, shows that the toe failure assumed for this analysis gives the lowest factor of safety provided that $\phi > 5^\circ$. The $\phi = 0$ analysis, involving failure below the toe of the slope through the base material has been discussed by Skempton[234] and by Bishop and Bjerrum[235] and is applicable to failures which occur during or after the rapid construction of a slope. Such conditions are unlikely to occur in typical mining operations.

The influence of various normal stress distributions upon the factor of safety of soil slopes has been examined by Frohlich[216] who found that a *lower bound* for all factors of safety which satisfy statics is given by the assumption that the normal stress is concentrated at a single point on the failure surface. Similarly, the *upper bound* is obtained by assuming that the normal load is concentrated at the two end points of the failure arc.

The unreal nature of these stress distributions is of no consequence since the object of the exercise, up to this point, is simply to determine the extremes between which the actual factor of safety of the slope must lie. In an example considered by Lambe and Whitman[212], the upper and lower bounds for the factor of safety of a particular slope corresponded to 1.62 and 1.27 respectively. Analysis of the same problem by Bishop's simplified method of slices gives a factor of safety of 1.30 which suggests that the actual factor of safety may lie reasonably close to the lower bound solution.

Further evidence that the lower bound solution is also a meaningful practical solution is provided by an examination of the analysis which assumes that the failure surface has the form of a logarithmic spiral[227]. In this case, the factor of safety is independent of the normal stress distribution and the upper and lower bounds coincide. Taylor[174] compared the results from a number of logarithmic spiral analyses with results of lower bound solutions* and found that the difference is negligible. On the basis of this comparison, Taylor concluded that the lower bound solution provides a value of the factor of safety which is sufficiently accurate for most practical problems involving simple circular failure of homogeneous slopes.

The authors have carried out similar checks to those carried out by Taylor and have reached the same conclusions. Hence, the charts presented in this chapter correspond to the lower bound solution for the factor of safety, obtained by assuming that the normal load is concentrated at a single point on the failure surface. These charts differ from those published by Taylor in 1948 in that they include the influence of a critical tension crack and of groundwater in the slope.

Groundwater flow assumptions

In order to calculate the uplift force due to water pressure acting on the failure surface and the force due to water in the tension crack, it is necessary to assume a set of groundwater flow patterns which coincide as closely as possible with those conditions which are believed to exist in the field.

In the analysis of rock slope failures, discussed in chapters 7 and 8, it was assumed that most of the water flow took place in discontinuities in the rock and that the rock itself was practically impermeable. In the case of slopes in soil or waste rock, the permeability of the mass of

---

*

The lower bound solution discussed in this chapter is usually known as the *Friction Circle Method* and was used by Taylor[174] for the derivation of his stability charts.

material is generally several orders of magnitude higher
than that of intact rock and, hence, a general flow pattern
will develop in the material behind the slope.

Figure 55a on page 137 shows that, within the soil mass,
the equipotentials are approximately perpendicular to the
phreatic surface. Consequently, the flow lines will be
approximately parallel to the phreatic surface for the
condition of steady state drawdown. Figure 104a shows that
this approximation has been used for the analysis of the
water pressure distribution in a slope under conditions of
normal drawdown. Note that the phreatic surface is assumed
to coincide with ground surface at a distance x, measured
in multiples of the slope height, behind the toe of the
slope. This may correspond to the position of a surface
water source such as a river or dam or it may simply be the
point where the phreatic surface is judged to intersect the
ground surface.

The phreatic surface itself has been obtained, for the range
of slope angles and values of x considered, by a computer
solution of the equations proposed by L. Casagrande[236],
discussed in the text book by Taylor[174].

For the case of a saturated slope subjected to heavy surface
recharge, the equipotentials and the associated flow lines
used in the stability analysis are based upon the work of
Han[237] who used an electrical resistance analogue method
for the study of groundwater flow patterns in isotropic
slopes.

Production of circular failure charts

The circular failure charts presented in this chapter were
produced by means of a Hewlett-Packard 9100 B calculator
with graph plotting facilities. This machine was programmed
to seek out the most critical combination of failure surface
and tension crack for each of a range of slope geometries
and groundwater conditions. Provision was made for the
tension crack to be located in either the upper surface of
the slope or in the face of the slope. Detailed checks
were carried out in the region surrounding the toe of the
slope where the curvature of the equipotentials results in
local flow which differs from that illustrated in Figure 104a.

The charts are numbered 1 to 5 to correspond with the
groundwater conditions defined in the table presented on
page 233.

Use of the circular failure charts

In order to use the charts to determine the factor of safety
of a particular slope, the steps outlined below and shown
in Figure 105 should be followed.

Step 1 :   Decide upon the groundwater conditions which are
           believed to exist in the slope and choose the
           chart which is closest to these conditions, using
           the table presented on page 233.

Step 2 :   Calculate the value of the dimensionless ratio

$$\frac{c}{\gamma H . \mathrm{Tan}\, \phi}$$

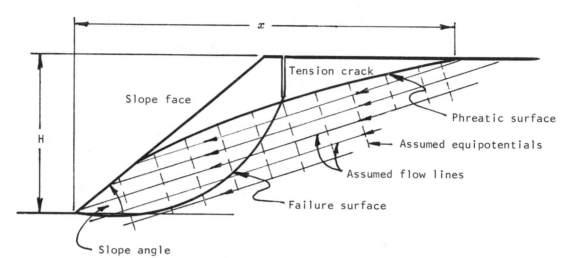

a. Groundwater flow pattern under steady state drawdown conditions where the phreatic surface coincides with the ground surface at a distance x behind the toe of the slope. The distance x is measured in multiples of the slope height H.

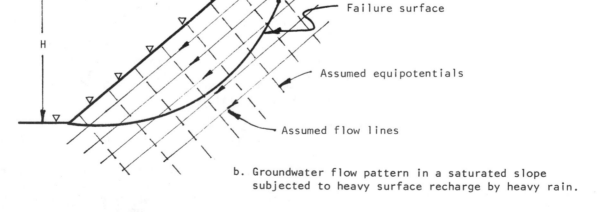

b. Groundwater flow pattern in a saturated slope subjected to heavy surface recharge by heavy rain.

Figure 104 : Definition of groundwater flow patterns used in circular failure analysis of soil and waste rock slopes.

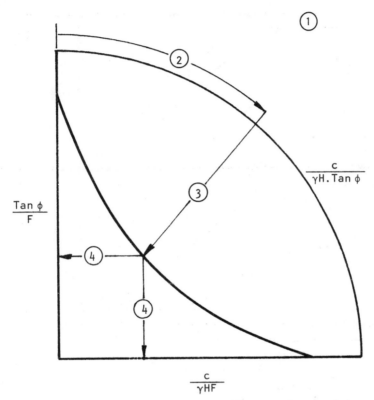

Figure 105 : Sequence of steps involved in using circular failure
charts to find the factor of safety of a slope.

Find this value on the outer circular scale of
the chart.

Step 3 : Follow the radial line from the value found in
step 2 to its intersection with the curve which
corresponds to the slope angle under consideration.

Step 4 : Find the corresponding value of $Tan\phi/F$ or $c/\gamma HF$,
depending upon which is more convenient, and
calculate the factor of safety.

Consider the following example :

A 50 foot high slope with a face angle of 40° is to be
excavated in overburden soil with a density γ = 100 lb/ft³,
a cohesive strength of 800 lb/ft² and a friction angle of
30°. Find the factor of safety of the slope, assuming that
there is a surface water source 200 feet behind the toe of
the slope.

The groundwater conditions indicate the use of chart No.3.
The value of $c/\gamma H.Tan\phi$ = 0.28 and the corresponding value
of $Tan\phi/F$, for a 40° slope, is 0.32. Hence, the factor of
safety of the slope is 1.80.

Because of the speed and simplicity of using these charts,
they are ideal for checking the sensitivity of the factor of
safety of a slope to a wide range of conditions and the
authors suggest that this should be their main use.

GROUNDWATER FLOW CONDITIONS	CHART NUMBER
FULLY DRAINED SLOPE	1
SURFACE WATER 8 x SLOPE HEIGHT BEHIND TOE OF SLOPE	2
SURFACE WATER 4 x SLOPE HEIGHT BEHIND TOE OF SLOPE	3
SURFACE WATER 2 x SLOPE HEIGHT BEHIND TOE OF SLOPE	4
SATURATED SLOPE SUBJECTED TO HEAVY SURFACE RECHARGE	5

CIRCULAR FAILURE CHART NUMBER 1

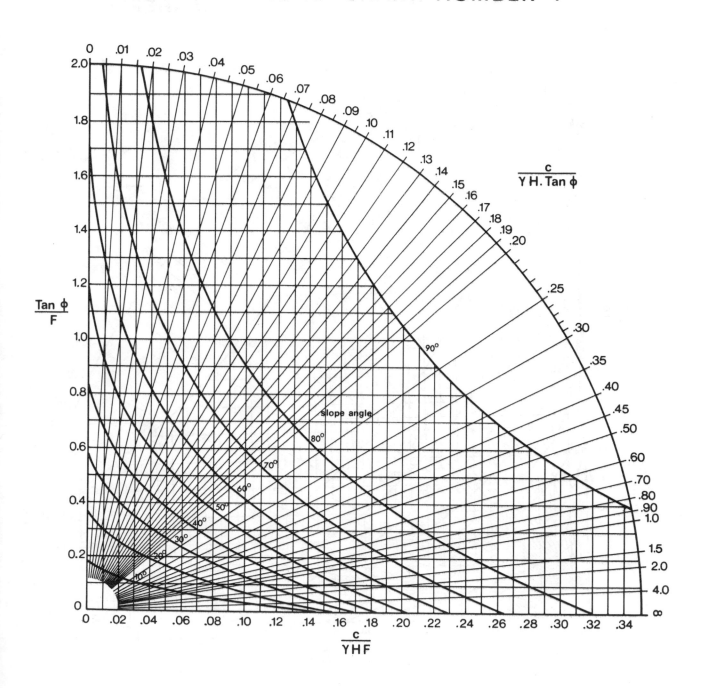

CIRCULAR FAILURE CHART NUMBER 2

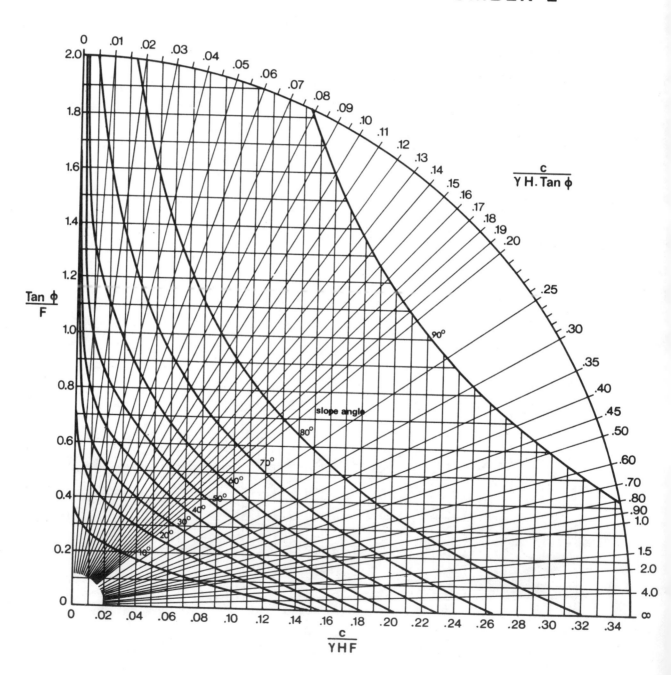

CIRCULAR FAILURE CHART NUMBER 3

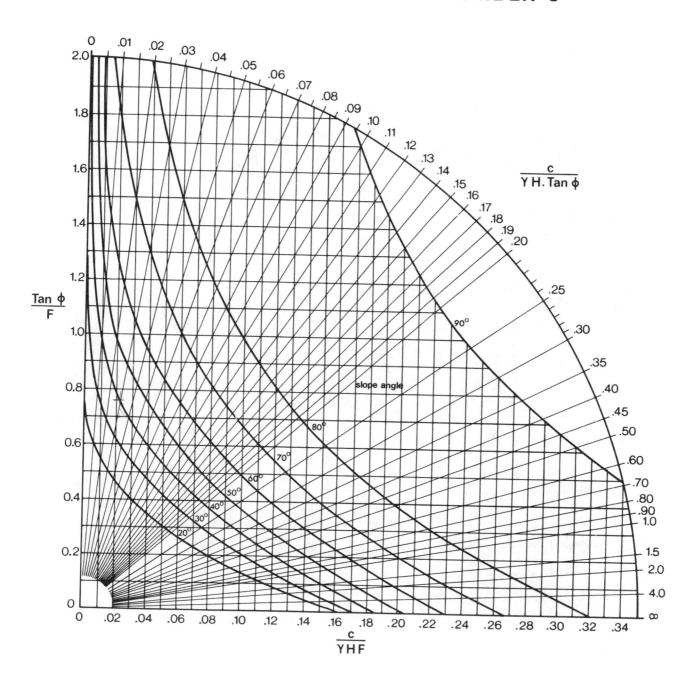

CIRCULAR FAILURE CHART NUMBER 4

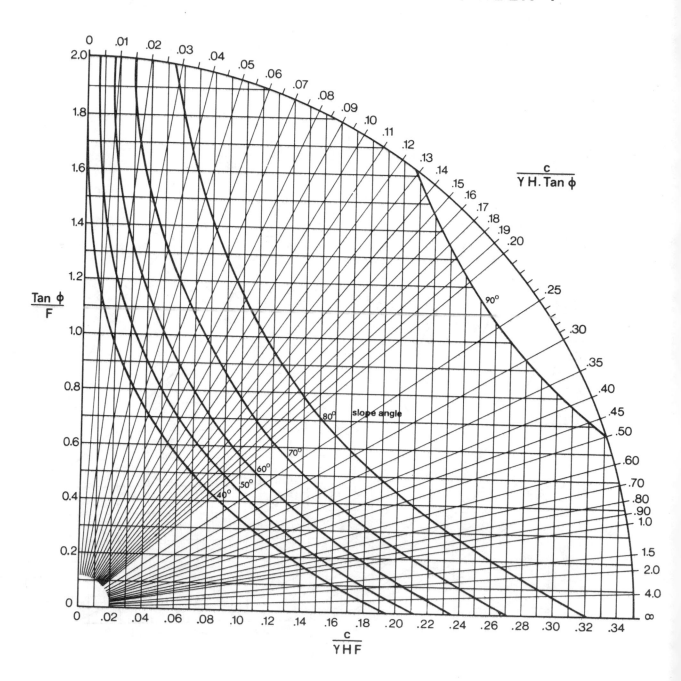

CIRCULAR FAILURE CHART NUMBER 5

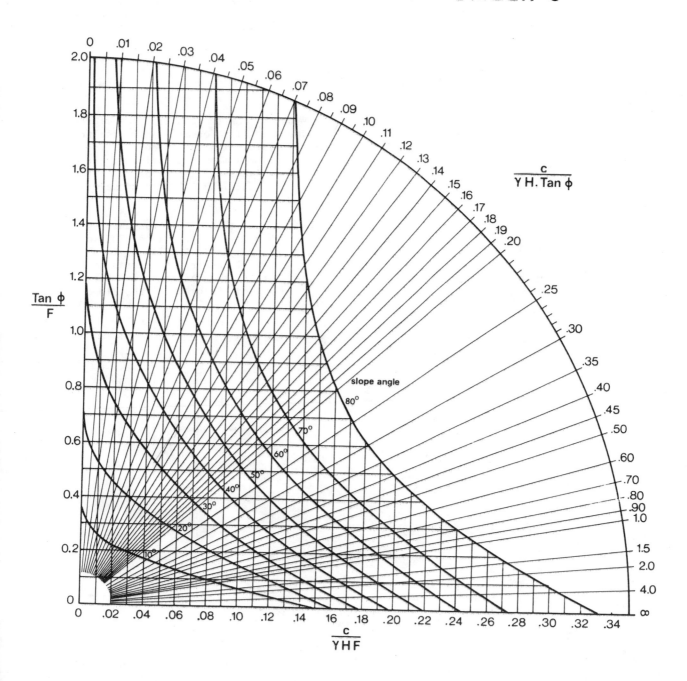

Location of critical failure circle and tension crack

During the production of the circular failure charts, presented on the previous pages, the locations of both the critical
failure circle and the critical tension crack for limiting
equilibrium (F = 1) were determined for each slope analysed.
These locations are presented, in the form of charts, in
Figures 106 and 107.

It was found that, once groundwater is present in the slope,
the locations of the critical circle and the tension crack
are not particularly sensitive to the position of the phreatic
surface and hence only one case, that for chart No. 3, has
been plotted. It will be noted that the location of the
critical circle centre given in Figure 107 differs significantly from that for the drained slope plotted in Figure
106.

These charts are useful for the construction of drawings
of potential slides and also for estimating the friction
angle when back-analysing existing circular slides. They
also provide a start for a more sophisticated circular
failure analysis in which the location of the circular
failure surface having the lowest factor of safety is found
by iterative methods.

As an example of the application of these charts, consider
the case of a slope having a face angle of 30° in a drained
soil with a friction angle of 20°. Figure 106 shows that
the critical failure circle centre is located at X = 0.2H
and Y = 1.85H and that the critical tension crack is at
a distance b = 0.1H behind the crest of the slope. These
dimensions are shown in Figure 108 below.

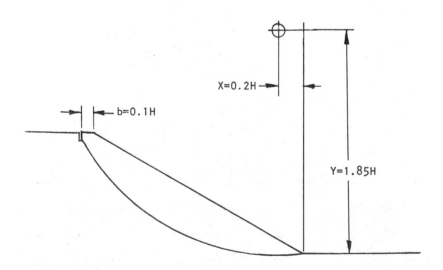

Figure 108 : Location of critical failure surface and
critical tension crack for a 30° slope in
drained soil with a friction angle of 20°.

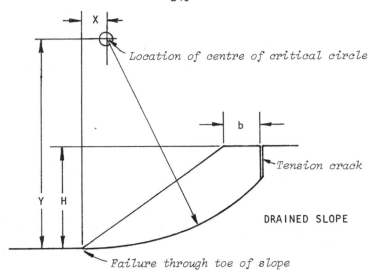

DRAINED SLOPE

Location of centre of critical circle

Tension crack

Failure through toe of slope

Distance X

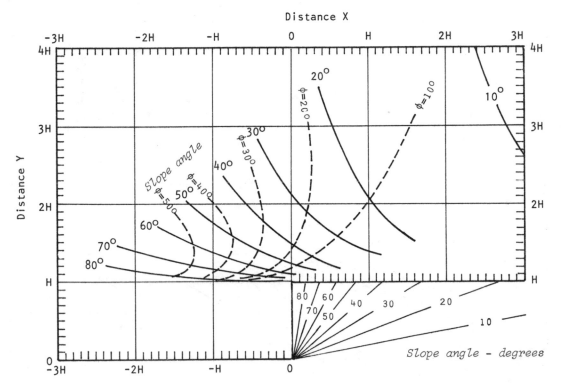

Location of centre of critical circle for failure through toe

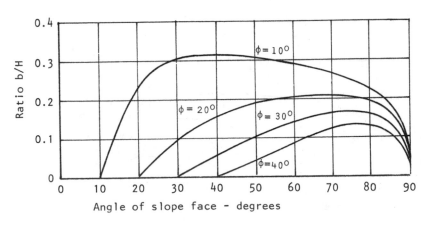

Location of critical tension crack position

Figure 106 : Location of critical failure surface and critical tension crack for drained slopes.

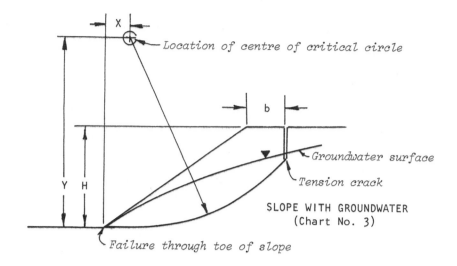

SLOPE WITH GROUNDWATER
(Chart No. 3)

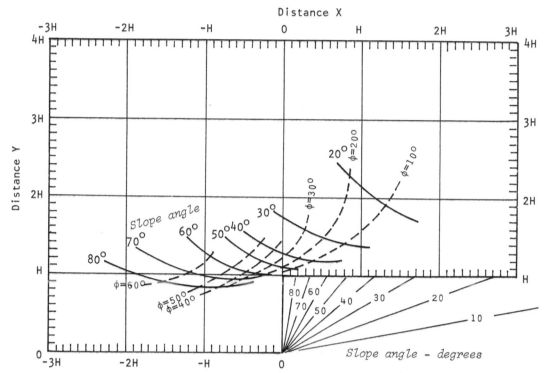

Location of centre of critical circle for failure through toe

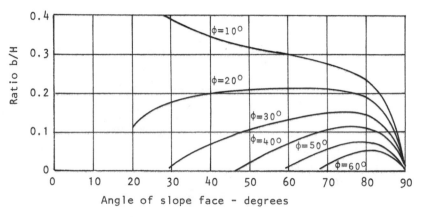

Location of critical tension crack position

Figure 107 : Location of critical failure surface and critical tension crack
for slopes with groundwater present.

Practical example number 1

China clay pit slope

Ley[153] has investigated the stability of a China clay pit slope which was considered to be potentially unstable. The slope profile is illustrated in Figure 108 below and the input data used for the analysis is included in this figure. The material, a heavily kaolinised granite, was carefully tested by Ley and the friction angle and cohesive strength are considered reliable for this particular slope.

Two piezometers in the slope and a known water source some distance behind the slope enabled Ley to postulate the phreatic surface shown in Figure 108. The chart which corresponds most closely to these groundwater conditions is considered to be chart number 2.

From the information given in Figure 108, the value of the ratio $c/\gamma H.\text{Tan}\phi = 0.0056$ and the corresponding value of $\text{Tan}\phi/F$, from chart number 2, is 0.76. Hence, the factor of safety of the slope is 1.01.

Ley also carried out a number of trial calculations using Janbu's method[238] and, for the critical slip circle shown in Figure 108, found a factor of safety of 1.03.

These factors of safety indicated that the stability of the slope was inadequate under the assumed conditions and steps were taken to deal with the problem.

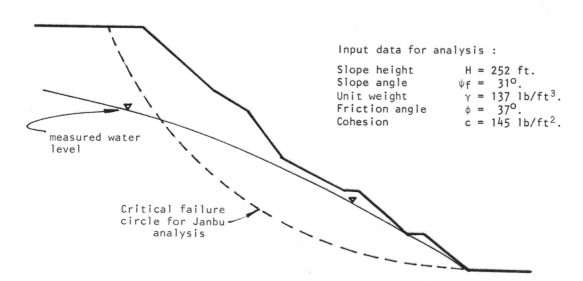

Input data for analysis :

Slope height	$H = 252$ ft.
Slope angle	$\psi f = 31°$.
Unit weight	$\gamma = 137$ lb/ft³.
Friction angle	$\phi = 37°$.
Cohesion	$c = 145$ lb/ft².

measured water level

Critical failure circle for Janbu analysis

Figure 108 : Slope profile of China clay pit slope considered in example number 1.

Practical example number 2

Projected open pit slope

An open pit mine plan calls for a slope on one side of the
pit to have an angle of 42°. The total height of the slope
will be 200 feet when completed and it is required to check
whether the slope will be stable. A site visit enables the
slope engineer to assess that the slope is in weathered and
altered material and that failure, if it occurs, will be of
a circular type. Insufficient time is available for groundwater levels to be accurately established or for shear tests
to be carried out. The stability analysis is carried out
as follows :

For the condition of limiting equilibrium, F = 1 and
Tanϕ/F = Tanϕ. For a range of friction angles, the values
of Tanϕ are used to find the values of c/γH.Tanϕ, for 42°,
by reversing the procedure outlined in Figure 104. The
value of the cohesion c which is mobilised at failure, for
a given friction angle, can then be calculated. This
analysis is carried out for dry slopes, using chart number 1,
and for saturated slopes, using chart number 5. The
resulting range of friction angles and cohesive strengths
which would be mobilised at failure are plotted in Figure 109.

The shaded circle included in Figure 109 indicates the
range of shear strengths which are considered probable for
the material under consideration, based upon the data
presented in Figure 70 on page 152. It is clear from this
figure that the available shear strength may not be adequate
to maintain stability in this slope, particularly when the
slope is saturated. Consequently, the slope engineer would
have to recommend that, either the slope should be flattened
or, that investigations into the groundwater conditions and
material properties should be undertaken in order to
establish whether the analysis presented in Figure 109 is
too pessimistic.

The effect of flattening the slope can be checked very
quickly by finding the value of c/γH.Tanϕ for a flatter
slope, say 30°, in the same way as it was found for the
42° slope. The dashed line in Figure 109 indicates the
shear strength which would be mobilised in a dry slope with
a face angle of 30°.

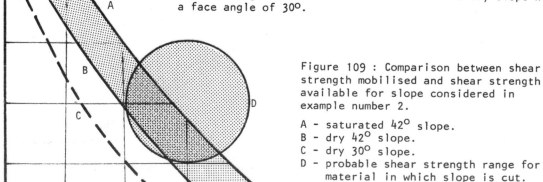

Figure 109 : Comparison between shear
strength mobilised and shear strength
available for slope considered in
example number 2.

A - saturated 42° slope.
B - dry 42° slope.
C - dry 30° slope.
D - probable shear strength range for
    material in which slope is cut.

Practical example number 3

Stability of waste dumps

As a result of the catastrophic slide in colliery waste
material at Aberfan in Wales on October 22, 1966, attention
was focussed on the potential danger associated with large
dumps of waste material from mining operations[239]. Since
1966, a number of excellent papers and handbooks dealing
with waste dump stability and with the disposal of finely
ground waste have become available [241-243] and the authors
do not feel that a detailed discussion on this subject
would be justified in this book. The purpose of this
example is to illustrate the application of the design
charts for circular failure, presented earlier in this
chapter, to waste dump stability problems.

McKechnie Thompson and Rodin[240] have shown that the relationship between shear strength and normal stress for colliery
waste material is usually non-linear as shown in Figure 110.
In view of the discussion on shear strength presented in
chapter 5, this finding is not particularly surprising and
the authors suspect that most waste materials exhibit this
non-linearity to a greater or lesser degree. Consequently,
the methods used in this example, although applied specifically to colliery waste, are believed to be equally applicable
to most rock waste dumps.

The method used in analysing the stability of a slope in
material which exhibits non-linear shear strength behaviour
was discussed on page 92 and illustrated in Figure 37. This
figure shows that a number of tangents are drawn to the
failure curve, giving values for apparent cohesion and angle
of friction for various normal stress levels. These values
are then used for stability analyses in exactly the same
way as the cohesion and friction values for a normal soil.

In the case of the failure curve for colliery waste, shown
in Figure 110, the apparent cohesion values and the friction
angles given by the three tangents are as follows :

Tangent number	Apparent cohesion kN/m^2	Friction angle degrees.
1	0	38
2	20	26
3	40	22

The relationship between slope height and slope angle for
the condition of limiting equilibrium, F = 1, will be
investigated for a dry dump (using chart No.1) and for a
dump with some groundwater flow (using chart No.3).

Tangent number 1

Since the cohesion intercept is zero for this tangent, the
value of the dimensionless ratio $c/\gamma H.Tan\phi = 0$ and hence,
the slope angle at which the face would repose is given
by the slope angle corresponding to the value of
$Tan 38° = 0.78$ on the $Tan\phi/F$ axis (noting that F = 1). From
chart No. 1, this intercept is 38° and for chart No.3 it is
approximately 25°.

Note that, for zero cohesion, the dump face angle would be

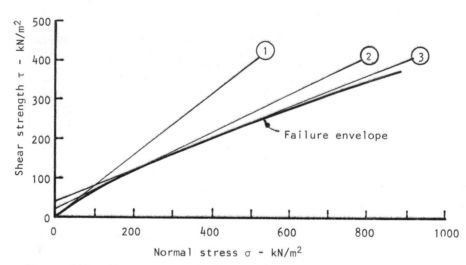

Figure 110 : Shear strength of typical colliery waste material.

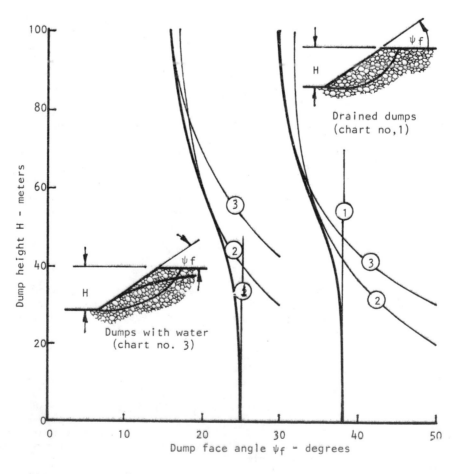

Figure 111 : Relationship between slope height and slope angle for a
typical colliery waste dump with different water conditions.

independent of the slope height. It is normally assumed that the angle of repose of a waste dump is independent of the height of the dump and is equal to the angle of friction of the material. Figure 111 shows that this assumption is only correct for a dry slope of limited height. Any build-up of water pressure within the dump causes a serious reduction in the stable face angle and, once the normal stress across the potential failure surface becomes high enough for the next tangent to become operative, the high initial friction angle no longer applies and the dump face assumes a flatter angle.

Tangent number 2

For $c = 20$ kN/m^2, $\gamma = 18$ kN/m^3 and $\phi = 26°$,

$$\frac{c}{\gamma H.\mathrm{Tan}\,\phi} = \frac{20}{18.\mathrm{Tan}\,26°.H} = \frac{2.28}{H}$$

and $\mathrm{Tan}\phi/F = \mathrm{Tan}\,26° = 0.49$

Hence

H meters	$\dfrac{c}{\gamma H.\mathrm{Tan}\phi}$	Slope angle° Chart 1	Chart 3
20	0.114	50	39
40	0.057	39	25
60	0.038	34	20
80	0.029	32	18
100	0.023	31	17

Plotting these values on Figure 111 gives the curves numbered 2 for the drained and the wet dumps.

Tangent number 3

$c = 40$ kN/m^2, $\gamma = 18$ kN/m^3 and $\phi = 22°$, hence

$c/\gamma H.\mathrm{Tan}\,\phi = 5.5/H$ and $\mathrm{Tan}\,\phi /F = 0.4$

H meters	$\dfrac{c}{\gamma H.\mathrm{Tan}\,\phi}$	Slope angle° Chart 1	Chart 3
20	0.275	61	56
40	0.138	43	31
60	0.092	34	22
80	0.069	31	18
100	0.055	30	16

The relationships between dump face angle and dump height, for both drained and wet dumps, are given by the envelopes to the curves derived from tangents 1, 2 and 3. These envelopes, shown in Figure 111, illustrate the danger in continuing to increase the height of a dump on the assumption that it will remain stable at an angle of repose equal to the friction angle. The dangers associated with poor dump drainage are also evident in this figure.

The reader who attempts this type of analysis for himself, and it is strongly recommended that he should, will find that the slope height versus slope angle relationship is extremely sensitive to the shape of the shear failure curve. This emphasises the need for reliable in situ shear test methods such as those described by McKechnie Thompson and Rodin[240] and Schultze and Horn[244] to be further developed for application to waste dump problems.

Janbu's non-circular failure analysis

When the properties of the soil or waste rock mass vary
throughout a slope or when the shape of the failure surface
is not circular, as a result of some structural feature
such as the soil/rock interface, the conditions assumed in
deriving the circular failure charts presented on the previous
pages no longer apply. Significant errors can arise from
the application of the circular failure charts in such
cases, particularly when low shear strength planar features
form part of the failure surface. Consequently, a more
elaborate form of analysis must be used when strength
variations or structural features are known to occur in
the material forming the slope.

Janbu's method of analysing non-circular failure in slopes[238]
is one of the most versatile of the available methods and
it is simple enough to permit the solution of problems by
hand. This method is described on the following pages *.

The failure surface is described by the parameters given in
Figure 112. Note that the slices into which the failing
mass is divided need not be of constant width. For design
purposes it is necessary to assume a factor of safety and
to iterate until the derived factor of safety agrees with
that from the previous iteration. In order to facilitate
the solution of problems by hand, values for the two functions
n_α and f_O are plotted in Figures 113 and 114 and the use
of these functions in the solution of a typical problem is
illustrated in the practical example given at the end of
this discussion. In the back-analysis of an existing failure,
no iteration is necessary as the cohesion c and angle of
friction ϕ mobilised in the failure are given by assuming
that the factor of safety F = 1.

The analysis is carried out in the following steps :

Step 1 : Slice parameters

> Divide the sliding mass into slices. The slice
> width should be selected to take account of changes
> in material properties, slope geometry and water
> pressure distribution. The calculations are sim-
> plified if equal slice widths can be used but the
> reader should not force the geometry into equal
> slices if conditions indicate that unequal slice
> widths are more appropriate. The inclination α of
> the centre of the base of each slice with respect
> to the horizontal and the width Δx of the slice
> are measured. Tabulate the values of α, Δx, c
> and Tan ϕ for each slice.

Step 2 : Weight parameters

> The weight of the slice ΔW and the average weight
> of the slice per unit area of base p are calculated.
> If the slice geometry is reasonably regular, p may
> be taken as γh_m where h_m is the midheight of the
> slice and $\Delta W = \gamma h_m \Delta x$. If the slice height is
> irregular, the weight of the slice may be calculated

---

* This discussion is based upon an internal company working
document prepared by Dr L.R.Richards of Golder, Hoek and
Associates of Maidenhead, England.

by measuring the slice area with a planimeter and multiplying this by the density of the material. In this case $p = \Delta W/\Delta x$. Tabulate p, h_m and ΔW.

Step 3 : Water pressure on failure surface

Calculate the average water pressure on the base of each slice as indicated in Figure 112 and enter this value into the calculation table. If there is a vertical tension crack at the rear of the slide, the horizontal water force Q due to water in the tension crack should be calculated.

Step 4 : Detailed calculations

Calculate $\Delta W.\tan\alpha$ and $X = \{c + (p - u)\tan\phi\}\Delta x$ for each slice and enter these values into the table.

Step 5 : Assume a factor of safety (usually taken as 1 for the first trial). Find the value of n_α for each slice from Figure 113a and b. Calculate X/n_α for each slice and enter this value in the table.

Step 6 : Determine the value of f_o from Figure 114 and calculate the value of the new factor of safety from

$$F = f_o.\frac{\Sigma X/n_\alpha}{\Sigma p.\tan\alpha}$$

Step 7 : If the factor of safety calculated in step 6 does not agree with that assumed in step 5, assume a new value of F (close to that calculated in step 6) and repeat steps 5 and 6 until the calculated factor of safety agrees with the assumed value. The calculations converge very rapidly and three iterations are generally adequate for the solution of the problem.

In the example given at the end of this chapter, the initial value of F was assumed to be 1.00 and the final value of $F = 1.95$ was obtained after three iterations.

Janbu's method of non-circular failure analysis can be applied to planar failure on one or more surfaces (but not to wedge failures) provided that the failure mode is kinematically admissible.

Janbu's method will generally be found to be adequate for most practical problems but it tends to become very tedious to use when a large number of sensitivity studies are required or when a large number of slopes have to be designed. In such cases, the use of a digital computer for the solution of the equations published by Bishop[213] or by Morgenstern and Price[222] is more practical. Most large university civil engineering departments and geotechnical consulting organisations have computer programs available for these types of studies and the interested reader is advised to contact one of these organisations for assistance if he or she becomes involved in a large number of complex slope designs.

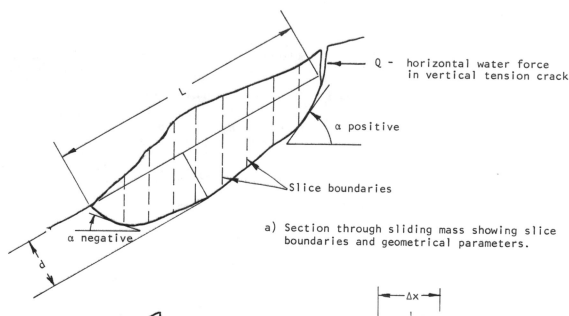

Q - horizontal water force in vertical tension crack

α positive

Slice boundaries

α negative

d

a) Section through sliding mass showing slice boundaries and geometrical parameters.

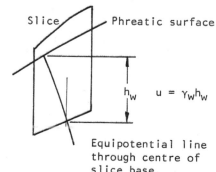

Slice   Phreatic surface

h_w $u = \gamma_w h_w$

Equipotential line through centre of slice base.

b) Calculation of average water pressure u on base of slice.

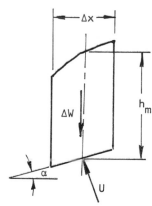

Δx

ΔW

h_m

α

U

c) Slice parameters used in the stability analysis.

Symbols :

f_o - correction factor (Figure 114)

n_α - geometrical functions (Figures 113a and 113b)

c - cohesive strength

ϕ - angle of friction

p - average weight per unit width of slice

u - average water pressure on base of slice

L - chord length of failure surface

d - depth of failure surface

Calculation of factor of safety :

$$F = f_o \frac{\sum \left\{ \dfrac{c + (p - u)\,\mathrm{Tan}\,\phi}{n_\alpha} \right\} \Delta x}{\sum \Delta W . \mathrm{Tan}\,\alpha + Q}$$

If Q is zero and Δx is constant

$$F = f_o \frac{\sum \left\{ \dfrac{c + (p - u)\,\mathrm{Tan}\,\phi}{n_\alpha} \right\}}{\sum p . \mathrm{Tan}\,\alpha}$$

Figure 112 : Definition of geometrical parameters and method of calculation for Janbu's non-circular failure analysis.

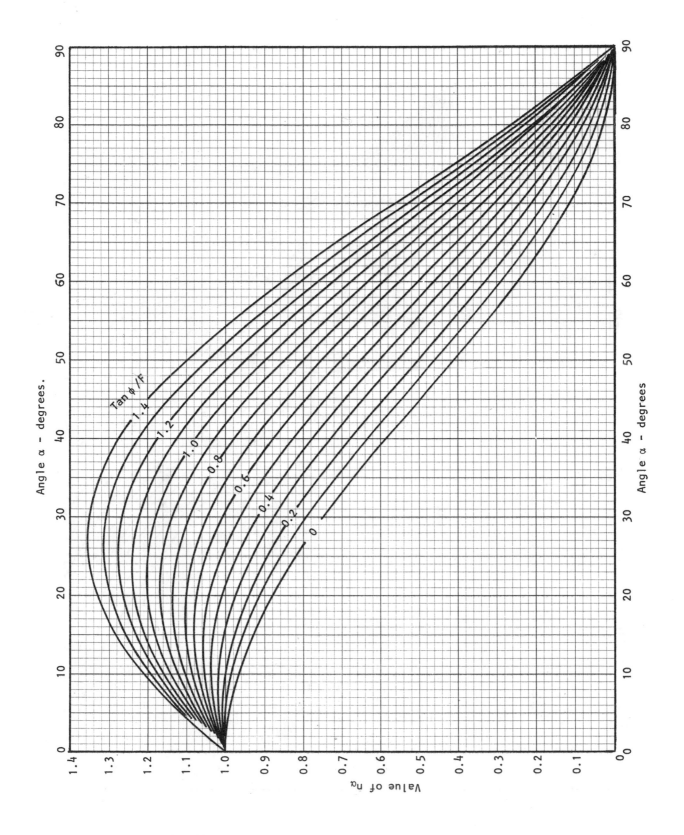

Figure 113 a : Determination of value of n_α for positive angles .

*Note that α is positive when the slope of the base of
the slice is in the same quadrant as the ground slope.*

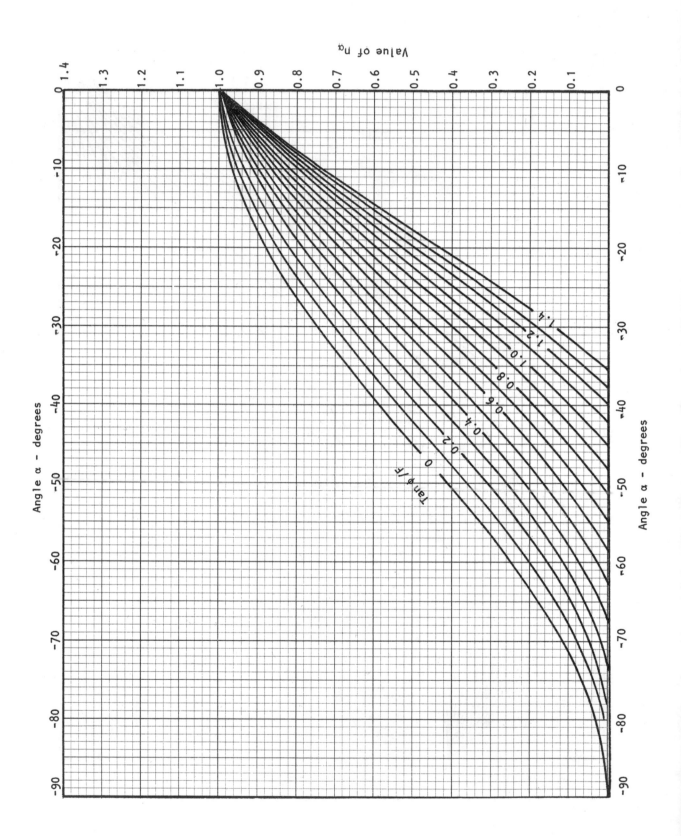

Figure 113 b : Determination of the value of n_α for negative angles.

Note that α is negative when the slope of the base of
the slice is in a different quadrant from the ground slope.

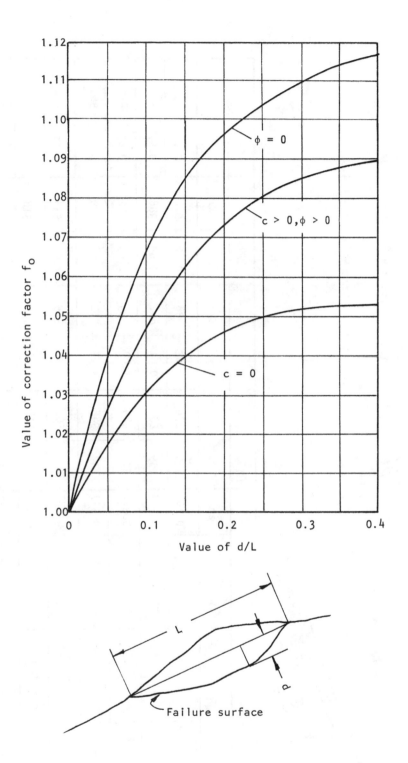

Figure 114 : Correction factor f_o to allow for inter-slice
forces in Janbu's non-circular failure analysis.

Sand

Tan φ = 0.76
c = 0
γ = 18 kN/m³

Clay

Tan φ = 0.577
c = 40 kPa
γ = 20 kn/m³

Shear strength of clay/rock interface as for clay

d/L = 0.17, hence f_0 = 1.065 from Figure 114.

$$F = f_0 \frac{\Sigma(X/n_\alpha)}{\Sigma\Delta W \cdot Tan\,\alpha} \qquad \text{where } X = \{c + (p - u)Tan\,\phi\}\Delta x$$

Trial 1 : Assume F = 1.00
Calculated F = (1.065 × 4400)/2648 = 1.77

Trial 2 : Assume F = 1.80
Calculated F = (1.065 × 4786)/2648 = 1.92

Trial 3 : Assume F = 1.95
Calculated F = (1.065 × 4848)/2648 = 1.95

Factor of safety for this failure surface = 1.95

Piezometric height on failure surface
Failure surface
Sand / Clay / Bedrock

	VALUES FROM SECTION								CALCULATIONS		TRIAL 1		TRIAL 2		TRIAL 3	
Slice	α	u	h_m	Δx	p	ΔW	c	Tan φ	ΔW Tanα	X	n_α	X/n_α	n_α	X/n_α	n_α	X/n_α
1	58	–	4.0	5.0	72	360	0	0.761	576	274	0.63	438	0.52	526	0.50	548
2	44	34.30	8.1	6.0	162	972	40	0.577	939	682	0.81	842	0.69	988	0.67	1018
3	29.5	49.05	7.5	6.0	150	900	40	0.577	509	589	1.00	589	0.89	662	0.89	662
4	20.5	58.86	7.4	6.0	148	888	40	0.577	332	548	1.07	512	0.98	559	0.97	565
5	15.0	49.05	7.1	6.0	142	852	40	0.577	228	562	1.07	525	1.01	556	1.01	556
6	9.5	36.79	6.0	6.0	120	720	40	0.577	120	528	1.06	498	1.02	518	1.02	518
7	3.5	19.62	4.5	6.0	90	540	40	0.577	33	483	1.03	469	1.02	474	1.01	478
8	-12	–	3.0	7.0	60	420	40	0.577	-89	448	0.85	527	0.89	503	0.89	503
									ΣΔWTanφ=2648		Σ(X/n_α)	4400		4786		4848

Chapter 9 references

210. GOLDER, H.Q. The stability of natural and man-made slopes in soil and rock. *Geotechnical Practice for Stability in Open Pit Mining*. Editors Brawner, C.O. and Milligan, V. AIME, New York, 1972, pages 79-85.

211. TERZAGHI, K. *Theoretical Soil Mechanics*. Wiley, New York, 1943.

212. LAMBE, W.T. and WHITMAN, R.V. *Soil Mechanics*. Wiley, New York, 1969, 553 pages.

213. BISHOP, A.W. The use of the slip circle in the stability analysis of earth slopes. *Geotechnique*. Vol.5, 1955, pages 7-17.

214. BISHOP, A.W. and MORGENSTERN, N. Stability coefficients for earth slopes. *Geotechnique*. Vol.10, 1960, pages 29-150.

215. FELLENIUS, W. Calculation of the stability of earth dams. *Trans. 2nd Congress on Large Dams*. Washington, Vol.4, 1936, page 445.

216. FROHLICH, O.K. General theory of the stability of slopes. *Geotechnique*. Vol.5, 1955, pages 37-47.

217. JANBU, N. Application of composite slip circles for stability analysis. *Proc. European Conference on Stability of Earth Slopes*. Stockholm, Vol.3, 1954, pages 43-49.

218. JANBU, N. Stability analysis of slopes with dimensionless parameters. *D.Sc. Thesis. Harvard Soil Mechanics Series*. No.46, 1954, 81 pages.

219. LOWE, J. Stability analysis of embankments. *J. Soil Mech. Foundation Div. ASCE*. Vol.93, No.SM4, 1967, pages 286-299.

220. MEYERHOF, G.G. The mechanism of flow slides in cohesive soils. *Geotechnique*. Vol.7, 1961, pages21-31.

221. MENCL, V. The influence of the stiffness of a sliding mass on the stability of slopes. *Rock Mechanics and Engineering Geology*. Vol.4, 1966, pages 127-131.

222. MORGENSTERN, N.R. and PRICE, V.E. The analysis of the stability of general slip surfaces. *Geotechnique*. Vol.15, 1965, pages 79-93.

223. MORGENSTERN, N.R. Stability charts for earth slopes during rapid drawdown. *Geotechnique*. Vol.13, 1963, pages 121-131.

224. NONVELLIER, E. The stability analysis of slopes with a slip surface of general shape. *Proc. 6th Intnl. Conf. Soil Mech. Foundation Engg*, Montreal. Vol.2, 1965, page 522.

225. PECK, R.B. Stability of natural slopes. *Proc. ASCE*. Vol.93, No.SM 4, 1967, pages 403-417.

226. RODRIGUEZ, A. Analysis of slope stability. *Proc. 5th Intnl. Conf. Soil Mech. Foundation Engg.* Paris. 1961, Vol.2, page 709.

227. SPENCER, E. Circular and logarithmic spiral slip surfaces. *J. Soil Mech. Foundation Div. ASCE.* Vol.95. No.SM 1, 1969, pages 227-234.

228. SPENCER, E. A method of analysis of the stability of embankments assuming parallel inter-slice forces. *Geotechnique.* Vol.17, 1967, pages 11-26.

229. TAYLOR, D.W. Stability of earth slopes. *J. Boston Soc. Civil Engineers.* Vol.24, 1937, page 197.

230. TERZAGHI, K. Critical height and factor of safety of slopes against sliding. *Proc. Intnl. Conf. Soil Mechanics.* Cambridge, Mass. 1936, Vol.1, pages 156-161.

231. VERGHESE, P.C. Investigations of a new procedure for analysing the stability of slopes. *M.Sc. Thesis, Harvard University.* 1949.

232. VISHER, D. Use of moment planimeter for analysis of slope stability. (in German). *Schwerz. Bauztg.* Vol.81, No.12, 1963, pages 183-187.

233. WHITMAN, R.V. and BAILEY, W.A. The use of computers for slope stability analysis. *J. Soil Mech. Foundation Engg, ASCE.* Vol.93, No.SM 4, 1967, pages 475-498.

234. SKEMPTON, A.W. The $\phi = 0$ analysis for stability and its theoretical basis. *Proc. 2nd Intnl. Conf. Soil Mech. Foundation Engg.* Rotterdam. Vol.1, 1948, page 72.

235. BISHOP, A.W. and BJERRUM, L. The relevance of the triaxial test to the solution of stability problems. *Proc. ASCE Conf. on shear strength of cohesive soils.* Boulder, Colorado, 1960, pages 437-501.

236. CASAGRANDE, L. Näherungsverfahren zur Ermittlung der Sickerung in geschütteten Dämmen auf underchlässiger Sohle, *Die Bautechnik.* Heft 15, 1934.

237. HAN. C. The technique for obtaining equipotential lines of groundwater flow in slopes using electrically conducting paper. *M.Sc. Thesis, London University (Imperial College),* 1972.

238. JANBU, N., BJERRUM, L and KJAERNSLI, B. Soil mechanics applied to some engineering problems. (in Norwegian with an English summary). *Norwegian Geotechnical Inst.* Publ. 16, 1956.

239. *Report of the tribunal appointed to enquire into the disaster at Aberfan on October 21st, 1966.* Her Majesty's Stationery Office, London, 1968.

240. MCKECHNIE THOMPSON, G. and RODIN, S. *Colliery spoil tips - after Aberfan.* Inst. Civil Engineers, London. 1972, Paper 7522, 60 pages.

241. NATIONAL COAL BOARD. *Technical handbook on spoil heaps and lagoons.* National Coal Board Mining Department. London. Final Draft 1970, 233 pages.

242. CHINA CLAY ASSOCIATION. *China Clay industry technical handbook for the disposal of waste materials.* China Clay Association. St. Austell, Cornwall, 1971, 82 pages.

243. CANADIAN DEPARTMENT OF ENERGY, MINES AND RESOURCES. *Tentative design guide for mine waste embankments in Canada.* Mining Research Centre, Canadian Department of Energy, Mines and Resources. Ottawa. Final Draft, May, 1971, 185 pages.

244. SCHULTZE, E. and HORN, A. The base friction for horizontally loaded footings in sand and gravel. *Geotechnique.* Vol.17, 1967, pages 329-347.

Chapter 10: Toppling failure

Suggested toppling failure mechanism of the north face of the Vajont slide. After Müller[245].

Deep wide tension cracks are associated with toppling failure in hard rock slopes. Photograph by R.E.Goodman.

Introduction

The failure modes discussed in previous Chapters of this book are all related to the sliding of a rock or soil mass along an existing or an induced failure surface. On page 31 brief mention was made of a different failure mode - that of toppling. Toppling failure involves rotation of columns or blocks of rock about some fixed base and the simple geometrical conditions governing the toppling of a single block on an inclined surface were defined in Figure 11 on page 32.

Toppling failures in hard rock slopes have only been described in literature during the past few years. One of the earliest references is by Müller[245] who suggested that block rotation or toppling may have been a contributory factor in the failure of the north face of the Vajont slide. Hofmann[246] carried out a number of model studies under Müller's direction to investigate block rotation. Similar model studies were carried out under the authors' direction at Imperial College by Ashby[247], Soto[248] and Whyte[249] while Cundall[250] made one of the earliest attempts to study the problem numerically. Burman[251], Byrne[252] and Hammett[253], all of the James Cook University of North Queensland in Australia, made significant contributions to the understanding of this problem and to the incorporation of rotational failure modes into computer analysis of rock mass behaviour. An excellent descriptive paper on toppling failures in the United Kingdom by de Freitas and Watters[254] is recommended reading and Heslop[255] has discussed an example of toppling associated with mining. Most of the discussion which follows in this chapter is based on a paper by Goodman and Bray[256] in which a formal mathematical solution to a simple toppling problem is attempted. This solution, which is reproduced here, is believed to represent a basis for the development of methods for designing rock slopes in which toppling is present. Several years of development work will be necessary before these methods can be used with the same degree of confidence as other methods of stability analyses described in this book.

Types of toppling failure

Goodman and Bray have described a number of different types of toppling failures which may be encountered in the field and each of these types is discussed briefly on the following pages.

Flexural toppling

The process of flexural toppling is illustrated in Figure 115 which shows that continuous columns of rock, which are separated by well developed steeply dipping discontinuities, break in flexure as they bend forward. An example of this type of failure is illustrated in the photograph in the margin on page 31 which shows a large flexural topple in the Dinorwic slate quarry in North Wales. The example illustrated in Figure 115 is in the Penn Rynn slate quarry in North Wales.

Sliding, undermining or erosion of the toe of the slope allows the toppling process to start and it retrogresses backwards into the rock mass with the formation of deep, wide tension cracks. The lower portion of the slope is

Figure 115 : Flexural toppling occurs in hard rock slopes
with well developed steeply dipping discontinuities.
Photograph by R.E.Goodman.

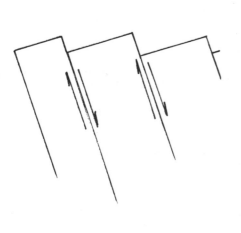

Figure 116 : Interlayer sliding between toppling columns
results in a series of back facing or
obsequent scarps in the upper surface of
the rock slope.
Photograph by R.E.Goodman.

covered with disoriented and disordered blocks and it is sometimes very difficult to recognise a toppling failure from the bottom of a slope.

The outward movement of each cantilevered column produces an interlayer slip and a portion of the upper surface of each plane is exposed in a series of back facing or obsequent scarps such as those illustrated in Figure 116.

Block toppling

As illustrated in Figure 117, block toppling occurs when individual columns of hard rock are divided by widely spaced orthogonal joints. The short columns forming the toe of the slope are pushed forward by the loads from the longer overturning columns behind and this sliding of the toe allows further toppling to develop higher up the slope. The base of the failure is better defined than that in a flexural topple and it generally consists of a stairway rising from one cross-joint to the next.

Block-flexure toppling

As shown in Figure 118, this type of toppling failure is characterised by pseudo-continuous flexure along long columns which are divided by numerous cross joints. Instead of the flexural failure of continuous columns, resulting in flexural toppling, the toppling of the columns in this case results from accumulated displacements on the cross joints. Because of the large number of small movements in this type of topple, there are fewer tension cracks than in flexural toppling and fewer edge-to-face contacts and voids than in block toppling.

Secondary toppling modes

Figure 119 illustrates a number of possible secondary toppling mechanisms suggested by Goodman and Bray. In general, these failures are initiated by some undercutting of the toe of the slope, either by natural agencies such as erosion or weathering or by the activities of man. In all cases, the primary failure mode involves sliding or physical breakdown of the rock and toppling is induced in some part of the slope as a result of this primary failure.

Analysis of toppling failure

Goodman[24] has published a detailed discussion on the base friction modelling technique which is an ideal tool for simple physical model studies of toppling phenomena. As illustrated in Figure 120, the apparatus consists of a base and frame to hold a pair of wide rollers over which a sanding belt runs. This sanding belt applies a friction force to the underside of the model resting on the belt and, if the base of the model is prevented from moving, the base friction forces will simulate the gravitational loads of the individual blocks which make up the model. Block toppling in models made from cork, plaster, plastic blocks or wooden blocks can be studied by means of this technique and the type of behaviour illustrated in the computer generated margin drawing can be simulated very easily.

While this method is ideal for demonstration and teaching purposes, its value as a design tool for rock slope engineer-

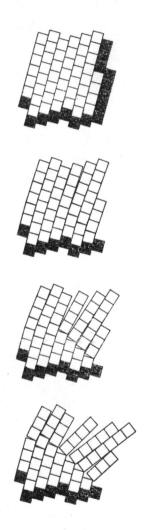

Computer generated model of toppling failure by Cundall[250]. Solid blocks are fixed in space.

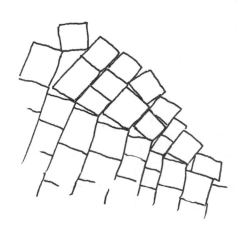

Figure 117 : Block toppling can occur in a hard rock mass with
widely spaced orthogonal joints.
Photograph by R.E.Goodman.

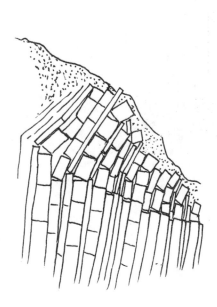

Figure 118 : Block flexure toppling is characterised by pseudo-
continuous flexure of long columns through accumulated motions along numerous cross joints.
Photograph by R.E.Goodman.

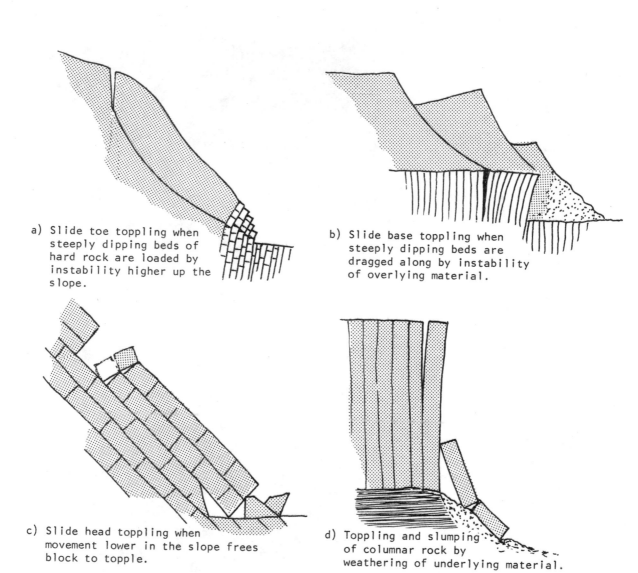

a) Slide toe toppling when steeply dipping beds of hard rock are loaded by instability higher up the slope.

b) Slide base toppling when steeply dipping beds are dragged along by instability of overlying material.

c) Slide head toppling when movement lower in the slope frees block to topple.

d) Toppling and slumping of columnar rock by weathering of underlying material.

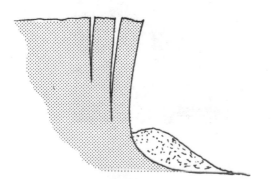

e) Tension crack toppling in cohesive materials.

Figure 119 : Secondary toppling mechanisms suggested by Goodman and Bray.

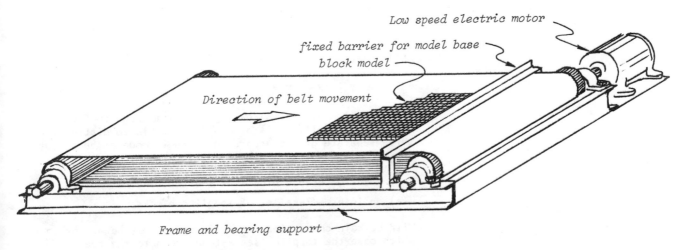

Figure 120 : Base friction model apparatus which can be used for demonstrating toppling effects in block models.

Incipient toppling failure in a steeply jointed hard rock slope. Photograph by R.E.Goodman.

ing is limited because studies on the sensitivity of the slope to small changes in geometry become very tedious. In addition, the range of physical properties which can be incorporated into the model is limited by available modelling materials.

Attempts to overcome these problems by modelling toppling processes numerically have been made by Cundall[250], Byrne[252] and Hammett[253] and, while the results of these studies have been very promising, the numerical techniques are demanding on computer storage and time and these methods are not yet suitable for general engineering use. The authors believe that these numerical methods will eventually become practical design tools, particularly the computer graphics techniques being explored by Cundall at the University of Minnesota.

The method of analysis described below utilises the same principles of limiting equilibrium which have been used throughout the remainder of this book and, while the solution is limited to a few simple cases of toppling failure, it should provide the reader with a basic understanding of the factors which are important in toppling situations.

Limit equilibrium analysis of toppling on a stepped base

Consider the regular system of blocks shown in Figure 121 in which a slope angle θ is excavated in a rock mass with layers dipping at $90 - \alpha$. The base is stepped upwards with an overall inclination β. The constants a_1, a_2 and b shown in the figure are given by

$$a_1 = \Delta x . \text{Tan}(\theta - \alpha) \tag{98}$$

$$a_2 = \Delta x . \text{Tan}\,\alpha \tag{99}$$

$$b = \Delta x . \text{Tan}(\beta - \alpha) \tag{100}$$

where Δx is the width of each block.

In this idealised model, the height of the nth block in a

position below the crest of the slope is

$$y_n = n(a_1 - b) \qquad (101)$$

while above the crest

$$y_n = y_{n-1} - a_2 - b \qquad (102)$$

When a system of blocks, having the form shown in Figure 121, commences to fail, it is generally possible to distinguish three separate groups according to their mode of behaviour :

   a) a set of sliding blocks in the toe region,

   b) a set of stable blocks at the top, and

   c) an intermediate set of toppling blocks.

With certain geometries, the sliding set may be absent in which case the toppling set extends down to the toe.

Figure 122a shown a typical block (n) with the forces developed on the base (R_n, S_n) and on the interfaces with adjacent blocks (P_n, Q_n, P_{n-1}, Q_{n-1}).

When the block is one of the toppling set, the points of application of all forces are known, as shown in Figure 122b.

If the nth block is below the slope crest :

$$M_n = y_n \qquad (103)$$

$$L_n = y_n - a_1 \qquad (104)$$

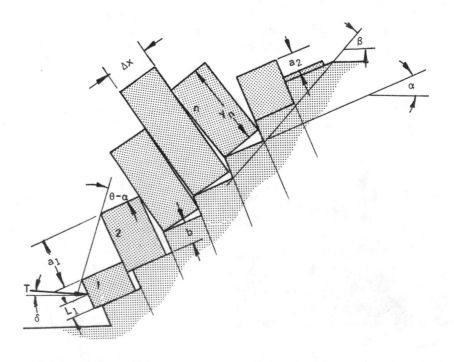

Figure 121 :   Model for limiting equilibrium analysis of toppling on a stepped base.

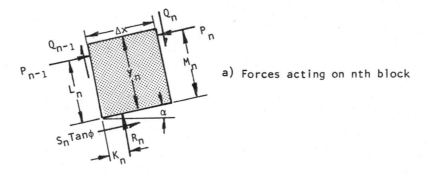

a) Forces acting on nth block

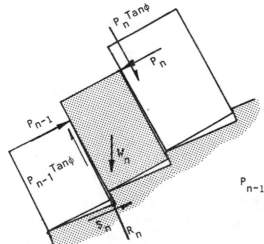

b) Toppling of nth block

$$P_{n-1} = \frac{P_n(M_n - \Delta x.Tan\phi) + (W_n/2)(y_n.Sin\alpha - \Delta x.Cos\alpha)}{L_n}$$

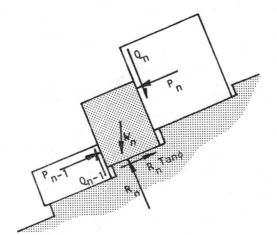

c) Sliding of nth block

$$P_{n-1} = P_n - \frac{W_n(Tan\phi.Cos\alpha - Sin\alpha)}{1 - Tan^2\phi}$$

Figure 122 : Limiting equilibrium conditions for toppling and for sliding of the nth block.

If the nth block is the crest block :

$$M_n = y_n - a_2 \tag{105}$$

$$L_n = y_n - a_1 \tag{106}$$

If the nth block is above the slope crest

$$M_n = y_n - a_2 \tag{107}$$

$$L_n = y_n \tag{108}$$

In all cases $K_n = 0$ $\hspace{4cm}$ (109)

For an irregular array of blocks, y_n, L_n and M_n can be determined graphically.

For limiting friction on the sides of the block :

$$Q_n = P_n.\text{Tan } \phi$$

$$Q_{n-1} = P_{n-1}.\text{Tan } \phi$$

By resolving perpendicular and parallel to the base,

$$R_n = W_n.\text{Cos } \alpha + (P_n - P_{n-1})\text{Tan } \phi \tag{110}$$

$$S_n = W_n.\text{Sin } \alpha + (P_n - P_{n-1}) \tag{111}$$

Considering rotational equilibrium, it is found that the force P_{n-1} which is just sufficient to prevent toppling has the value

$$P_{n-1,t} = \frac{P_n(M_n - \Delta x.\text{Tan } \phi) + (W_n/2)(y_n.\text{Sin } \alpha - \Delta x.\text{Cos } \alpha)}{L_n} \tag{112}$$

When the block under consideration is one of the sliding set,

$$S_n = R_n.\text{Tan } \phi \tag{113}$$

However, the magnitudes and points of application of all the forces applied to the sides and base of the block are unknown. The procedure suggested here is to assume that, as in the toppling case, conditions of limiting equilibrium are established on the side faces so that equations (110) and (111) apply. Taken in conjunction with (113), these show that the force P_{n-1} which is just sufficient to prevent sliding has the value

$$P_{n-1,s} = P_n - \frac{W_n(\text{Tan } \phi.\text{Cos } \alpha - \text{Sin } \alpha)}{1 - \text{Tan}^2\phi} \tag{114}$$

The assumption introduced here is quite arbitrary, but a little consideration will show that it has no effect on calculations of the overall stability of the slope. Any other reasonable assumption would produce the same results.

Calculation procedure

Let n_1 = uppermost block of the toppling set,

$\hspace{1cm} n_2$ = uppermost block of the sliding set.

a) To determine the value of ϕ for limiting equilibrium.

1. Assume a reasonable value of ϕ, such that $\phi > \alpha$.

2. Establish n_1 by determining the uppermost block of the whole group which satisfies the condition

$$y_n/\Delta x \ > \ Cot\,\alpha$$

3. Starting with this block, determine the lateral forces $P_{n-1,t}$ required to prevent toppling and $P_{n-1,s}$ to prevent sliding.

   If $P_{n-1,t} > P_{n-1,s}$, the block is on the point of toppling and P_{n-1} is set equal to $P_{n-1,t}$.

   If $P_{n-1,s} > P_{n-1,t}$, the block is on the point of sliding and P_{n-1} is set equal to $P_{n-1,s}$.

   For this particular block, and all other tall blocks of the system, it will be found that the toppling mode is critical, and this check is purely a matter of routine. It is required at a later stage to determine n_2 which defines the upper limit of the sliding section. Further checks should be carried out to ensure that

$$R_n > 0$$
$$|S_n| < R_n.Tan\,\phi$$

4. The next lower block (n_1-1) and all the lower blocks are treated in succession, using the same procedure.

5. Eventually a block may be reached for which $P_{n-1,s} > P_{n-1,t}$. This establishes block n_2, and for this and all lower blocks, the critical state is one of sliding. If the condition $P_{n-1,s} > P_{n-1,t}$ is not met for any of the blocks, the sliding set is absent and toppling extends down to block 1.

6. Considering the toe block 1 :

   If $P_o > 0$, the slope is unstable for the assumed value of ϕ. It is necessary to repeat the calculations for an increased value of ϕ.

   If $P_o < 0$, repeat the calculations with a reduced value of ϕ.

   When P_o is sufficiently small, the corresponding value of ϕ can be taken as that for limiting equilibrium.

b) To determine the cable force required to stabilise a slope.

   Suppose that a cable is installed through block 1 at a distance L_1 above its base. The cable is inclined at an angle δ degrees below horizontal and anchored a safe distance below the base. The tension in the cable required to prevent toppling of block 1 is

$$T_t = \frac{(W_1/2)(y_1.Sin\,\alpha - \Delta x.Cos\,\alpha) + P_1(y_1 - \Delta x.Tan\,\phi)}{L_1.Cos(\alpha + \delta)} \qquad (115)$$

   while the tension in the cable to prevent sliding is

$$T_s = \frac{P_1(1 - Tan^2\phi) - W_n(Tan\,\phi.Cos\,\alpha - Sin\,\alpha)}{Tan\phi.Sin(\alpha + \delta) + Cos(\alpha + \delta)} \qquad (116)$$

The normal and shear force on the base of the block are respectively :

$$R_1 = P_1.\text{Tan}\,\phi + T.\text{Sin}(\alpha + \delta) + W_1.\text{Cos}\,\alpha \quad (117)$$

$$S_1 = P_1 - T.\text{Cos}(\alpha + \delta) + W_1.\text{Sin}\,\alpha \quad (118)$$

The procedure in this case is identical to that described above apart from the calculations relating to block 1. The required tension is the greater of T_t and T_s defined by equations (115) and (116).

Example

An idealised example is illustrated in Figure 123. A rock slope 92.5m high is cut on a 56.6° slope in a layered rock mass dipping at 60° into the hill. A regular system of 16 blocks is shown on a base stepped at 1m in every 5 (angle $\beta - \alpha = 5.8°$). The constants are $a_1 = 5.0m$, $a_2 = 5.2m$, $b = 1.0m$, $\Delta x = 10.0m$ and $\gamma = 25$ kN/m^3. Block 10 is at the crest which rises 4° above the horizontal. Since $\text{Cot}\,\alpha = 1.78$, blocks 16, 15 and 14 comprise a stable zone for all cases in which $\phi > 30°$ ($\text{Tan}\,\phi > 0.577$).

In this example, $\text{Tan}\,\phi$ is set as 0.7855, P_{13} is then equal to 0 and P_{12} calculated as the greater of $P_{12.t}$ and $P_{12.s}$ given by equations (112) and (114) respectively. As shown in the table given on page 268, $P_{n-1,t}$ turns out to be the larger until a value of $n = 3$, whereupon $P_{n-1.s}$ remains larger. Thus blocks 4 to 13 constitute the potential toppling zone and blocks 1 to 3 constitute a sliding zone.

The force required to prevent sliding in block 1 tends to zero which indicates that the slope is very close to limiting equilibrium. The installed tension required to stabilise block 1 is 0.5 kN per meter of slope crest length, as compared with the maximum value of P (in block 5) equal to 4837 kN/m.

If $\text{Tan}\,\phi$ is reduced to 0.650, it will be found that blocks 1 to 4 in the toe region will slide while blocks 5 to 13 will topple. The tension in a bolt or cable installed horizontally through block 1, required to restore equilibrium, is found to be 2013 kN/meter of slope crest. This is not a large number, demonstrating that support of the "keystone" is remarkably effective in increasing the degree of stability. Conversely, removing or weakening the keystone of a slope near failure as a result of toppling can have serious consequences. The support force required to stabilise a slope from which the first n toe blocks have been removed can be calculated from equations (115) and (116), substituting P_{n+1} for P_1.

Now that the distribution of P forces has been defined in the toppling region, the forces R_n and S_n on the base of the columns can be calculated using equations (110) and (111); and assuming $Q_{n-1} = P_{n-1}.\text{Tan}\,\phi$, R_n and S_n can also be calculated for the sliding region. Figure 123 shows the distribution of these forces throughout the slope. The conditions defined by $R_n > 0$ and $|S_n| < R_n.\text{Tan}$ are satisfied everywhere.

n	y_n	$y_n/\Delta x$	M_n	L_n	$P_{n.t}$	$P_{n.s}$	P_n	R_n	S_n	S_n/R_n	Mode
16	4.0	0.4			0	0	0	866	500	0.577	
15	10.0	1.0			0	0	0	2165	1250	0.577	STABLE
14	16.0	1.6			0	0	0	3463	2000	0.577	
13	22.0	2.2	17	22	0	0	0	4533.4	2457.5	0.542	
12	28.0	2.8	23	28	292.5	-2588.7	292.5	5643.3	2966.8	0.526	
11	34.0	3.4	29	34	825.7	-3003.2	825.7	6787.6	3520.0	0.519	T
10	40.0	4.0	35	35	1556.0	-3175.0	1556.0	7662.1	3729.3	0.487	O
9	36.0	3.6	36	31	2826.7	-3150.8	2826.7	6933.8	3404.6	0.491	P
8	32.0	3.2	32	27	3922.1	-1409.4	3922.1	6399.8	3327.3	0.520	P
7	28.0	2.8	28	23	4594.8	156.8	4594.8	5872.0	3257.8	0.555	L
6	24.0	2.4	24	19	4837.0	1300.1	4837.0	5352.9	3199.5	0.598	I
5	20.0	2.0	20	15	4637.5	2013.0	4637.5	4848.1	3159.4	0.652	N
4	16.0	1.6	16	11	3978.1	2284.1	3978.1	4369.4	3152.5	0.722	G
3	12.0	1.2	12	7	2825.6	2095.4	2825.6	3707.3	2912.1	0.7855	
2	8.0	0.8	8	3	1103.1	1413.5	1413.5	2471.4	1941.3	0.7855	SLIDING
1	4.0	0.4	4	-	-1485.1	472.2	472.2	1237.1	971.8	0.7855	

CALCULATION OF FORCES FOR EXAMPLE SHOWN IN FIGURE 123.

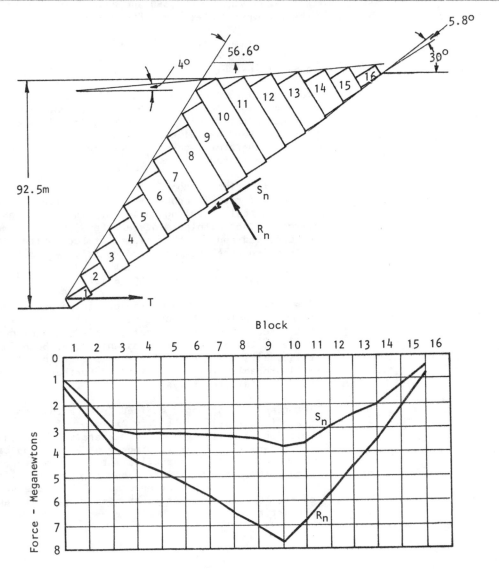

Figure 123 : Limiting equilibrium of a toppling slope with Tanϕ = 0.7855.

Factor of safety for limiting equilibrium analysis of toppling failures

The factor of safety for toppling can be defined by dividing the tangent of the friction angle believed to apply to the rock layers ($Tan\phi_{available}$) by the tangent of the friction angle required for equilibrium with a given support force T ($Tan\phi_{required}$).

$$F = \frac{Tan\phi_{available}}{Tan\phi_{required}} \qquad (119)$$

If, for example, the best estimate of $Tan\phi = 0.800$ for the rock surfaces sliding on one another, the factor of safety in the example, with $Tan\phi_{required} = 0.7855$ and with a 0.5 kN support force in block 1, is equal to $0.800/0.7855 = 1.02$. With $Tan\phi_{required} = 0.650$ and a support force of 2013 kN, the factor of safety is $0.800/0.650 = 1.23$.

Once a column overturns by a small amount, the friction required to prevent further rotation increases. Hence, a slope just at limiting equilibrium is meta-stable. However, rotation equal to $2(\beta - \alpha)$ will convert the edge to face contacts along the sides of the columns into continuous face contacts and the friction angle required to prevent further rotation will drop sharply, possibly even below that required for initial equilibrium. The choice of factor of safety, therefore, depends on whether or not some deformation can be tolerated.

The restoration of continuous face to face contact of toppled columns of rock is probably a very important arrest mechanism in large scale toppling failures. In many cases in the field, large surface displacements and tension crack formation can be observed and yet the volumes of rock which detatch themselves from the rock mass are relatively modest.

General comments on toppling failure

The analysis presented on the preceding pages can be applied to a few special cases of toppling failure and it is obviously not a rock slope design tool at this stage of development. However, the basic principles which have been included in this analysis are generally true and, with suitable additions, will probably provide a basis for further developments of toppling failure analysis.

The reader is strongly advised to work through the example given for himself and to try examples of his own since this can be a very instructive exercise. The calculations are relatively simple to program on a desk top calculator or a computer and the availability of such a program will enable the user to explore a number of possibilities, thereby gaining a better understanding of the sensitivity of the toppling process to changes in geometry and material properties.

Chapter 10 references

245. MULLER, L. New considerations of the Vajont slide. *Felsmechanik und engenieurgeologie.* Vol. 6, No. 1, 1968, pages 1-91.

246. HOFMANN, H. Kinematische Modellstudien zum Boschungs-problem in regelmassig geklüfteten Medien. *Veröffent-lichungen des Institutes für Bodenmechanik und Fels-mechanik.* Kalsruhe, Hetf 54, 1972.

247. ASHBY, J. Sliding and toppling modes of failure in models and jointed rock slopes. *M.Sc Thesis.* London University, Imperial College, 1971.

248. SOTO, C. A comparative study of slope modelling techniques for fractured ground. *M.Sc Thesis.* London University, Imperial College, 1974.

249. WHYTE, R.J. A study of progressive hanging wall caving at Chambishi copper mine in Zambia using the base friction model concept. *M.Sc Thesis.* London University, Imperial College, 1973.

250. CUNDALL, P. A computer model for simulating progres-sive, large scale movements in blocky rock systems. *Proc. Intnl. Symposium on Rock Fracture.* Nancy, France, 1971, Paper 11-8.

251. BURMAN, B.C. Some aspects of the mechanics of slope and discontinuous media. *Ph.D Thesis.* James Cook University of North Queensland, Australia, 1971.

252. BYRNE, R.J. Physical and numerical models in rock and soil slope stability. *Ph.D Thesis.* James Cook University of North Queensland, Australia, 1974.

253. HAMMETT, R.D. A study of the behaviour of discontinuous rock masses. *Ph.D Thesis.* James Cook University of North Queensland, Australia, 1974.

254. DE FREITAS, M.H and WATTERS, R.J. Some field examples of toppling failure. *Geotechnique.* Vol. 23, No. 4, 1973, pages 495-514.

255. HESLOP, F.G. Failure by overturning in ground adjacent to cave mining, Havelock Mine, Swaziland. *Proc. Third Congress Intnl. Soc. Rock Mechanics.* Denver. Vol. 2B. 1974, pages 1085-1089.

256. GOODMAN, R.E and BRAY, J.W. Toppling of rock slopes. *Proc. Speciality Conference on Rock Engineering for Foundations and Slopes.* Boulder, Colorado, ASCE, Vol.2, 1976.

Chapter 11: Blasting

Introduction

The excavation of rock slopes generally involves blasting and it is appropriate that the subject should receive some attention in a book on rock slope engineering. The fragmentation of rock by means of explosives is a major subject in its own right and the fundamentals of the subject have been dealt with in a number of excellent text books[257-260]. Most of the practical aspects of the subject have been described in handbooks published by manufacturers of drilling equipment and explosives[261-264] and no attempt will be made to summarise all this published material in this chapter.

This chapter will be devoted to those aspects of blasting which are particularly relevant to slope stability in open pit mining *.

Production blasting

Open pit mining is a process of controlled destruction of the rock mass in order that the ore may be extracted from it. The open pit blasting engineer is faced with the conflicting requirements of providing large quantities of well fragmented rock for the processing plant and of minimising the amount of damage inflicted upon the rock slopes left behind. A reasonable compromise between these two conflicting demands can only be achieved if the blasting engineer has a very sound understanding of the factors which control rock fragmentation and damage.

The economic pressures on the open pit blasting engineer to produce well fragmented rock are illustrated in the margin sketch which was taken from a paper by Harries and Mercer[265]. Production of a well fragmented and loosely packed muck pile facilitates loading and subsequent handling and crushing of the rock and the blasting engineer may vary each of the following parameters in order to achieve an optimum result:

1. Type, weight and distribution of explosive

2. Blasthole diameter

3. Effective burden

4. Effective spacing

5. Sub-drill depth

6. Blasthole inclination

7. Stemming

8. Initiation sequence for detonation of explosives

9. Delays between successive hole or row firing.

Factors 2 to 7 are defined in Figure 124.

Each of the factors listed above will be considered in relation to its influence upon the effectiveness of the production blast and its influence upon the amount of damage inflicted upon the remaining rock.

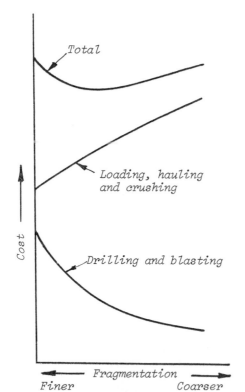

Effect of fragmentation on the cost of drilling, blasting, loading, hauling and crushing in an open pit operation.

---

*Much of the material presented in this chapter is based upon a company working document prepared by John Ashby of Golder, Brawner and Associates of Vancouver, Canada.

A production blast in a large open pit copper mine.

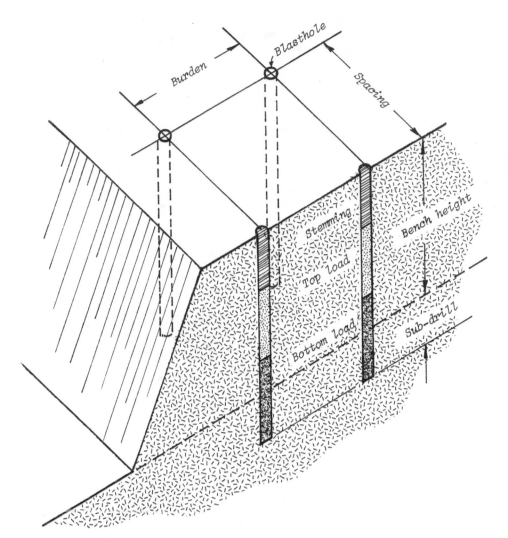

Figure 124 : Definition of bench blasting terms.

1. Type, weight and distribution of explosive

The strength of an explosive is a measure of the work done by a certain weight or volume of explosive . This strength can be expressed in absolute units or as a ratio relative to a standard explosive such as Gelignite or ANFO (Ammonium Nitrate/ Fuel Oil). Comparison of explosive strengths to that of ANFO is preferred because of the very wide use of this explosive in open pit production blasting.

The weight strengths of some commonly used explosives are compared with that of ANFO in Table VI. These weight strengths are useful when rationalising blasting designs when various explosives of different strengths are used. These figures are also useful in comparing costs since explosives are sold by weight. The volume or bulk strength is related to weight strength by specific gravity and this figure is important in calculating the volume of blasthole required to contain a given amount of explosive energy. A high bulk strength is obviously advantageous since less blasthole capacity is required to contain a required charge.

The sensitivity of an explosive is a characteristic which determines the method by which a charge may be detonated, the minimum diameter of the charge and the safety with which

TABLE VI - STRENGTHS OF EXPLOSIVES COMPARED TO ANFO			
Explosive	*Weight strength % ANFO*	*Bulk Strength % ANFO*	*Specific Gravity*
ANFO (Gravity loaded)	100	100	0.82
ANFO (Pressure loaded)	100	109	0.92
A.N. Gelatine Dynamite '75'	114	195	1.40
A.N. Gelignite '60'	95	174	1.50
A.N. 'Ligdyn 40'	85	149	1.43
A.N. 'Ligdyn 25'	68	119	1.42
'Anzite' Blue	114	193	1.40
'Anzite' Red	114	193	1.40
'Anzite' Yellow	97	165	1.43
'Aquamex'	100	170	1.39
Blasting Gelatine	127	233	1.50
'Exactex'	90	107	0.96
'Geophex'	85	163	1.55
'Hydrogel'	111	205	1.50
'Hydromex' M1	95	124	1.50
'Hydromex' M2	127	233	1.50
'Hydromex' M4	152	279	1.50
'Molanal' A	82	140	1.3-1.4
'Molonal' D	114	195	1.3-1.4
'Molonal' DQ	114	195	1.3-1.4
'Monograin'	90	107	0.90
'Plastergel'	95	174	1.50
'Quarigel'	101	186	1.50
Quarry 'Monobel'	100	121	0.98
'Rollex' 60	97	174	1.45
'Roxite'	63	121	1.65
'Seismex'	101	174	1.10
'Seismex' (Aluminised)	113	151	1.10
S.N. Gelignite 50%	89	163	1.50
Semigel	106	226	1.20
Semigel No. 2	99	135	1.12
'Ajax'	71	135	1.50
'Dynagex'	57	84	1.39
'Dynobel' No. 2	81	109	1.10
'Morcol'	80	116	1.20
'Polar' A3 'Monobel'	71	86	0.98
'Sunderite'	113	156	1.13

* *Adapted from ICI Handbook of Blasting Tables*[266]

the explosive may be handled. As a granular explosive such
as ANFO is compressed to higher densities in order to
develop a higher bulk strength, it becomes less sensitive
and finally becomes totally insensitive , a condition which
is known as 'dead pressed'.

Highly sensitive explosives will detonate when used in
smaller diameter charges and, as the sensitivity of the
explosive is decreased, the diameter of the charge must be
increased. Explosive slurries can be sensitised by the
provision of small chemically generated gas bubbles which,
when compressed adiabatically by the detonation of the
primer, are heated sufficiently to cause secondary detonation
which propagates through the charge at high speed.

The addition of aluminium to blasting slurries has become
common in recent years and, although expensive, the aluminium
will liberate considerably more energy than a fuel oil.
A typical fuel oil slurry with a weight strength of 82%
that of ANFO, a bulk strength of 116% that of ANFO and
a specific gravity of 1.15 will have the following characteristics when 11% aluminium is added : weight strength 123%
of that of ANFO, bulk strength 181% of that of ANFO and
specific gravity of 1.20. The addition of aluminium is
useful when a high energy bottom load is required to move
a heavy toe burden. Where drilling costs are high, the
use of high density aluminised slurries can result in a
substantial overall cost saving.

2. Blasthole diameter

Table VII is reproduced from a paper by Persson[267] and it
illustrates the cost benefit to be gained by increasing the
diameter of the blasthole. The use of 9 and 12 inch diameter
blastholes is common in open pit mining and many mines are
using 15 inch diameter holes and have made provision for
17 inch diameter hole drilling. The cost per litre of hole
shown in Table VII suggests that there is further benefit to
be gained by going to even larger holes but Persson points
out that there are some limitations to this increase.

Experience has shown that an effective utilisation of the
explosive charge is given by using a burden (see Figure 124)
of approximately 40 times the blasthole diameter. As the
blasthole diameter is increased, the burden approaches the
dimensions of the bench height and the explosion becomes
less efficient. Consequently, Persson suggests that the
blasthole diameter should be limited by the relationship :

$$\text{Blasthole diameter } d \leqq \text{ Bench height}/40 \qquad (120)$$

The use of large diameter blastholes also increases the
flyrock and air blast problems which will be discussed
later in this chapter. The use of large blastholes can
also give rise to excessive fracturing of the remaining
rock, a problem which will be discussed under the heading
of blasting delays.

The blasting engineer has to compromise between the potential
cost savings of large blastholes (always a major attraction
to mine management) and the problems resulting from the use
of large blast holes. This engineer would be well advised
to investigate the matter extremely thoroughly and to

TABLE VII - BLASTHOLE DIAMETER, VOLUME AND COST

| Hole diameter | | Hole volume | Hole cost (in granite) | |
Inches	Millimeters	litres/m	$/m	$/litres
1	25.4	0.51	1	2
2	50.8	2.03	2	1
3	76.2	4.56	3.1	0.67
4	101.6	8.11	4.1	0.50
6	152.4	18.2	6.1	0.33
10	254	50.7	10.1	0.20
15	381	114	15.2	0.13
20	508	203	20.3	0.10

discuss his recommendations with blasting engineers in open pit mines with similar rock types and production requirements before committing his management to the high capital investment of what may turn out to be the wrong drill-rig.

3. Effective burden

In order to understand the influence of the effective burden (the distance between the row of holes under consideration and the nearest free face), it is necessary to consider the sequence of rock breakage surrounding a blasthole. The following discussion is based upon a paper by McIntyre and Hagan[268].

When a charge confined in a blasthole is detonated, a longitudinal compressive strain wave is propagated outwards into the surrounding rock. The tangential strain associated with this wave creates a radial crack pattern around the blasthole. These cracks extend a radial distance of approximately four blasthole diameters into the rock.

When the compressive strain wave encounters a free face, created by the previous blast or by the detonation of the row of holes ahead of the one under consideration, a reflected tensile wave is generated and this is propagated back into the rock mass. This tensile strain wave tends to open those radial cracks around the blasthole which are tangential to the wave front, i.e. parallel or nearly parallel to the free face. The high-pressure explosion gases can penetrate these open cracks more easily than is the case for the other cracks and this generates a force acting outwards towards the free face. In a well designed blast with a correctly chosen burden, the outward force resulting from the explosion gas pressure will displace the burden rock towards the free face, fragmenting it and creating a good muck pile.

Too small a burden will allow the radial cracks to extend to the free face and this will give rise to venting of the explosion gases with a consequent loss of efficiency and the generation of flyrock and air blast problems.

Too large a burden will choke the blast and will give rise

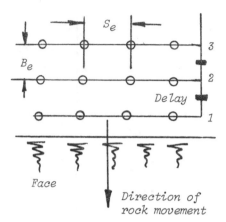

Effective burden and spacing for a square blasting pattern

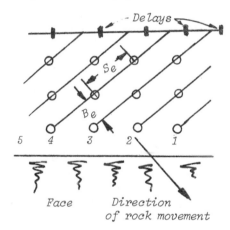

Effective burden and spacing for an en echelon blasting pattern

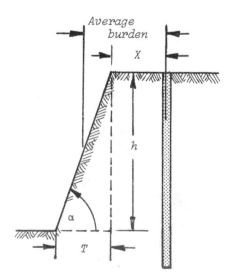

Average front row burden =

$X + \frac{1}{2}T = X + \frac{1}{2}h \cdot \cot\alpha$

to very poor fragmentation and a general loss of efficiency in the blast.

The effective burden B_e and the effective spacing S_e depend not only upon the blasthole pattern but also upon the sequence of firing. As illustrated in the margin sketch, a square blasthole pattern which is fired row by row from the face gives an effective burden equal to the spacing between successive rows parallel to the face. On the other hand, an identical pattern of blastholes can be fired en echelon resulting in completely different burdens and spacings as shown in the central margin sketch.

One of the most important questions to be considered in designing a blast is the choice of the front row burden. If vertical blastholes are used and the bench face is inclined as a result of the digging angle of the shovel in clearing the previous blast, the front row burden will not be constant but will vary with depth as illustrated in the margin sketch. Allowance can be made for this variation by using a higher energy bottom load in the front row of holes, e.g. increasing the aluminium content of a slurry for the bottom load as discussed in section 1. Alternatively, the blasthole can be inclined to give a more uniform burden as will be discussed later in this section. When the free face is uneven, the use of easer holes to reduce the burden to acceptable limits is advisable (see Figure 125).

Since the effectiveness of the fragmentation process depends upon the creation of a free face from which a tensile strain wave can be generated and to which the burden rock can move, the design of the front row blast is critical. Once this row has been detonated and effectively broken, a new free face is created for the next row and so on until the last row is fired.

4. Effective spacing

When cracks are opened parallel to the free face as a result of the reflected tensile strain wave, gas pressure entering these cracks exerts an outward force which fragments the rock and heaves it out onto a muck pile. Obviously, the lateral extent to which this gas can penetrate is limited by the size of the crack and the volume of gas available and a stage will be reached when the force generated is no longer large enough to fragment and move the rock. If the effect of a single blasthole is reinforced by holes on either side at an effective spacing S_e, the total force acting on the strip of burden material will be evened out and uniform fragmentation of this rock will result.

Experience suggests that an effective spacing of 1.25 times the effective burden gives good results. However, work by Lundborg of Nitro Nobel in Sweden , mentioned in the paper by Persson[267] shows that improved fragmentation may be obtained by increasing the spacing to burden ratio to as much as 4, 6 or even 8. This finding has been incorporated into the "Swedish" blasthole pattern illustrated in Figure 125. The reader should exercise caution is applying this pattern since its success depends upon rock of good quality. Joints running across a line of holes in a row could allow explosion gases to vent and reduce the effectiveness of the blast.

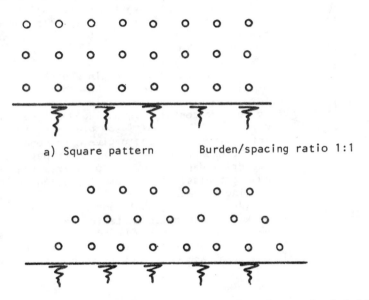

a) Square pattern   Burden/spacing ratio 1:1

b) Staggered pattern   Burden/spacing ratio 1:1.15

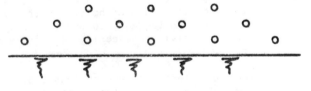

c) Swedish pattern   Burden/spacing ratio 1:4

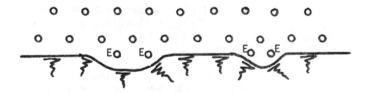

d) Use of easer holes (E) to move front
   row burden.

Figure 125 : Various blasthole patterns used in open pit
production blasting.

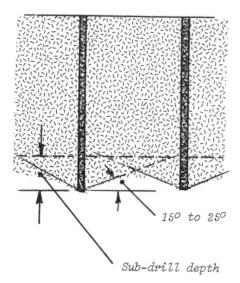

15⁰ to 25⁰

Sub-drill depth

Rock breakage at the bottom of a blasthole

5. *Sub-drill depth*

Sub-drilling or drilling to a depth below the toe of the bench is necessary in order to break the rock on the floor of the bench in such a way that the shovel can dig it to the required level. Poor fragmentation at this level can lead to very expensive shovel operation due to delays and break-downs. Excessive fragmentation probably means that the rock forming the lower bench is damaged and this means a reduction in stability.

As illustrated in the margin sketch, breakage of the rock usually projects from the base of the bottom load in the form of an inverted cone with sides inclined at 15° to 25° to the horizontal , depending upon the strength and structure of the rock. In multi-row blasting, the breakage cones interact and link up to give a reasonably even transition from broken to undamaged rock. Experience has shown that a sub-drill depth of 0.2 to 0.3 times the distance between adjacent blastholes is usually adequate to ensure effective digging to bench grade. It is particularly important that sub-drill depths should not be exceeded in the front and back rows otherwise unstable crest and toe conditions can be created in the new bench. In fact, there is good justi- fication for reducing or even eliminating sub-drilling in the front and back rows if bench stability is critical.

6. *Blasthole inclination*

As pointed out on page 277, the front row burden varies with depth if vertical blastholes are used and the bench faces are inclined. Many drill rigs can only be set up to drill vertical holes and, in such cases, effective front row blasting can only be achieved by varying the charge strength with depth of hole. On the other hand, there is an increasing tendency to use inclined blastholes , partic- ularly in civil engineering blasting where smaller rigs are used.

Inclined blastholes are obviously advantageous for the front row and, by drilling the blastholes parallel to the bench face, a constant front row burden is achieved. In order to maintain a constant burden with depth for the remainder of the blast, it follows that all the blastholes should be inclined. Some blasting engineers would argue that the use of blastholes drilled at between 10° and 30° to the vertical will give better fragmentation [269], greater dis- placement and reduced back-break problems[270].

7. *Stemming*

The use of stemming consisting of drill cuttings is a gener- ally accepted procedure for directing explosive effort into the rock mass. The same arguments as were used in the discussion on burden apply in the case of stemming. Too little stemming will allow the explosion gases to vent and will generate flyrock and air blast problems as well as reducing the effectiveness of the blast. Too much stemming will give poor fragmentation of the rock above the top load.

Hagan[270] points out that, although drill cuttings are usually the most convenient and cheapest materials to use for stemming, they are not necessarily the best. Dry and

well graded angular materials, e.g 10-15 mm crushed rock,
are more effective than wet materials. The use of materials
other than drill cuttings depends upon the availability and
cost of such materials.

The optimum stemming length depends upon the properties of
the rock and can vary between 0.67 and 2 times the burden
width [270] . Stemming columns shorter than two thirds of the
burden width normally cause flyrock, airblast and backbreak
problems. Excessively long stemming columns cause poor
fragmentation of the rock above the top of the charge.

If unacceptably large blocks are obtained from the top of
the bench, even when the minimum stemming column consistent
with flyrock and airblast problems is used, fragmentation
can be improved by locating a small 'pocket' charge centrally
within the stemming[270].

8. Initiation sequence for detonation of explosives

Having drilled and charged a blast it is then necessary to
tie-up the pattern. This involves laying out detonating
cord along the 'rows' to form trunk lines which are then
tied to the down-line of each charge. The rows are normally
parallel to the free face but, as shown in the margin sketch
on page 277, may be inclined to it. Safety lines are used
in large patterns to ensure complete detonation and reduce
the risk of cut-offs. A perimeter or 'ring' line is then
tied around the pattern to provide a further safe-guard.

The firing or initiating line will normally be connected to
the middle of the front row trunk line. Other firing sequences
are illustrated in Figure 126. The blasting sequence, after
the initiation of the first row, is controlled by the use
of delays as discussed in the next section.

9. Delays between successive hole or row firing

In a typical large open pit mine, the total amount of rock
handled per day is between 100,000 and 300,000 tons. Breakage
of this rock requires the detonation of 20 to 75 tons of
explosive. Simultaneous detonation of this quantity of
explosive generates accelerations in the pit slopes equivalent to those which would be associated with a reasonable
earthquake and which would certainly generate slides in
those slopes close to limiting equilibrium. In order to
prevent such instability and also to improve the effectiveness of the blast, the blast is broken down into a number
of sequential detonations by means of delays.

When the front row of a blast is detonated and moves away
from the rock mass to create a new free face, it is important
that time should be allowed for this new face to be established before the next row is detonated. Typically, delay
intervals of 1 to 2 milliseconds per foot of burden (3 to
6 milliseconds per metre) are used in production blasting.
A typical blast with a burden of 25 feet may be delayed
as follows :

            Front row -   instantaneous
            Row 2   -   35 milliseconds delay
            Row 3   -   70 milliseconds delay
            Row 4   -   105 milliseconds delay

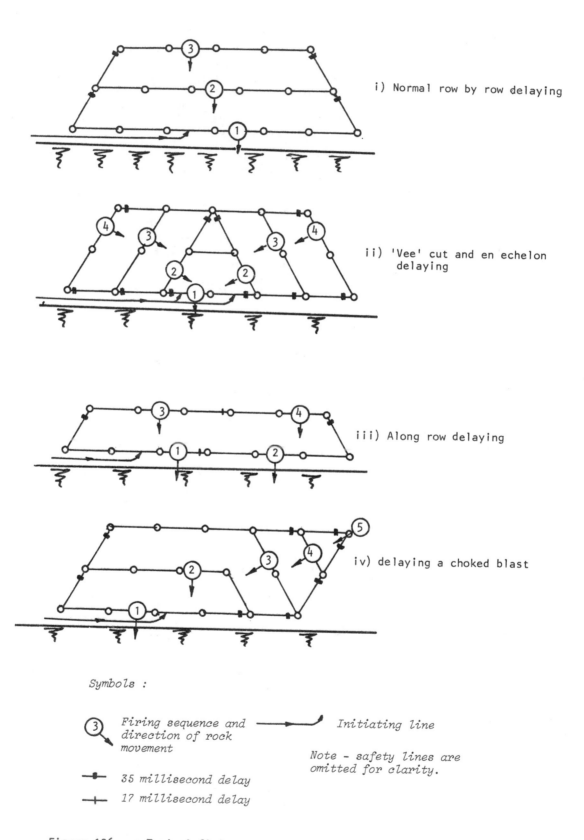

i) Normal row by row delaying

ii) 'Vee' cut and en echelon delaying

iii) Along row delaying

iv) delaying a choked blast

Symbols :

③ Firing sequence and direction of rock movement

35 millisecond delay

17 millisecond delay

Initiating line

Note - safety lines are omitted for clarity.

Figure 126 : Typical firing sequences.

Normal row by row delaying is the simplest and generally
the most satisfactory firing sequence. Delay patterns can
become quite complex and should be planned and checked
carefully. The number of rows should not exceed 4 to 6
as choking occurs with deeper blasts and vertical craters
may be formed above the back rows which do not have sufficient
room to move laterally.

An alternative is to use en echelon delaying and to initiate
the firing sequence with a vee cut to create the first free
face. This type of firing sequence can be useful when
blasting in strongly jointed rock where near vertical joints
strike across the bench at an angle to the face. Some
blasting engineers suggest that the row line should bisect
the angle between the strike of the joint and the face or the
strike of two joint sets. There do not appear to be very
strong theoretical grounds to support this suggestion
and the optimum blasting direction is usually established
by carefully controlled trials. McIntyre and Hagan[268] describe the use of en echelon blasting patterns in place of
normal row by row firing sequences to improve the performance
of blasts and the stability of the benches in a large strip
coal mine.

The importance of blasting to a free face has already been
stressed and it is equally important to plan the blast so
that suitable free faces are created for the next blast.
When free faces are not available, e.g. when the face
changes direction, there is a danger that the blast may
become choked and it may be necessary to use delays in
order to work in such a situation. A typical firing sequence
for a choked blast situation is illustrated in Figure 126 iv.

The use of delays in a blast is one of the most powerful
weapons in the fight against excessive blast damage and
the instability of benches in open pit mines. This
question will be discussed more fully in a later section
of this chapter.

Production blasting design

Nine factors which influence the effectiveness of a production
blast have been discussed on the preceding pages. Random
variation of each of these nine factors in order to achieve
good blasting results is likely to lead to total confusion
and it is essential that the blasting engineer should
follow a systematic path in attempting to achieve an optimum
result. One of the most useful elements in such a systematic
path is the use of the *powder factor* concept to link the
first four factors together. This concept deserves careful
attention.

Powder factor design

A powder factor or specific charge is a measure of the weight
of explosive required to break a unit volume or weight of
rock. The powder factor may be expressed as lb/short ton,
kg/tonne, lb/yd^3, kg/m^3 or by the reciprocals of these values
i.e. short ton/lb., tonne/kg., etc. and the reader should
pay particular attention to the quoted units when reading
tables of powder factors. It is also necessary to relate
the powder factor to the type of explosive used because
the amount of energy for a given weight of explosive varies

with explosive type as discussed on page 273. Unless otherwise stated, all powder factors quoted on the following pages are referred to gravity loaded ANFO.

Choosing a powder factor (deciding just how much explosive to place in a particular rock mass) is the first and most critical step in a blasting design. Unfortunately, there are very few simple rules or relationships which can be used to decide upon an optimum powder factor for a given rock mass. The evaluation of full scale production blasts is the most direct and usually the most successful method of optimising a powder factor but this can be very expensive unless the blasting engineer has a reasonable starting point and some sound criteria for judging the effectiveness of the trial blast.

Broadbent[271] has reported attempts to correlate the in situ seismic velocity to required powder factor and these efforts may eventually lead to a reasonable method of making a first choice of powder factor. The correlation reported by Broadbent is shown in Figure 127.

An empirical relationship between the required powder factor and the properties of the rock mass (characterised by the effective friction angle $(\phi + i)$ and the fracture frequency) was developed at the Bougainville Copper Ltd. mine by Ashby. This relationship is presented graphically in Figure 128. This type of relationship is particularly important because it takes the properties of the rock mass into account and does not rely on small scale laboratory tests on intact rock specimens. The authors feel that it should be possible to develop such relationships further in order to provide a reliable estimate of the required powder factor.

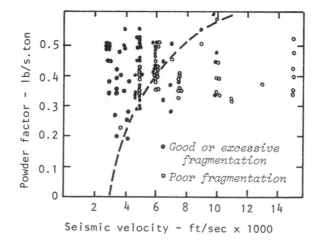

Figure 127 : Correlation between in situ seismic velocity and required powder factor . After Broadbent[271].

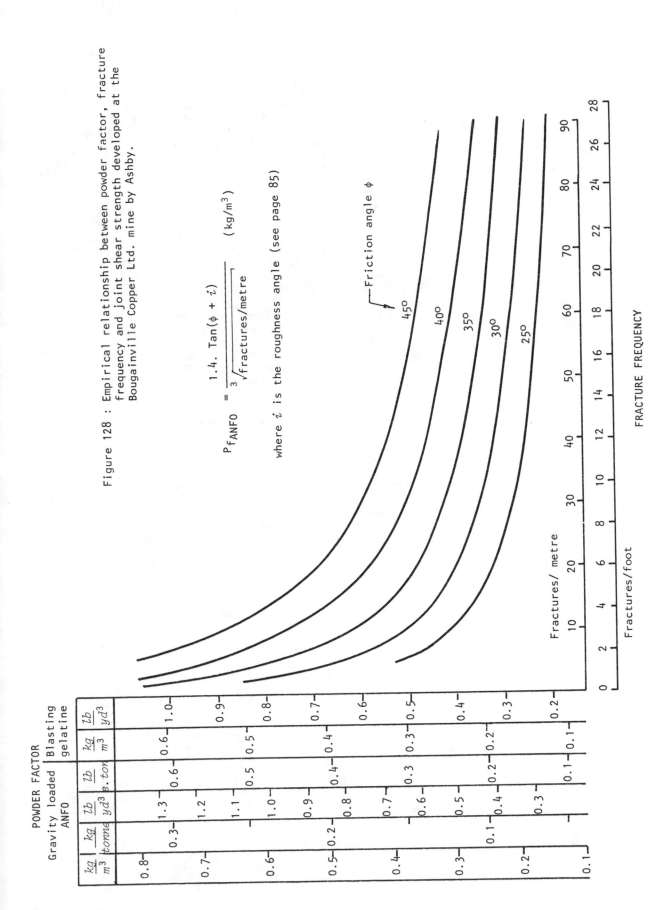

Figure 128 : Empirical relationship between powder factor, fracture frequency and joint shear strength developed at the Bougainville Copper Ltd. mine by Ashby.

$$Pf_{ANFO} = \frac{1.4 . \ Tan(\phi + i)}{\sqrt[3]{\text{fractures/metre}}} \quad (kg/m^3)$$

where i is the roughness angle (see page 85)

Optimum charge

Having chosen a powder factor, either from experience or by using a relationship such as that given in Figure 128, the blasting engineer has then to determine the optimum charge to be placed in each blasthole. This charge is calculated from the relationship :

$$\text{Optimum charge} = \text{powder factor} \times \text{burden volume} \qquad (121)$$

where

$$\text{Burden volume} = \text{burden} \times \text{spacing} \times \text{bench height} \qquad (122)$$

For example, assume a burden of 26 feet, a spacing of 30 feet and a bench height of 40 feet, the burden volume is $26 \times 30 \times 40 = 31200$ ft^3 = 1155.5 yd^3. If the powder factor has been chosen as 0.8 lb/yd^3, the optimum charge is found to be 925 lb.

In a typical open pit mine, the bench height is fixed by the dimensions of the loading equipment, the blasthole diameter is fixed by the equipment available and the sub-drill length and stemming column length variations are restricted by fragmentation requirements. Consequently, the blasthole volume is defined within reasonably close limits . The volume of a blasthole of diameter d is :

$$\text{Blasthole volume} = \tfrac{1}{4}\pi d^2 \times (\text{bench height} + \text{sub-drill length} - \text{stemming}) \qquad (123)$$

If the blasthole volume does not match the volume of explosive which makes up the optimum charge, the simplest means of correcting this mismatch is by changing the optimum charge by altering the burden and spacing. Most open pit blasting engineers prepare blasting charge tables which give the optimum charge for a range of burdens, spacings and powder factors for the bench heights and blasthole diameters on their particular mine. Such charge tables greatly facilitate the layout of blasthole patterns. A typical charge table is reproduced below.

CHARGE TABLE - POUNDS ANFO   SPACING = 30 FEET
Bench height = 40 feet, Rock density = 12 cu.ft/ton

BURDEN (FT) POWDER FACTOR (LB/TON)	20	22	24	26	28	30	32	34
0.25	500	550	600	650	700	750	800	850
0.30	600	650	700	800	850	900	950	1000
0.35	700	750	850	900	1000	1050	1100	1200
0.40	800	900	950	1050	1100	1200	1300	1350
0.45	900	1000	1100	1150	1250	1350		
0.50	1000	1100	1200	1300	1400			
0.55	1100	1200	1300					
0.60	1200	1300		Maximum charge 1400 lb.				
0.65	1300							
0.70	1400							
0.75								

SUMMARY OF OPEN PIT PRODUCTION BLASTING DESIGN PROCEDURE

Variables	Typical values	Determining factors and comments
BENCH HEIGHT h	35 to 50 feet 10 to 15 metres	Loading equipment dimensions
BLASTHOLE DIAMETER d	$9\frac{7}{8}$, 12, 15 in. 250 to 380 mm.	Blasthole drilling equipment capacity. Typically d = h/40 for high powder factors. For low powder factors, d = h/66 may be adequate. Cost/hole volume generally decreases with hole diameter.
EXPLOSIVE TYPE	ANFO (Ammonium Nitrate/ Fuel Oil)	Holes must be dry or lined. Low cost, simple to mix, safe to handle components. Fairly safe in use. *Low bulk strength*. Good for low powder factor situations. May be expensive in high powder factor situations owing to high drilling costs.
	ANFO Slurry	Used in dry or wet holes. Usually mixed and pumped by contractors. Higher bulk strength hence lower drilling costs than for ANFO, but lower weight strength and higher cost than ANFO.
	Metal fuel (Replacement or partial replace- of fuel oil with aluminium)	May be added to ANFO or slurry. Cost is high on equivalent weight strength basis, lower on bulk strength basis. Use may significantly reduce drilling costs, particularly in high powder factor situations.
DESIGN POWDER FACTOR	0.4 to 2.0 lb/yd^3. 0.25 to 1.2 kg/m^3. 0.2 to 1.0 lb/s.ton 0.1 to 0.4 kg/tonne	Controlled by rock conditions. May be determined by crater tests or by empirical relationship but ultimate test is full scale blast. Design powder factor should be expressed in standard dimensions of weight, length and standard explosive.
ACTUAL POWDER FACTOR	as above	Measured for whole blast or total production - less than design powder factor due to back-break and free digging.
BURDEN VOLUME Burden x spacing x bench height.	600 to 2600 yd^3 500 to 2000 m^3	Cross sectional area of blasthole x charged length x design powder factor x explosive density.
SUB-DRILL DEPTH	3 to 7 feet 1 to 2 metre	Determined approximately by crater testing and hole spacing. 0.2 to 0.3 x burden.
STEMMING COLUMN LENGTH	20 to 25 feet 6.6 to 8.3 m.	Determined approximately by crater testing and burden. 0.67 to 2 x burden.

Variables	Typical values	Determining factors and comments
MAIN CHARGE WEIGHT	Minimum Maximum 800 to 1600 lb 350 to 700 kg	Density of explosive x available hole volume (cross sectional area of hole x (bench height + sub-drill - stemming))
BURDEN:SPACING RATIO	Square 1:1 Staggered 1:1.15 Swedish 1:>1 to 1:<8	Square pattern - easy to lay out, poor charge distribution, more difficult to delay. Staggered pattern - even charge distribution, easy delaying. Good in large patterns for low powder factors. Swedish pattern - good for high powder factor in good rock.
BURDEN	12 to 30 feet 4 to 10 metres	Approximately 45 x hole diameter
SPACING	18 to 33 feet 6 to 11 metres	Determined by burden:spacing ratio
MINIMUM ALLOWABLE DISTANCE FROM CREST TO FRONT ROW	10 feet 3 metres	Determined by safety requirements. Drill dimensions and bench crest conditions determine safe distance from front row to bench crest.
FRONT ROW MEAN BURDEN	12 to 30 feet 4 to 10 metres	Crest to front row distance + $\frac{1}{2}$ bench height x Cotangent face angle.
FRONT ROW BURDEN VOLUME		= mean from row burden x spacing x height
MAXIMUM DISTANCE FROM CREST TO FRONT ROW	10 to 30 feet 3 to 10 metres	$\dfrac{\text{Maximum charge}}{\text{Powder factor x spacing x bench height}}$ - $\frac{1}{2}$ maximum toe distance. If maximum crest distance for a pattern is less than the minimum allowable crest to front row distance the pattern must be closed in. Note that this is a common fault in widely spaced patterns.
DELAY INTERVALS	17, 25, 35, 45 milliseconds.	1 to 2 milliseconds/ft (3 to 6 m.s/m) of effective burden. Place delays 3 ft. (1 m) from delayed holes.
DELAY SEQUENCE		Delay to free face. Delays may be used to avoid choked blast. Delays may be used to modify burden: spacing ratios. Delays can be used to reduce maximum instantaneous charge levels.
MAXIMUM INSTANTANEOUS CHARGE	1000 to 10000 lb. 450 to 4500 kg	Loosely referred to as charge/delay depends upon risk of damage and distance to surrounding structures.
BACK-BREAK	25 to 45 feet 8 to 15 metres	Diggable back-break approximately equal to burden. Varies with rock conditons, explosive strength and powder factor of back row.
BACK ROW TO FINAL WALL DISTANCE	25 to 45 feet 8 to 15 metres	Should be equal to anticipated back-break.

Evaluation of a blast

Once the dust has settled and the fumes have dispersed after a blast, an inspection of the area should be carried out. The main features of a satisfactory blast are illustrated in Figure 129.

The front row should have moved out evenly but not too far. Excessive throw is unnecessary and very expensive to clean up. The heights of most open pit benches are designed for efficient shovel operation. Low muck piles, due to excessive front row movement, represent low productivity excavation volumes.

The main charge should have lifted evenly and cratering should, at worst, be an occasional occurrence. Flat or wrinkled areas are indicative of misfires or of poor delaying.

The back of the blast should be characterised by a drop, indicating a good forward movement of the free face. Tension cracks should be visible in front of the final diglines. Excessive cracking behind the final digline represents damage to the slopes and wastage of powder.

The quality of a blast has a significant influence on later stages of the materials handling process of the mining system. Secondary drilling and blasting, digging rate, the condition of the mine haul roads, crushing and grinding are all influenced by the efficiency of the blast. For this reason, critical evaluation of the fragmentation and digging conditions of each blast is an essential part of any blasting operation. Although this evaluation is time consuming and therefore expensive, the cost of this evaluation is usually justified in the development of an efficient blasting system.

The amount of secondary drilling and blasting is easily quantified and contributes directly to the drilling and blasting budget. Efforts should be made to compare modifications to the primary blasting system to reductions in secondary drilling and blasting costs.

Secondary drilling of over-size fragments in a Swedish mine.

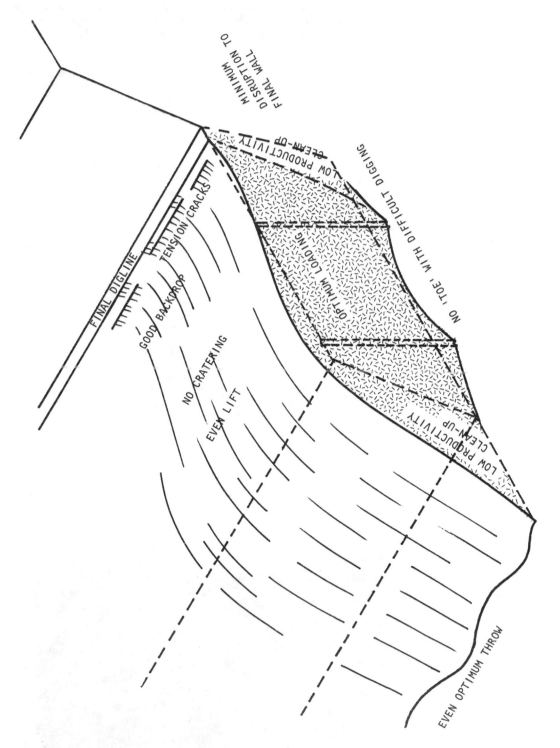

Figure 129 : Features of a satisfactory production blast.

Oversized fragments, hard toes , tight areas and low muck piles (caused by excessive throw) have the most significant detrimental effect on the digging rate and digging conditions. A study of shovel behaviour and of complaints from shovel operators helps to maintain an awareness of these problems amongst blasting personnel. An attempt should be made to measure digging rate by noting the time required to fill trucks or by comparing average daily production rates. Similarly, shovel wear and tear should be noted since this may reflect difficult digging conditions.

Poor fragmentation of the toe due to an excessive toe burden can lead to poor digging and can also lead to uneven haul road conditions at a later stage of the pit development. Uneven haul roads lead to suspension wear on the trucks and can also cause spillage which can give rise to high tyre wear. An attempt to correct this problem by additional sub-drilling rather than by correcting the front row charge can lead to excessive sub-break which can give rise to blasthole instability during a later drilling phase and also to poor bench crest conditions when the face is moved back.

Modification of blasting methods

When it is evident that unsatisfactory results are being obtained from a particular blasting method and that the method should be modified, the blasting engineer may have to embark on a series of trials in order to arrive at an optimum design. As with any trials, careful documentation of each blast is essential and, whenever possible, only one variable at a time should be changed. The following sequence of test work is an illustration of the type of experiment which would be carried out to evaluate the cost effectiveness of using a higher energy explosive. Similar test sequences could be carried for each of the other factors which are relevant in a particular situation.

Rationalisation
a. Document present powder factors on an equivalent energy basis using the weight strengths of various explosives compared to that of the explosive in current use. Weight strength data should be obtained from the explosives manufacturer if these are not already available.

Evaluation
b. For a blast with the explosive currently in use, document the behaviour of the blast during initiation and the condition of the resulting muck pile.
c. Document rate and conditions of digging.
d. Document fragmentation based upon the ratio of oversized material requiring secondary blasting to the total blast tonnage.
e. Document drilling and blasting costs.

Experimentation
f. Select a similar area of ground and carry out a blast with a higher powder factor which is obtained by using a higher energy explosive e.g. by increasing the aluminium content of a slurry.
Evaluation
g. Document the results as for steps b to e.
h. Carry out a cost-benefit study.

i. Repeat the experiment before preparing a statement to management suggesting a modification or a retention of the existing method.

Blasting damage and its control

In a review of factors affecting blasting damage, carried out by Ladegaard-Pedersen and Dally[272], four types of damage are identified :

1. Structural damage due to vibrations induced in the rock mass

2. Damage due to flyrock or boulders ejected from the blast area

3. Damage due to airblast

4. Damage due to noise

In most open pit mines, only the first two types of damage are of concern but, with increasing mineral demands and population densities, airblast and noise damage can no longer be ignored. In the USA, recent legislation has imposed a restriction of 140 decibels on noise from blasting and it is probable that such legislation will become general in years to come.

Space in this book does not permit a complete review of blast damage mechanisms and the interested reader is referred to the body of excellent literature on this subject [257,258, 260,272-282]. A brief review of those aspects of the subject which are most relevant in open pit blasting is given on the following pages.

Structural damage

The fragmentation of rock by the detonation of an explosive charge depends upon the effects of both strain induced in the rock and upon the gas pressure generated by the burning of the explosive. Structural damage resulting from vibration is dependent upon the strains induced in the rock.

When an explosive charge is detonated near a free surface, two body waves and one surface wave are generated as a result of the elastic response of the rock. The faster of the two waves propagated within the rock is called the primary or P wave while the slower type is known as the secondary or S wave. The surface wave, which is slower than either the P or S wave, is named after Rayleigh who proved its existence and is known as the R wave. Ladegaard-Pedersen and Dally[272] suggest that, in terms of vibration damage, the R wave is the most important since it propagates along the surface of the earth and because its amplitude decays more slowly with distance travelled than the P or S waves. While this may be true for damage to surface structures, it should also be remembered that damage to the rock remaining in the final slopes is of prime concern to the open pit engineer and hence the effects of all three waves may be of concern.

In the review by Ladegaard-Pedersen and Dally it is concluded that the wide variations in geometrical and geological conditions on typical blasting sites preclude the solution of ground vibration problems by means of elastodynamic equations and that the most reliable predictions are given by empirical relationships developed as a result of observations of actual blasts. Of the many empirical relationships which have been postulated, the most reliable appears to be that relating particle velocity to scaled distance.

The scaled distance is defined by the function $R/\sqrt{W}$ where R is the radial distance from the point of detonation and W is the weight of explosive detonated per delay. The US Bureau of Mines has established that the maximum particle velocity V is related to the scaled distance by the following relationship :

$$V = k(R/\sqrt{W})^{\beta} \tag{124}$$

where k and β are constants which have to be determined by measurements on each particular mine or blasting site.

Equation 124 plots as a straight line on log-log paper and the value of k is given by the V intercept at a scaled distance of unity while the constant β is given by the slope of the line. A hypothetical example of such a plot is given in Figure 130.

In order to obtain data for the construction of a plot such as that given in Figure 130, some form of vibration measuring instrument must be available to the open pit blasting engineer. Duvall[283] has drawn up specifications for two types of velocity seismograph for the measurement of blasting vibrations and these specifications are reproduced in Table VIII. Ladegaard-Pedersen and Dally have listed the specifications of a number of commercially available seismographs which meet these requirements and their list is summarised in Table IX.

Langefors and Kihlström[257], Duvall and Fogelson[278] and others have examined the relationship between maximum particle velocity and structural damage and the following threshold values have been suggested :

Particle velocity V		Damage
In/sec	*mm/sec*	
2	51	Limit below which risk of damage or structures, even old buildings, is very slight (less than 5%)
5	127	Minor damage, cracking of plaster, serious complaints.
12	305	Rockfalls in unlined tunnels
25	635	Onset of cracking of rock
100	2540	Breakage of rock

These threshold values have been plotted in Figure 131, for different combinations of distance R and weight of explosive charge W. In plotting this figure, values of k = 200 and β = -1.5 have been used in solving equation 124. These values are based upon an average range derived from a paper by Oriard[284] and this plot should only be used for very general guidance. When damage is a serious potential problem on an open pit mine, values of k and β should be determined from a plot of measured particle velocities, such as that presented in Figure 130, and a chart similar to that presented in Figure 131 plotted for

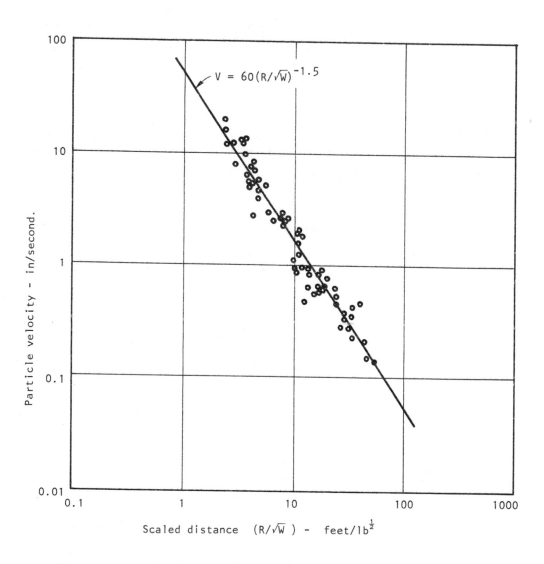

Figure 130 : Hypothetical plot of measured particle velocity versus
scaled distance from blast.

Typical values of k and β quoted by Oriard[284] are :

Down hole blasting   :   k = 26 to 260,   β ≈ -1.6
Coyote blasting   :   k = 5 to 20,   β ≈ -1.1
Pre-splitting   :   k ≈   800,   β ≈ -1.6

TABLE VIII - SPECIFICATIONS FOR VELOCITY SEISMOGRAPHS BY DUVALL[283]			
Property		Type 1	Type 2
Resonant frequency	cycles/sec	1-5	1-5
Useful frequence range	cycles/sec	10-125	6-500
Sensitivity Inches deflection per in/sec		0.5	0.1-100
Particle velocity range	in/sec	0.1-2.0	0.1-10
Attenuation range		1	1-100
Allowable tilt angle	degrees	2	2
Internal impedance of coil	ohms	<1000	<1000
Maximum acceleration without instrument damage		10g	100g
Maximum displacement without distortion	in	0.1	0.3
Record paper speed	in/sec	5-10	5-50
Power requirement*	volt x amp	<100	<100
Total weight of instrument	lb	<50	<50

* The power requirement is based upon the assumption that the instrument
  will be operated from an automobile storage battery.

TABLE IX - COMMERCIALLY AVAILABLE VELOCITY SEISMOGRAPHS		
Manufacturer	Type	Comments
W.F.Sprengnether Instruments	VS-1200 Engineering and Research Seismograph.	A highly sophisticated multi-purpose instrument which exceeds Duvall's type 2 specifications on almost all counts.
W.F.Sprengnether Instruments	VS-1100 Engineering Seismograph	A three-component package which can also be used for seismic refraction work.
W.F.Sprengnether Instruments	VS-4000 Portable Blast and Vibration Seismograph.	Self-contained three-component seismograph suitable for open pit work.
VME Nitro Consult, Inc.	Velocity Recorder Model E	Portable seismograph suitable for blast monitoring. Additional options on models F and G.
VME Nitro Consult, Inc.	Velocity Seistector	Responsive to ground motion in 3 dimensions. Can record for 30 days on 115 volt AC. Suitable for monitoring blasts.
VME Nitro Consult, Inc.	Seismolog	Self-contained portable instrument suitable for open pit work.
VME Nitro Consult, Inc.	Ampligraph	Clockwork instrument records motions in a single plane for 1 week.
Dallas Instruments, Inc.	BR-2-3 Blast Monitor	Peak velocity recorder which can operate for 30 days for monitoring.
Dallas Instruments, Inc.	3B-2 Blast Monitor	Three component instrument measures peak particle velocity for monitoring.
Dallas Instruments, Inc.	BR-2 Blast Monitor	Measured vertical peak particle velocity only.
Dallas Instruments, Inc.	BT-4 Blast Monitor Tape System	Tape recording of three veolocity components and over-pressure value.
Slope Indicator Co.	S-2 Vibration Monitor	Records three velocity components at each of two stations up to 2000 feet apart.
Slope Indicator Co.	S-3 Vibration Monitor	As for S-2 but can record displacement, velocity and acceleration.
Nimbus Instruments	ES-6 Engineering Seismograph	Portable instrument suitable for blast monitoring and seismic refraction.

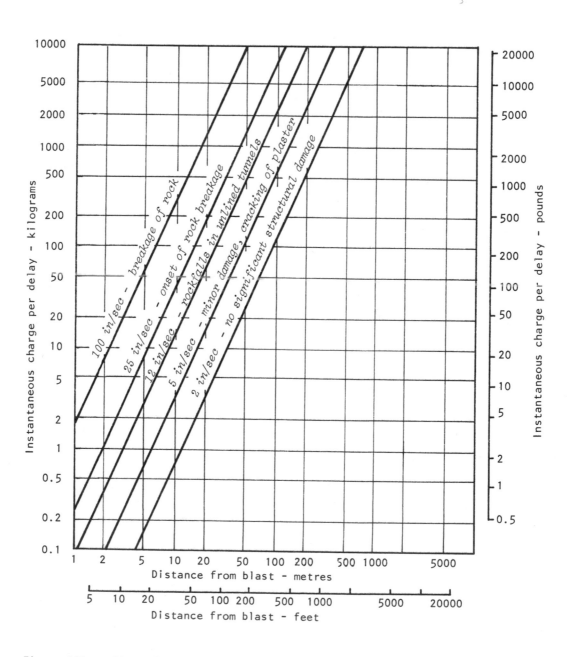

Figure 131 : Plot of particle velocities induced at given distances
by particular charges.

these values.

While the open pit engineer may feel that this problem is of little concern to him, particularly if there are no old buildings or residential dwellings close to his mine, he would do well to remember that the rock forming the final walls of the open pit should be damaged as little as possible if their stability is to be retained. Figure 131 shows that the detonation of 1000 pounds of explosive per delay will cause rock breakage up to 100 feet from the blasthole while the detonation of 10000 pounds per delay will extend this distance to approximately 400 feet.

It is obvious, from this example, that the use of delays to limit the amount of explosive detonated at any one time is an extremely important method of limiting damage to the rock which is to remain in the slope after a bench has been blasted. This is particularly important when the ultimate pit slope is being approached and when the minimum possible charge per delay should be used.

Figure 132 illustrates the damage to rock in the immediate vicinity of a blasthole. It will be noted that the breakage is severe enough to ignore potential weakness planes in the rock and to cause fracturing of the intact rock material. The stability of a slope in which such blast damage has been induced will be reduced to a point where it may be impossible to maintain working benches.

Figure 132 : Rock breakage in the immediate vicinity of a blasthole.

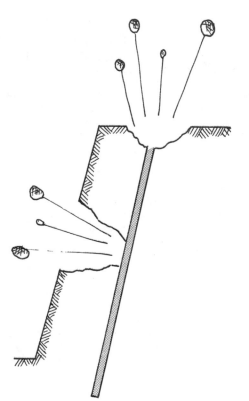

Flyrock problems are caused by cratering as a result of inadequate stemming or too small a front row burden.

Control of flyrock

When the front row burden is inadequate or when the stemming column is too short, a crater is formed as illustrated in the margin sketch. Under these conditions, rock is ejected from the crater and it may be thrown a considerable distance. In a study by the Swedish Detonic Research Foundation[282], the maximum distance which boulders were thrown was studied for a range of powder factors and the results are plotted in Figure 133. This plot shows that, for the particular rock mass and blasting geometry tested , the flyrock problem could be eliminated by reducing the powder factor to 0.2 kg/m[3]. As shown in Figure 128, a low powder factor such as that required to eliminate flyrock may not give adequate fragmentation and hence a compromise solution may be required.

An alternative to changing the powder factor would be to increase the front row burden and/or the length of the stemming column but, as pointed out earlier in this chapter, this could give rise to choking the blast and to poor fragmentation of the rock above the top load. A stemming column length of 40 blasthole diameters is recommended by the Swedish Detonic Research Foundation for the control of flyrock and this is in line with the optimum stemming column length of 0.67 to 2 times the burden which is recommended by Hagan[270].

No absolute rules can be given for the optimum relationship between powder factor, front row burden and stemming column length since these will depend upon the characteristics of the rock mass being blasted. The open pit blasting engineer should start with the values suggested earlier in this chapter and experiment until a reasonable compromise between flyrock control and good fragmentation is achieved.

A relationship between the blasthole diameter and the throw distance for a boulder of given size was established by the Swedish Detonic Research Foundation[282] and is plotted in Figure 134. From this figure is can be seen that a 1 metre granite boulder would be thrown 1000 metres by the cratering of a 254 mm (10 inch) diameter charge. This result should provide ample inducement for the open pit blasting engineer to either do something about controlling flyrock or else stand a long way from the blast.

Airblast and noise problems associated with production blasts

These two problems are taken together because they both stem from the same cause. Airblast, which occurs close to the blast itself, can cause structural damage such as the breaking of windows. Noise, into which the airblast degenerates with distance from the blast, can cause discomfort and will almost certainly give rise to complaints from those living close to an open pit mine.

Factors contributing to the development of an airblast and to noise include overcharged blastholes, poor stemming, uncovered detonating cord, venting of developing cracks in the rock and the use of inadequate burdens giving rise to cratering. The propagation of the pressure wave depends upon atmospheric conditions including temperature, wind and the pressure-altitude relationship. Cloud cover can also cause reflection of the pressure wave back to ground level at some distance from the blast.

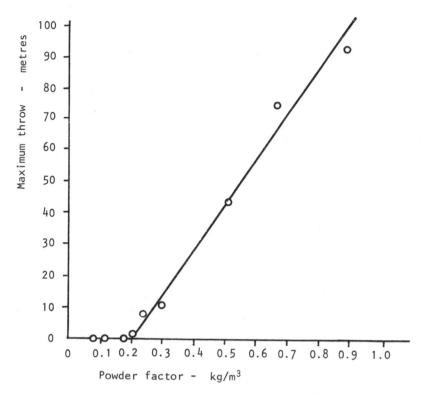

Figure 133 : Maximum throw of flyrock as a function of powder
factor in tests by Swedish Detonic Research Foundation.

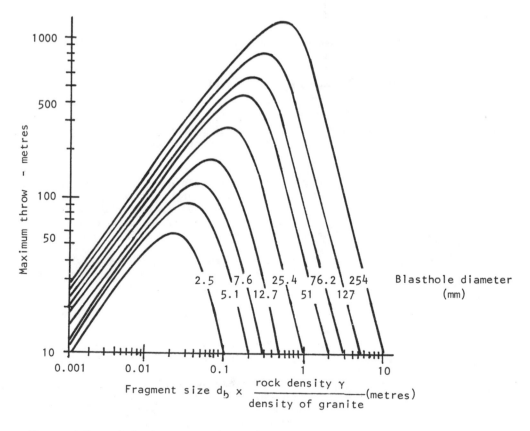

Figure 134 : Relationship between fragment size and maximum throw
established by Swedish Detonic Research Foundation.

HUMAN AND STRUCTURAL RESPONSE TO SOUND PRESSURE LEVEL

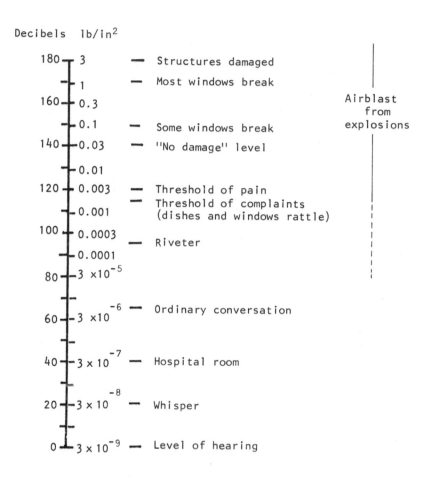

Figure 135 : Human and structural response to sound pressure level.
After Ladegaard-Pedersen and Dally[272]

Figure 135 gives a useful guide to the response of structures
and humans to sound pressure level. As mentioned earlier,
legislation in the USA now restricts blast noise to 140 dB
which corresponds to the "no damage" threshold shown in
Figure 135.

The decrease of sound pressure level with distance can be
predicted by means of cube root scaling. The scaling
factor with distance K_R is given by :

$$K_R = R/\sqrt[3]{W} \tag{125}$$

where   R is the radial distance from the explosion
         W is the weight of charge detonated.

Figure 136 gives the results of pressure measurements
carried out by the US Bureau of Mines in a number of quarries
(reported by Ladegaard-Pedersen and Dally[272]). The burden B
was varied and the length of stemming was 2.6 feet per inch
diameter of borehole.

300

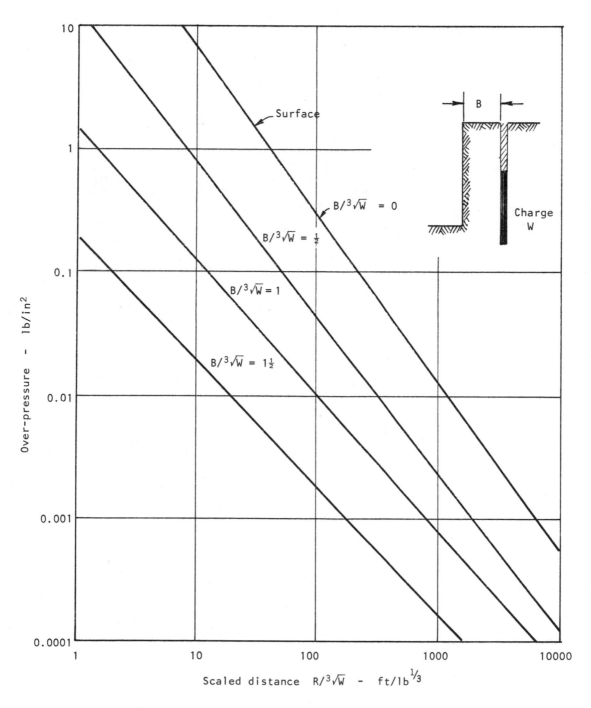

Figure 136 : Over-pressure as a function of scaled distance for bench blasting.

Consider the example of a 1000 lb charge fired in a bench with a burden of 10 feet. The scaled quantity $B/^3\sqrt{W} = 1$ and, from Figure 136, the over-pressure is approximately 0.01 lb/in^2 for a scaled distance of 100. Substituting W = 1000 gives a distance of R = 1000 feet. Hence, at a distance of 1000 feet, a 1000 lb charge confined by a burden of 10 feet and with adequate stemming will produce a pressure of 0.01 lb/in^2. From Figure 135, this pressure is seen to correspond to a noise level of 130 dB which is below the "no damage" threshold defined by current legislation in the USA.

If cratering occurs during the firing of a 1000 lb charge, $B/^3\sqrt{W}$ is reduced to zero and the over-pressure is increased to approximately 0.3 lb/in^2 (160 dB) which will give minor structural damage at a distance of 1000 feet. This again emphasises the need for adequate stemming and a correct front row burden.

These estimates are very crude since, as already pointed out, the actual pressure or noise level at any point depends upon the atmospheric conditions as well as the geometry of the blast. There are cases where blasting is not permitted on cloudy days because noise is reflected back to ground level at distances of several miles from the blast and can cause minor structural damage and give rise to serious complaints from home owners. However, in spite of these wide variations, Figures 135 and 136 provide a reasonable starting point for the estimation of airblast and noise problems. Such estimates can be supplemented by noise or pressure measurements if it is anticipated that these problems could give rise to damage or to serious complaints.

Special blasting techniques for improving slope stability

Based upon the assumption that the damage caused by a blast increases in proportion to the weight of explosive used, it follows that any reduction in explosive consumption will lead to a reduction in damage to the rock. Since damage to the rock behind the final row will lead to slope stability problems in the new bench face, it is evident that optimisation of the production blast is the most important step in controlling slope damage.

The following conditions should be satisfied if the production blast is to be optimised and damage to the rock behind the final row minimised :
   a. Choke blasting into excessive burden or broken muck piles should be avoided.
   b. The front row charge should be adequately designed to move the front row burden
   c. The main charge and blasthole patterns should be optimised to give the best possible fragmentation and digging conditions for the minimum powder factor
   d. Adequate delays should be used to ensure good movement towards free faces and the creation of new free faces for following rows
   e. Delays should be used to control the maximum instantaneous charge to ensure that rock breakage does not occur in the rock mass which is supposed to remain intact (see Figure 131)
   f. Back row holes should be drilled at an optimum distance from the final digline to permit free digging and yet minimise damage to the wall. Experience can be used

to adjust the back row positions and charges to achieve this result.

If all of these conditions have been satisfied and a bench instability problem due to over-break still exists, consideration should be given to the use of special blasting techniques such as buffer blasting, pre-splitting and smooth-wall blasting.

Buffer blasting

Buffer or cushion blasting involves increasing the distance between the back row charges and the final digline. Obviously, there is a limit to the amount this distance can be increased before unacceptable digging conditions are created at the final digline. The burden and spacing in the back row can be decreased to approximately one half that of the main charge and the holes can be charged with a lower strength explosive than that used in the main blast. The buffer holes are fired in a normal blasting sequence, i.e. they are fired last with a delay of 1 to 2 milliseconds per foot of burden.

Pre-split blasting

Pre-splitting or pre-shearing is a technique which is used very extensively and very successfully in civil engineering excavations in hard rock. Its use in mining, particularly with large diameter blastholes, is less common but the technique merits serious consideration by open pit engineers.

A row of closely spaced and usually small diameter holes is drilled along the line of the final face. These holes are lightly charged and the charge is de-coupled from the rock by leaving an air space between the charge and the walls of the blasthole. This is usually achieved by placing the charge in a plastic tube of smaller diameter than the hole and centering this tube in the blasthole with some form of spacer.

The row is fired *before* the main charge and the reinforcing effect of the closely spaced holes together with the very large burden results in the formation of a clean fracture running from one hole to the next. A good pre-split face is characterised by a clean fracture running between the parallel half barrels of the blastholes as illustrated in the margin photograph.

Table X gives recommended dimensions for blasthole diameter and spacing together with the charge per unit length of hole. These recommendations are based upon Swedish experience in hard rocks and may require considerable modification to suit local geological conditions.

When only large diameter blasthole rigs are available , pre-splitting can still be carried out but some experimentation may be required in order to achieve an acceptable result. The charge, usually a high energy explosive such as aluminised ANFO or slurry can be poured into plastic or cardboard tubes which are centred in the blasthole and the hole back-filled with drill cuttings. The detonating line is taped to the explosive and the entire pre-split line is fired simultaneously. A stemming column length of 7 to 10 blasthole diameters is usually adequate.

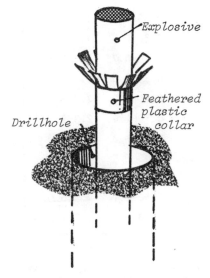

Centering a pre-split charge in a drillhole

Pre-split face in a limestone quarry in South Africa.
Photograph provided by
Mr O.K.H. Steffen of Steffen,
Robertson and Kirsten.

TABLE X - RECOMMENDED DIMENSIONS FOR SMOOTH-WALL AND PRE-SPLIT BLASTING											
Drillhole diameter		Charge diameter		Explosive*		SMOOTH-WALL BLASTING Spacing		Burden		PRE-SPLIT BLASTING Spacing**	
mm	in	mm	in	kg/m	lb/ft	m	ft	m	ft	m	ft
30	1.25	11	0.5	0.07	0.05	0.5	1.6	0.7	2.3	0.25-0.3	0.8-1.0
37	1.5	17	0.63	0.12	0.08	0.6	2.0	0.9	3.0	0.30-0.5	1.0-1.6
44	1.75	17	0.63	0.17	0.11	0.6	2.0	0.9	3.0	0.30-0.5	1.0-1.6
51	2.0	22	0.88	0.25	0.17	0.8	2.6	1.1	3.6	0.45-0.75	1.5-2.5
62	2.38	22	0.88	0.35	0.23	1.0	3.3	1.3	4.2	0.55-0.8	1.8-2.6
75	3.0	25	1.0	0.50	0.34	1.2	4.0	1.6	5.2	0.60-0.9	2.0-3.0
87	3.5	25	1.0	0.70	0.47	1.4	4.6	1.9	6.2	0.70-1.0	2.3-3.3
100	4.0	29	1.13	0.90	0.60	1.6	5.2	2.1	6.9	0.80-1.2	2.6-4.0
125	5.0	40	1.63	1.40	0.94	2.0	6.6	2.7	8.8	1.00-1.5	3.3-4.9
150	6.0	50	2.0	2.00	1.34	2.4	7.9	3.2	10.5	1.20-1.8	4.0-5.9
200	8.0	52	2.0	3.00	2.02	3.0	9.8	4.0	13.0	1.50-2.1	4.9-6.9
250	10.0	65	2.5	3.38	2.27	3.4	11.2	4.5	14.8	1.80-2.4	5.9-7.9

*Based on Nitro Nobel's Dynamex B explosive, charge per unit length of hole.

** The burden is assumed to be infinite since the pre-split charge is fired before the main charge.

Pre-split holes should normally be drilled at approximately 70° to the vertical but, if the available blasthole rigs can only drill vertical holes, the holes should be drilled along the line of the toe of the proposed bench. With large diameter blastholes, some break-back will occur and clean faces such as that illustrated in the margin photograph on page 302 will not be achieved. Never-the-less, a considerable improvement in back-break will be achieved and, when bench stability is critical, the additional expense is usually justified.

A common misconception is that a pre-split will protect the rock behind it from vibrations induced by the main blast. Measurements carried out in recent years have shown that this is not the case and this is hardly surprising since these vibrations are induced by a compressive strain wave radiating outwards from the main blast. Such a compressive wave would not see the small gap created by the pre-spilt fracture any more than it would see a closed joint in the rock. Such gaps will provide a vent path for the explosion gases however and this venting will prevent propagation of radial cracks from the main blast across the pre-split line.

Pre-split blasting is not usually successful in well jointed hard rocks, particularly where the joints are open and are inclined to the pre-split line. These open joints allow the explosion gases to vent and fracturing follows the joints rather than the intended pre-split line. There is very little that can be done to remedy this problem other than to change the direction of the face.

An 18m high rock face created by pre-split blasting. Note the clean fracture running between parallel holes in the face, in spite of the variability of the rock through which these holes have been drilled.

Photograph reproduced with permission of Atlas Copco, Sweden.

Slopes excavated after pre-splitting the final faces on a site for a hydro-electric project in Austria.

Photograph reproduced with permission of Atlas Copco, Sweden.

Smooth-wall blasting

Smooth-wall or post-split blasting is similar to pre-split blasting except that the line of holes of fired *after* the main blast . This means that a free face exists close to the line of charged holes and hence a burden and spacing design has to be specified for this blast. Recommended values for burden, spacing and charge are given in Table X on page 303.

Smooth-wall blasting is sometimes used as a clean-up operation to minimise the danger of rockfalls from a face which has been heavily blasted or where jointing has created loose blocky conditions on the face.

Drilling blastholes for smooth-wall blasting on a civil engineering site.

Photograph reproduced with permission of Atlas Copco, Sweden.

Drilling smooth-wall blast holes with a hand-held drill rig on a dam site in Tasmania. Drilling accuracy cannot be maintained beyond about 10 feet (3 metres) with hand held machines and hence the blasts must be carried out in steps.

Use of smooth-wall blasting to control the stability of bench faces adjacent to the haul road of an open pit mine in Ireland.

Chapter 11 references

257. LANGEFORS, U and KIHLSTROM, B. *The modern technique of rock blasting*. John Wiley and Sons, New York, Second edition, 1973, 405 pages.

258. GUSTAFSSON, R. *Swedish blasting technique*. Published by SPI, Gotherburg, Sweden, 1973, 378 pages.

259. COOK, M.A. *The science of high explosives*. Reinhold Book Corp., New York, 1958.

260. JOHANSSON, C.H and PERSSON, P.A. *Detonics of high explosives*. Academic Press, London, 1970.

261. DU PONT DE MEMOURS & CO., INC. *Blaster's handbook*. du Pont, Wilmington, Delaware, 15th edition, 1966.

262. FRAENKEL, K.H. *Manual on rock blasting*. Atlas Copco AB and Sandvikens Jernverks AB, Stockholm, Sweden, Second edition, 1963.

263. C.I.L. *Blaster's handbook*. Canadian Industries Limited, Montreal, Quebec, 1959.

264. I.C.I. *Blasting practice*. Imperial Chemical Industries Limited, Glasgow, 1975

265. HARRIES,G and MERCER, J.K. The science of blasting and its use to minimise costs. *Proc. Australian Inst. Min. Metall. Annual Conf*. Adelaide, Part B, 1975, pages 387-399.

266. I.C.I. *Handbook of blasting tables*. Nobel(Australia) Pty., Ltd., 1971.

267. PERSSON, P.A. Bench drilling - an important first step in the rock fragmentation process. *Atlas Copco Bench Drilling Symposium*. Stockholm, 1975.

268. MC INTYRE, J.S and HAGAN, T.N. The design of over-burden blasts to promote highwall stability at a large strip mine. *Proc. 11th Canadian Rock Mechanics Symposium*. Vancouver, October 1976, *In press*.

269. ANTILL, J.M. Modern blasting techniques for construction engineering. *Australian Civil Engineering and Construction*. November 1964, page 17.

270. HAGAN, T.N. Blasting physics - what the operator can use in 1975. *Proc. Australian Inst. Min. Metall. Annual Conf*. Adelaide. Part B, 1975, pages 369-386.

271. BROADBENT, C.D. Predictable blasting with in situ seismic surveys. *Mining Engineering*. S.M.E., April 1974, pages 37-41.

272. LADEGAARD-PEDERSEN, A and DALLY, J.W. A review of factors effecting damage in blasting. *Report to the National Science Foundation*. Mechanical Engineering Department, University of Maryland. January 1975, 170 pages.

273. BOLLINGER, G.A. *Blast vibration analysis*. Feffer and Simons, Inc. London and Amsterdam, 1971.

274. CRANDELL, F.J. Ground vibrations due to blasting and its effects upon structures. *J. Boston Civil Engineers*. Vol. 36, No. 2, 1949, pages 222-245.

275. DEVINE, J.F., BECK, R.H., MEYER, A.V.C and DUVALL,W.I. Vibration levels transmitted across a presplit failure plane. *US Bureau of Mines Report of Investigations*. 6695, 1966, 29 pages.

276. DEVINE, J.F., BECK, R.H., MEYER, A.V.C and DUVALL, W.I. Effect of charge weight on the vibration levels in quarry blasting. *US Bureau of Mines Report of Investigations*. 6774, 1966, 37 pages.

277. DUVALL, W.I., DEVINE, J.F., JOHNSON, C.F & MEYER,A.V.C. Vibration from blasting at Iowa Limestone Quarries. *US Bureau of Mines Report of Investigations*. 6270, 1963, 28 pages.

278. DUVALL, W.I and FOGELSON, D.E. Review of criteria for estimating damage to residences from blasting vibrations. *US Bureau of Mines Report of Investigations* 5968, 1962, 19 pages.

279. DUVALL,W.I., JOHNSON, C.F., MEYER, A.V.C & DEVINE, J.F. Vibrations from instantaneous and millisecond-delayed quarry blasts. *US Bureau of Mines Report of Investigations*. 6151, 1963, 34 pages.

280. EDWARDS, A.T and NORTHWOOD, T.D. Experimental studies of the effects of blasting on structures. *The Engineer*. Vol. 210, September 30, 1960, pages 538-546.

281. LEET, L.D. *Vibrations from blasting rock*. Harvard University Press. Cambridge, Mass., 1960.

282. LUNDBORG,N., PERSSON,A., LADEGAARD-PEDERSEN, A and HOLMBERG, R. Keeping the lid on flyrock in open pit blasting. *Engineering and Mining Journal*. May 1975, pages 95-100.

283. DUVALL,W.I. Design requirements for instrumentation to record vibrations produced by blasting. *US Bureau of Mines Report of Investigations*. 6487, 1965. 5 pages.

284. ORIARD, L.L. Blasting effects and their control in open pit mining. *Proc. 2nd Intnl. Conf. on Stability in Open Pit Mining*. Vancouver 1971. Published by AIME, New York, 1972, pages 197-222.

Chapter 12: Miscellaneous topics

Introduction

The theoretical models presented in the previous chapters are intended to provide a rational basis for the design of rock slopes. It should, however, be clear to the reader that the successful design of a slope depends not only on the choice of the correct theoretical model but also upon a number of other considerations which cannot conveniently be quantified. Some of these considerations are discussed in this chapter.

Influence of slope curvature upon stability

All methods of stability analysis currently in use treat slopes two-dimensionally, i.e. it is assumed that the section of slope under consideration is part of an infinitely long straight slope. It is also generally assumed that the slope is planar from crest to toe, i.e. the slope face is not curved in section and it can be defined by a single slope angle.

Actual slopes are invariably curved in plan and there is no reason why they should not also be curved in section. Moffitt, Friese-Greene and Lillico[1] have shown that the pit slope geometry, in both plan and elevation, is important in the economics of an open pit mining operation. It is the task of the rock slope engineer to define a slope geometry which provides the steepest possible slope angles, within the operational and safety constraints imposed by the geometry of the ore body, and he should be free to incorporate any factor which is likely to improve the stability of the slopes.

Figure 137 : An extension of an open pit to follow a branch of the ore body results in the formation of convex slopes. These "noses" are invariably less stable than concave slopes.

Figure 138 : Straight clean slopes at the Palabora Mining Company's copper open pit in South Africa.

Figure 137 shows a small secondary open pit which was created to exploit a branch of the main ore body at a mine in Tasmania. The two noses at the entrance to this secondary pit showed serious signs of instability whereas the remainder of the slopes were stable. This suggests that convex slopes are less stable than concave slopes.

If one considers the lateral restraint provided by the material on either side of a potential failure, it is clear that this restraint will be greater if the slope is concave than it would be if the potential failure is situated in a nose which has freedom to expand laterally. A crude analogy can be found in the case of a wedge failure where the stability increases with a decrease in the included angle of the wedge (see Figure 96 on page 204). This analogy shows that the lateral forces acting on the wedge have a very important influence upon its stability.

The only serious attempt to study the influence of slope curvature was made by Jenike and Yen[285]. This study, although giving useful indications of the improvement of stability with decreasing plan radius, was based upon assumptions which the authors do not consider to be applicable to rock slope design. Piteau and Jennings[286] studied the influence of plan curvature upon the stability of the slopes in five large diamond mines in South Africa. As a result of the mining method used, these slopes are all in a condition approaching that of limiting equilibrium. The average slope height for these mines is 320 feet and Piteau and Jennings found that the average slope angle for slopes with a radius of curvature of 200 feet was $39.5° \pm 9°$

as compared with 27.3° ± 5° for slopes with a radius of 1000 feet.

These studies by Piteau and Jennings confirm the general conclusions reached by the authors as a result of slope design experience in many parts of the world. This experience can be summed up in the following suggested design procedure :

When the radius of curvature of a *concave* slope is less than the height of the slope, the slope angle can be 10° *steeper* than the angle suggested by a conventional stability analysis. When considering a *convex* slope with a radius of curvature smaller than the slope height, the slope should be 10° *flatter* than the angle predicted by stability analysis. As the radius of curvature increases to a value greater than the slope height, these corrections should be decreased for either concave or convex slopes. For radii of curvature in excess of twice the slope height, the slope angle given by a conventional stability analysis should be used for the pit design.

Obviously, care must be exercised in the application of these suggestions since major changes in geology or in the drainage characteristics of the rock mass can invalidate the assumptions upon which these suggestions are based.

Note that this discussion also applies to waste dump design. A very common source of instability in waste dumps is the creation of "noses" by uncontrolled end-dumping. Wherever possible, slopes and waste dump faces should be straight and convex changes of direction should be located in places where minor instability can be tolerated.

In considering the effect of the slope profile upon stability it is necessary to examine the influence of cohesion. The resisting force due to cohesion is a product of the cohesive strength and the area of the failure surface which is proportional to the square of the slope height. On the other hand, the driving force due to the weight of material is proportional to the cube of the slope height. Hence, the slope angle must decrease with depth below surface if the same factor of safety is to be maintained with depth.

A very simple study of the influence of slope curvature upon the factor of safety of a slope, in which circular failure is assumed to occur, is presented in Figure 139. This shows that a significant increase in the factor of safety can be achieved by making the slope face concave in section.

Rana and Bullock[287] exploited this concept in the proposed design for open pit slopes for the Iron Ore Company of Canada. Their design is presented in Figure 140.

It should be noted that the design of concave slope profiles depends upon the assumption that the materials forming the slope do not vary with depth below surface. In many cases it may be necessary to *flatten* rather than steepen slopes in the upper benches as a result of deterioration of the quality of the rock due to weathering.

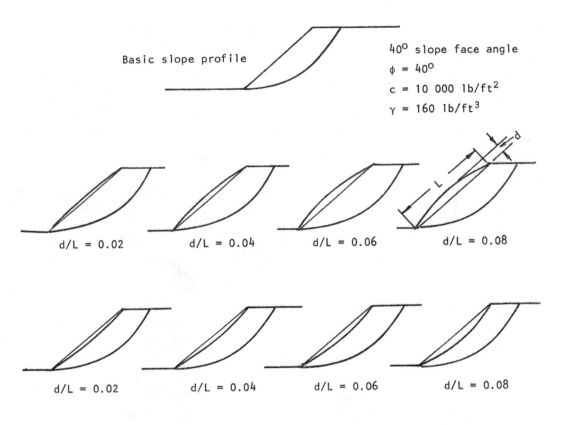

Basic slope profile

40° slope face angle
$\phi = 40^\circ$
$c = 10\ 000\ \text{lb/ft}^2$
$\gamma = 160\ \text{lb/ft}^3$

d/L = 0.02   d/L = 0.04   d/L = 0.06   d/L = 0.08

d/L = 0.02   d/L = 0.04   d/L = 0.06   d/L = 0.08

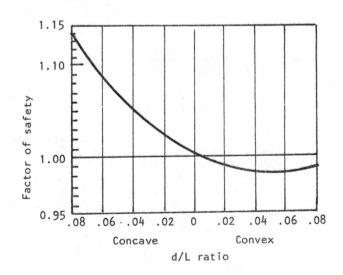

Concave   Convex

d/L ratio

Figure 139 :   Variation of factor of safety with change in
slope profile for a homogeneous slope in which
circular failure is assumed to occur. Analysis
carried out using Janbu method of analysis (see
pages 247 - 253.)

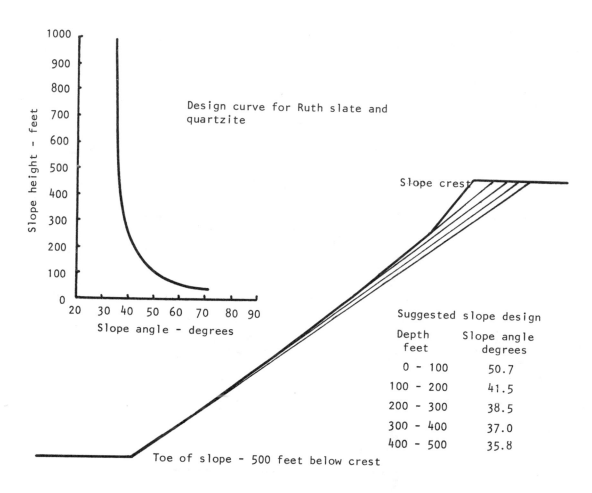

Figure 140 : Rana and Bullock's proposed slope design for the
Iron Ore Company of Canada.

Slope de-pressurisation

The very important influence of water pressure upon slope
stability has been emphasised throughout this book. Since
water pressure decreases the stability of a slope, it
follows that reduction of this water pressure will increase
the stability of the slope. Since it is water *pressure*
rather than the quantity of water which matters, this
section has been headed slope de-pressurisation rather
than slope drainage.

Three basic principles have to be kept in mind when considering slope de-pressurisation :

1. Prevent surface water from entering the slope through
   open tension cracks and fissures.

2. Reduce water pressure in the vicinity of the potential
   failure surface by selective surface and sub-surface
   drainage.

3. Position the drainage so that it reduces the water
   pressure in the immediate vicinity of the slope -
   there is no point in draining the country-side for

314

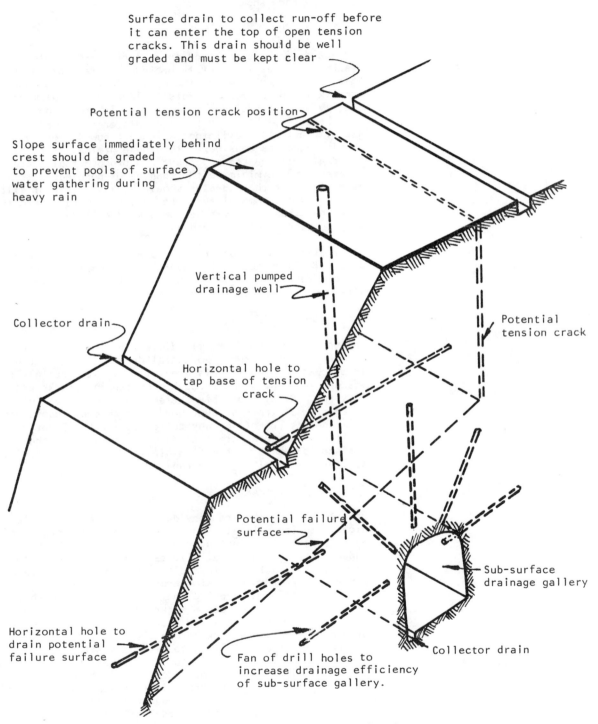

Surface drain to collect run-off before
it can enter the top of open tension
cracks. This drain should be well
graded and must be kept clear

Potential tension crack position

Slope surface immediately behind
crest should be graded
to prevent pools of surface
water gathering during
heavy rain

Vertical pumped
drainage well

Potential
tension crack

Collector drain

Horizontal hole to
tap base of tension
crack

Potential failure
surface

Sub-surface
drainage gallery

Horizontal hole to
drain potential
failure surface

Fan of drill holes to
increase drainage efficiency
of sub-surface gallery.

Collector drain

Figure 141 : Slope drainage and de-pressurisation measures.

miles around the pit.

The most common drainage and de-pressurisation methods are illustrated in Figure 141 and the following comments refer to the methods shown :

Surface drains

Surface drains are designed to collect run-off before it reaches the area immediately behind the crest of the slope. This is the area in which the most dangerous tension cracks are likely to occur. Attempts to line these drains are usually unsuccessful due to movements in the slope and also because of heavy traffic in this area. However, the use of heavy plastic linings which are now commercially available can be considered for these drains. Alternatively, the drains can be steeply graded to promote rapid water movement and to minimise the chances of ponding. It is essential that these drains should be kept clear of silt and debris and some of the most successful drains are those which have been cut by a narrow-bladed bull-dozer which is also used to clean the drains.

River diversion and other major drainage schemes are not dealt with here but it is important to remember that these could also feed water into the slope if they are incorrectly sited and improperly sealed.

Upper slope surface

The upper slope surface, immediately behind the crest, is an area of considerable potential danger since water which is allowed to pond in this area will almost certainly find its way into the slope through open tension cracks and fissures. Grading of this surface and the removal of piles of waste rock or over-burden which could cause damming will enhance run-off of any collected water. Sometimes additional measures are considered necessary and the slope surface is sealed with a flexible layer such as laterite soil or even a plastic membrane.

Open tension cracks

Open tension cracks are very dangerous in areas liable to high intensity rainfall since the water forces generated by a water-filled tension crack can be very high and can induce very sudden slope failures (see practical example on pages 174 to 179). In addition to diverting surface water away from open tension cracks, it is advisable to prevent water from entering the cracks by sealing them with a flexible impermeable material such as clay. When the crack is more than a few inches wide it should be filled with gravel or waste rock before the flexible seal is placed. The purpose of this fill is to allow any water which does find its way into the crack to flow out again as freely as possible. Under no circumstances should the crack be filled with concrete or grout since this would result in the creation of an impermeable dam which could cause the build up of high water pressures in the slope.

Horizontal drains

Horizontal drain holes drilled into the slope face can be very effective in reducing water pressures near the base

A horizontal drain in an altered rock slope. A plastic pipe has been inserted into the hole to prevent caving. The pipe is drilled or slotted as required to allow water to drain into it.

of a suspected tension crack or along a potential failure surface. The spacing and positioning of these holes depends upon the geometry of the slope and upon the structural discontinuities within the rock mass. In a hard rock slope, water is generally transmitted along joints and horizontal drains can only be effective if they intersect such features. In the case of a soft rock or soil slope, the holes can be regularly spaced but a certain amount of trial and error is necessary in order to determine the optimum spacing. In either case, the installation of piezometers *before* the drilling of the horizontal holes is strongly recommended since, without an indication of the change in water level, the rock slope engineer will have no idea of the effectiveness of the de-pressurisation measures which he has implemented.

Collector drains

Water which is drained out of the rock mass should be lead away in collector drains otherwise this water will simply find its way into the next bench down and the problem will have been transferred from one level to the next.

Vertical drainage wells

Vertical drainage wells drilled from the slope surface and fitted with down-hole pumps can be effective in slope drainage and de-pressurisation. These wells have one major advantage in that they can be brought into operation *before* the slope is excavated and can play an important part in keeping the effective stresses in the rock mass high and in preventing the onset of slope movement. In rock or soil masses of low permeability, a period of a year or more may be required to depress the water level to that required for the slope design and vertical pumped wells are very effective in this type of application.

The disadvantage of vertical pumped wells is that the pumps have to be kept running for the system to remain effective. Electrical and mechanical breakdowns which could occur during periods of heavy rain are particularly dangerous if these are of sufficient duration to permit the groundwater levels in the slope to rise. In some cases, vertical pumped wells installed during the early stages of a slope stabilisation programme have later been connected to a sub-surface drainage gallery so that they become gravity drains.

Drainage galleries

Drainage galleries, with or without fans of radial holes, are probably the most effective means of sub-surface drainage. They are also the most expensive form of drainage and can only be considered in critical situations or where the economic benefits of steepening the slope will more than cover the cost of the drainage gallery.

In many open pit mining situations, such galleries already exist in the form of old underground mine workings or old exploration adits. In other cases, the excavation of an exploration adit which could later be used as a drainage gallery is justified.

The optimum location of drainage galleries has been discussed

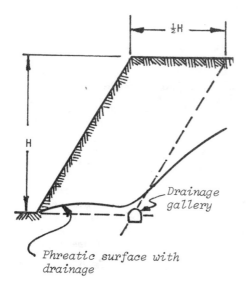

Drainage gallery

Phreatic surface with drainage

Optimum sub-surface drainage gallery location in a slope

by Sharp, Hoek and Brawner[288]. In very general terms, the optimum gallery location is at the corner of a parallelogram defined in the margin sketch. A sub-surface drainage gallery will effectively drain about 200 feet of over-lying material and, hence, for very large slopes, two or more levels of drainage galleries may be required.

The effectiveness of any slope drainage or de-pressurisation scheme is very difficult to gauge. Piezometers installed in the slope before the drainage system is brought into operation can give very valuable information on the reduction of water pressure which is achieved. Knowledge of overall groundwater flow patterns is very important in planning the most effective drainage measures. One positive aspect of drainage or de-pressurisation is that it can never do any harm - some drainage, however inefficient, is better than no drainage. Since most drainage measures, with the exception of drainage galleries, can be installed very quickly and at reasonable cost, these are usually the remedial measures which are taken first when attempting to control slope instability. It is encouraging that such measures are often effective in stopping or at least slowing down slides.

Surface protection of slopes

Slopes in soft rock or soil are prone to serious erosion during heavy rain and some rock slopes suffer from deterioration due to weathering when exposed. The protection of the suface of such slopes can be a serious problem and the following comments are offered as general guidance. Local conditions and the availability of materials will generally determine the measures which are taken on any particular site.

Vegetation is almost certainly the best form of slope protection, particularly against erosion of soil slopes. A grass mat covering the slope will not only bind the surface material together but it will also tend to inhibit the entry of water into the slope. Establishing the grass or other vegetation is the most difficult problem since the rain which is necessary to promote the growth of the young plants is also the agent which will remove these plants by erosion.

Figure 142 illustrates a scheme used in the state of Washington for the surface protection of soil slopes. Very small benches, two or three feet high and two or three feet wide, are cut into the slope and these benches are seeded and fertilised. Provided that the slope is not subjected to excessively heavy rains, the benches will last long enough to permit the grass to establish a strong growth and to take over the protection of the surface.

In some cases, grass seed and fertiliser pellets with a latex coating have been sprayed onto slopes and these will adhere to the surface for long enough to allow the grass to take root. In other cases, metal trays are suspended on steel cables from the top of the slope and these prevent erosion for long enough to permit vegetation to become established.

Where it is felt that vegetation will not provide sufficient surface protection, more positive mechanical measures can

Figure 142 : Use of very small benches in a soil slope to provide a stable base for the establishment of grass growth. Photographed in the state of Washington, USA.

be considered.

In Hong Kong, which suffers from slope stability problems in deeply-weathered granitic materials as well as violent rain storms, most cut slopes are covered with a hand-applied mortar known as *Chunam* . This slope protection system, illustrated in Figure 143, is very effective except when drainage of the layer has been forgotten or when the drains become blocked. Since the Chuman keeps the water out of the slope it can also keep it in and it is very important that weep holes be provided at regular intervals to allow water pressure to dissipate.

The use of pneumatically applied mortar , know as "shotcrete" or "gunite", is becoming increasingy common in the protection of slopes in materials such as shales which are prone to rapid weathering and breakdown upon exposure. The same warning about drainage applies to such protection schemes as to Chunam.

More elaborate (and more expensive) methods of slope protection are sometimes necessary. One such method, used in Hong Kong, is illustrated in Figure 144. Interlocking precast concrete members form an open framework into which a layer of porous no-fines mortar is placed. This supports a layer of top soil which is then seeded with grass seed.

Figure 143 : Chunam covered slope in Hong Kong. Small dots on slope are drain pipes and a stepped drainage channel can be seen running down the centre of the slope face.

Figure 144 : Protection of the surface of
a large soil slope in Hong Kong. Precast
concrete members interlock to form a framework into which layers of no-fines porous
mortar are placed. These layers support
top soil which is then seeded to produce
a grass-covered slope.

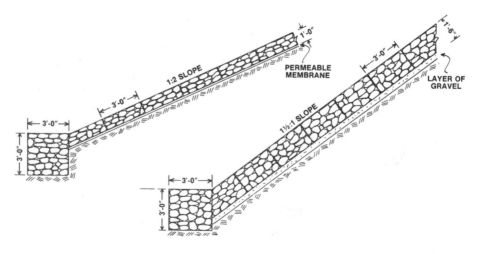

Figure 145 : The use of gabions for the stabilisation and surface
protection of slopes. Technical literature on gabions is available
from River and Sea Gabions (London) Ltd., Princes Street, London W1R 8SQ
and from Terra Aqua Conservation Ltd., 4930 Energy Way, Reno, Nevada 89502.

The use of gabions can also be considered for slope protection. Gabions are rock-filled wire baskets which are strong, heavy, flexible and permeable. Figures 145 to 148 show a variety of ways in which gabions can be used to improve the stability of slopes and to prevent erosion of river banks.

Figure 146 : The use of gabions to control river bank erosion in a high rainfall area.

Figure 147 : A gabion wall to control the flow of a stream and to protect a major highway slope.

Figure 148 : The use of gabions to buttress the abutment of an old bridge.

Control of rockfalls

One of the dangers associated with rock slopes is that of falls of loose rocks and boulders from the face of the slope. Such boulders can fall, bounce or roll down a slope as shown in Figure 149 and, unless steps are taken to dissipate the energy which has been acquired by the boulder, considerable damage can be caused. This danger is particularly acute in the case of highway slopes and a study carried out by Ritchie in 1963[289] was aimed at minimising this hazard.

Figure 149, adapted from a drawing in Ritchie's paper, summarises the main recommendations for the control of rockfalls. A ditch at the foot of the slope will contain much of the energy of the fall and a chain link fence on the shoulder of the trench will prevent the rock from bouncing onto the roadway. Unfortunately, to be fully effective, the ditch should be as much as 25 feet wide and 6 feet deep for a 100 foot high slope. The authors suggest that it may be possible to reduce these dimensions by placing a thick layer of gravel in the base of the ditch. Recent work has shown that such a layer of gravel can be very effective in dissipating the energy of a runaway airplane or truck. In most open pit mining situations such rockfall protection measures can only be considered for important haul roads or for public roads which pass through the mine area.

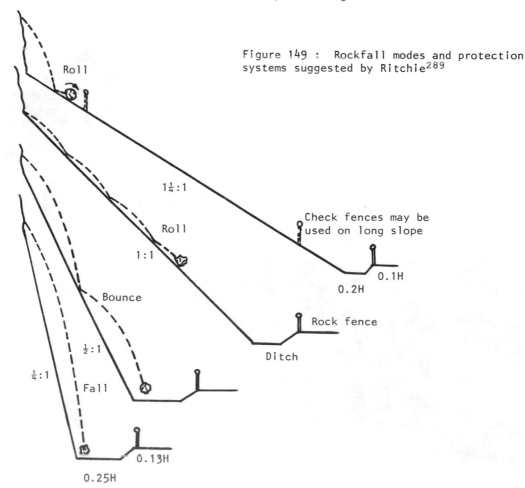

Figure 149 : Rockfall modes and protection systems suggested by Ritchie[289]

Protective canopies built over highways and railways are common in avalanche prone mountain regions and a typical canopy is illustrated in Figure 150. Such canopies are also used to deflect falling rock and, in order to dissipate the energy of the fall, waste rock is uaually placed on the top of the canopy.

For small critical slopes in civil engineering projects, wire mesh is sometimes draped over the face as illustrated in Figure 151. Any rocks which break loose from the face are contained between the rock face and the mesh and are thereby prevented from gaining momentum in a free fall.

Figure 150 : Concrete avalanche canopy over a major highway in Washington state, USA.

Figure 151 : Wire mesh draped over a steep rock face to prevent rockfalls in a critical area of a civil engineering project.

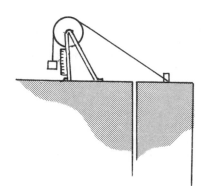

A simple tension crack monitoring system.

An electro-optical distance measuring instrument set up to monitor displacements in a quarry slope.

Monitoring and interpretation of slope displacements

When, in spite of all his efforts to design a stable slope or to improve the stability of a potentially dangerous slope, the engineer is forced to accept that a major failure is inevitable what course of action, other than catching the next 'plane, remains open to him ? Such an engineer should take courage from the fact that slopes seldom fail without giving ample warning that they are about to do so and, provided that these warnings are heeded, the slope may still fulfil its function before it finally fails.

As discussed earlier in this book, the opening of tension cracks is usually the first sign that a slope is in distress. The monitoring of the movement of such cracks will usually give a very good indication of the overall behaviour of the slope. Figure 152 illustrates the installation of an accurate tension crack monitoring system which could be used to monitor the response of the rock mass to changes in groundwater levels and even to the effects of blasting. Less sophisticated systems are in operation on many open pit mines and the layout of each system depends upon the nature of the slope and upon the ingenuity of the rock slope engineer. A common system utilises a wire anchored to the front block and running over a pulley which is fixed on the block of rock behind the tension crack. The movement of a weight suspended on the end of the wire will give an indication of tension crack movement.

Slope displacement observations should not be limited to the tension crack area. A convenient monitoring system involves the installation of a number of corner-cube reflecting targets on part of a moving rock slope and the monitoring of the movement of these targets with an electro-optical distance measuring device. A large number of such devices are now available commercially and full technical details can be obtained from the manufacturers. If the targets have been well placed and if they are monitored regularly, a contour plan of slope displacement rates can be prepared which will assist the rock slope engineer in defining the extent and the critical portions of the slide. A careful examination of the vectors of movement will often provide a valuable insight into the mechanism of failure.

One of the most spectacular predictions of slope failure based upon displacement monitoring was carried out at the Chuquicamata mine in Chile [9,10] *. Figure 153 shows the eastern slopes of the Chuquicamata pit. The overall height of the slope in the failure region (in the centre of the photograph) was 248 metres and the overall face angle was approximately 43°. The main rock type in the slide area was unaltered porphyritic granodiorite.

Tension cracks were first noticed in this slope in August 1966 and a simple monitoring system was established. Movements were found to be very small and eventually ceased so that monitoring was discontinued. An earthquake on December 20, 1967, of magnitude 5 on the Richter scale, was apparently responsible for reactivating the slide. Incidentally, the mine is in a desert area and hence groundwater is not a factor in this slide.

---

* The authors claim no credit for this example which was taken from papers by Kennedy et al.

a. Four pegs are set into holes drilled into the rock on either side of the tension crack. Epoxy resin is used to cement the pegs into the holes. The measuring heads are protected by grease-filled caps screwed onto the pegs. The pegs are left unpainted so that they will not attract the attention of vandals.

b. A large vernier caliper can be adapted to measure the displacement across a tension crack of up to 5 feet wide. The attachments on the caliper are cone seatings which are centred onto balls attached to the top of the measuring pegs.

c. A precision level placed along the caliper bar can be used to determine the changes in level of the pegs. Note that measurements are made across the diagonals as well as along the sides of the measuring square in order to detect and shear movement along the tension crack.

d. Precise measurement of movements across a narrow tension crack can be made by means of a mechanical extensometer. Measurements of this accuracy have proved useful for correlation with daily rainfall records and blasting records.

Figure 152 : Measurement of movements across a tension crack.

Figure 153 : Eastern slopes of the Chuquicamata mine in Chile in late 1968. The area on the top of the slope to the left of the two main smelter stacks is where 4½ million tons of material was removed during the unloading programme.

Figure 154 : Plot of slope displacement versus time used for the prediction of the Chuquicamata slide.

A. Plot of fastest moving target on slope face.
B. Plot of slowest moving target on slope face.
C. Prediction of slope failure date made on the basis of existing data on January 13.
D. Predicted and actual failure date - February 18, 1969.

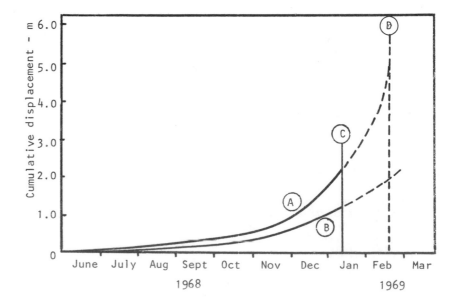

Displacement monitoring was re-commenced in June 1968 when
it was evident that movements were taking place in the slope.
The monitoring systems were basically very simple and con-
sisted of tension crack measurements, survey measurements
and some extensometer measurements. A three channel, short
period seismograph was also installed on the site and, although the results produced by this instrument were not used
in the failure prediction, it is interesting to compare these
results with displacement records[10].

In an effort to stabilise the slope, an unloading programme
was started in August 1968 and a total of 4.5 million tons
of material was eventually stripped from the top of the slope.
This stripping is visible on the top of the slope, to the
left of the two smelter stacks, in Figure 153. Although
the amount of material finally deposited on the pit floor
was probably reduced by this stripping programme, it is
unlikely that the unloading of the slope had any significant
influence on the rate of movement of the slide. This
confirms previous discussions in this book in which it was
pointed out that significant reductions in slope height have
to be made before the factor of safety of the slope is
improved to the point where a slide would stop.

By late 1968 it was evident that a major slope failure was
inevitable and steps were taken to re-route the haul road
system and to stockpile material for the mill. On January
13, 1969, a projection of displacement data, plotted in
Figure 154, was made. The earliest predicted failure date,
based upon the fastest moving target on the slope, was
given as February 18, 1969.

The failure itself, illustrated in the photograph reproduced
in Figure 155, occurred at 6.58 pm on February 18 and it
involved a movement of approximately 12 million tons of
material. Figure 156 shows the failure, as seen from the air,
and the relocated haul road can be seen in the lower right
hand side of the photograph.

Full production was resumed on February 19 after a shut-
down of the pit of 65 hours duration. The mill continued
working throughout this period on stockpiled material.

The spectacular accuracy of this prediction is not particularly relevant to this discussion since the overall result
would have been the same had the prediction been a few days
or even a few weeks out. The point of this example is that,
by knowing what to look for and by making full use of the
available data, a set of sound engineering decisions could
be made and the serious consequences which could have
resulted from a serious failure were avoided.

A look into the future

In order to present the basic principles of rock slope
stability as clearly as possible, the authors have restricted
the discussion in this book to a number of relatively simple
cases of sliding and toppling. Experience in the field
shows that more complex failure modes can and do exist and
that the development of methods for analysing the stability
of slopes in which such complex failures occur is becoming
increasingly important. Some of these failure modes are
reviewed on the following pages.

Figure 155 :   Failure of the Chuquicamata mine east slope at 6.58 pm.,
February 18, 1969.

Figure 156 :   The Chuquicamata slide seen from the air. Note the new
haul road location in the bottom right hand side of the photograph.

All photographs reproduced in Figures 153, 155 and 156 are prints of colour slides
made available to Imperial College by the Anaconda Company.

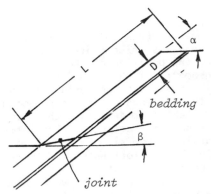

Failure model used by Brawner, Pentz and Sharp[290].

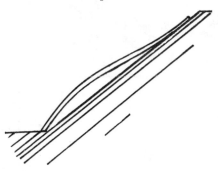

Buckling of a thin slab resting on a low-friction surface.

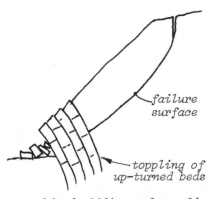

Combined sliding and toppling.

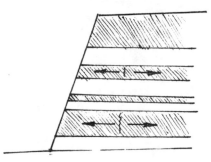

Tensile stressed induced in hard layers (shaded) by "extrusion" of soft layers.

Failure controlled by two surfaces

Brawner, Pentz and Sharp[290] have discussed a possible failure mechanism in a layered coal deposit involving the onset of sliding on two surfaces simultaneously. The failure model, illustrated in the upper margin sketch, assumes a slab of thickness D and length L resting on a bedding plane inclined at an angle α to the horizontal. Joints, inclined at an angle β to the horizontal, cut across the slab as shown. If the shear strengths of both the bedding surface and the joint are exceeded simultaneously, the toe of the slab will kick out and the slab will slide down the bedding surface. By considering different combinations of D, L, α and β and different strength properties for the two surfaces, these authors were able to derive a set of slope design charts which were applied to a particular slope problem.

Buckling of slabs

Buckling failure can occur in thin slabs resting on a base plane having a low friction angle. This type of failure occurs in layered coal deposits where very low friction materials such as mylonites [114] are present between the layers. Preliminary studies carried out at Imperial College by Bray and Walton have shown that such buckling can occur for certain combinations of D, L and φ. Further work on this analysis is required before it can be published as a slope design tool.

Sliding and toppling

The margin sketch illustrates a failure similar to that defined as slide toe toppling in Figure 119a on page 261. This type of failure, which the authors have encountered in coal mines, is extremely difficult to analyse theoretically. A useful understanding of the types of movement which can occur under different geometrical conditions can be gained from base friction model studies (see page 262). Further development of the toppling analysis presented in chapter 10 may eventually provide an analytical slope design method.

Induced stresses in slopes

Studies of the stress distribution in horizontally bedded slopes in which alternate layers of hard and soft materials occur have indicated that tensile stresses can be induced in the hard layers as a result of "extrusion" of the soft layers. Since rock is very weak in tension, this means that vertical joints will be opened up in the hard rock and this opening, assisted by water pressure, may be sufficient to induce near surface failure in the slope. This process is very poorly understood at present and a great deal more work will be required before practical design tools are available for this type of problem.

Progressive failure in slopes

The fact that movements of the rock mass forming a slope can occur for many years before the slope finally collapses suggests that the failure process is progressive rather than instantaneous as is assumed in most forms of stability analyses. Time-dependent phenomena such as weathering and creep obviously play an important role in this progressive failure process but it is probable that the geometrical

shapes of the interlocking blocks within the failing mass
also play a part in this process. Once collapse of the
slope starts, the dynamic processes which control the
violence of the collapse and the position of the muck pile
certainly depend upon the geometrical shapes of the interlocking blocks in the failing mass. The monitoring of slope
movements, as described on the previous pages, is the only
practical tool currently available for dealing with slopes
in which progressive failure has been detected. The interpretation of the results obtained from slope movement monitoring is based upon experience and, fortunately, a growing
body of such experience is being accumulated. There is
little doubt, however, that this interpretation would benefit
greatly from a sounder understanding of progressive failure
mechanisms.

Chapter 12 : references

285. JENIKE, A.and YEN, B. Slope stability in axial symmetry. *Proc. 5th Rock Mechanics Symposium.* University of Minnesota. 1962. Published by Pergamon Press 1963, pages 689-711.

286. PITEAU, D.R.and JENNINGS, J.E. The effects of plan geometry on the stability of natural slopes in rock in the Kimberley area of South Africa. *Proc. 2nd Congress of the International Society of Rock Mechanics.* Belgrade, Vol. 3, 1970, paper 7-4.

287. RANA, M.H and BULLOCK,W.D. The design of open pit mine slopes. *Canadian Mining Journal.* August 1969, pages 58 - 66.

288. SHARP, J.C., HOEK,E and BRAWNER, C.O. Influence of groundwater on the stability of rock masses - drainage systems for increasing the stability of slopes. *Trans. Inst. Mining and Metallurgy. London.* Section A, Vol. 81, No. 788, 1972, pages 113-120.

289. RITCHIE, A.M. The evaluation of rockfall and its control. *Highway Record.* Vol. 17, 1963, pages 13-28.

290. BRAWNER, C.O., PENTZ, D.L. and SHARP, J.C. Stability studies of a footwall slope in layered coal deposit. *Proc. 13th Symposium on Rock Mechanics.* Urbana, Illinois, 1971, Published by ASCE, New York, 1972, pages 329-366.

Appendices

Appendices 1 and 2 which follow both deal with the
analysis of the sliding of a wedge along the line
of intersection of two planes. These analyses occupy
62 pages and the reader would be justified in asking
why so much attention has been devoted to this one
problem. The answer is that four different types of
analysis have been presented in the hope that one of
these will suit the reader. Each of the analyses
represents a different phase in the process which
the authors went through in order fully to understand
this problem and it was felt that the reader may
benefit from being able to follow the same learning
process.

Appendix 1: Wedge failure analysis

Introduction

A problem frequently encountered in rock slope engineering is that involving the sliding of a wedge on two intersecting discontinuity planes. The problem studied hereunder includes an inclined upper slope surface, a tension crack running behind the slope crest and the influence of water pressure in the tension crack and along the sliding surfaces. The stabilisation of the slope by means of cable anchors or rockbolts is also investigated.

In part I, the problem is analysed by an engineering graphics method, in part II using spherical projections and, in part III, an analytical solution is presented.

The following assumptions are made:

a. The wedge remains in contact with both discontinuity surfaces during sliding.

b. The influence of moments is neglected, i.e. it is assumed that neither toppling nor rotational slip can occur.

c. It is assumed that the shear strength of the sliding surfaces is defined by a linear relationship $\tau = c + \sigma \text{Tan}\phi$, where c is the cohesive strength and ϕ is the angle of friction of the surface.

d. Sliding of the wedge is kinematically possible, i.e. the line of intersection of the two planes on which sliding occurs daylights in the face of the slope.

Problem

The face of a slope has a dip of 65° and a dip direction of 185°. The upper surface of the slope has a dip of 12° and a dip direction of 195°. This slope is intersected by two planes A and B and by a tension crack having the dips, dip directions and properties listed in Table A1. The line of intersection of planes A and B daylights in the slope at a point 100 feet vertically below the point of intersection of plane A with the crest line of the slope. The tension crack intersects the trace of plane A on the upper slope surface at a distance of 40 feet (measured along the trace) behind the point of intersection of plane A with the crest line of the slope.

It is required to determine the factor of safety of the slope for the following conditions :

a. Assuming that there is no tension crack and the slope is dry.

b. A dry slope with the tension crack present.

c. With the slope surface flooded with water which can enter the top of the tension crack.

d. With an external force due to cable anchors or rockbolts with dip at 30° in a dip direction of 355°. The total tension T is that required to ensure that the factor of safety of the slope is 1.5 with the slope surface flooded and it is required to determine the magnitude of T.

Part I - Engineering Graphics Solution

The basic contruction required for the engineering graphics solution to this problem is that which is necessary to produce a true plan view of the wedge. From this plan view, the direction of sliding and the forces acting on the wedge can be determined.

Table A1 - Geometry and properties of planes				
Plane A	Dip - degrees	Dip direction degrees	Strike (Dip dir. -90°)	Location and properties
Plane A	45	105	15	Cohesion c_A = 500 lb/ft^2 Friction angle ϕ_A = 20°
Plane B	70	235	145	Cohesion c_B = 1000 lb/ft^2 Friction angle ϕ_B = 30°
Slope face	65	185	95	Intersection of plane A and slope crest 100 feet vertically above toe of wedge.
Upper slope	12	195	105	
External force	30	355	265	Magnitude T to be determined
Tension crack	70	165	75	40' along trace of plane A.

There are many ways in which a true plan view of the wedge can be constructed but the method presented hereunder has been chosen to reduce errors and to ensure that the minimum number of steps has to be carried out in arriving at a complete solution to the problem. This method is to construct a set of contours along each of the surfaces being considered and, from the intersection of contours of the same elevation, to determine the traces of these planes on each other. In constructing these contours, it is convenient to express their elevations relative to the lowest point of the wedge which is generally the point at which the line of intersection of planes A and B daylights in the slope face.

The scale chosen for the construction depends upon the overall size of the drawing board available and several attempts may be required before a satisfactory scale is achieved. It is therefore recommended that a preliminary construction be carried out on a piece of rough paper in order to determine how much room is required for the complete construction. The scale used for the original drawings from which the figures in this appendix were prepared was 1 inch to 25 feet and it is suggested that this is the minimum scale which should be used for acceptable accuracy.

Note that this construction may be carried out directly on a contoured topographic map.

Details of plan view construction

Step 1. Views along the strikes of planes A and B are constructed as shown in Figure A1. The point of intersection of the 100 foot contours (marked a in the construction) is fixed near the centre of the paper and the line of intersection $a0$ is determined from the contour intersections. Note that the points A and B are not known at this stage in the construction but because it is known that the intersection of plane A with the crest line of the slope is 100 feet vertically above the toe of the wedge, the position of this toe is determined by the intersection of the zero elevation contours.

Step 2. A zero elevation contour line is drawn through point 0, along the strike of the slope face, and the dip of the slope face is set off on this line. The 25, 50, 75 and 100 foot contours on the slope face are now constructed and the points A and b are fixed at the intersections of the 100 foot contours on planes A and B and the slope face.

Step 3. The point A has been specified as a point on the crest line of the slope and hence a 100 foot contour along the strike of the upper slope surface can be drawn through this point as shown. A 112.5 foot contour (chosen to suit the geometry of this particular drawing) is drawn along the strike of the upper slope and the intersection of this contour with the 112.5 foot contours on planes A and B (marked c and d on the drawing) are points on the traces of these planes on the upper slope surface.

Step 4. The line Ac is drawn until it intersects the line of intersection of planes A and B at point C and the line AC is the visible trace of plane A on the upper surface of the slope. The trace of plane B is constructed by joining points C and d and projecting this line towards the slope crest (as yet undetermined).

Step 5. From the intersections of the contours on plane B and the slope face, the trace 0b is established and this trace is projected to meet the trace of plane B on the upper surface. The intersection of these traces defines the point B on the crest of the slope and the true plan view of the wedge AOBC can now be completed.

Step 6. At this stage it is worth determining the elevations of points C and B since these elevations will be required at a later stage in the analysis. These elevations are found in the views along the strike of plane A and along the strike of the slope face as shown and are 130.0' and 104.5' respectively.

Step 7. The tension crack is located by a point T which is 40 feet from point A, measured along the trace AC. In order to establish the point T on the plan view, the horizontal distance between A and T must be determined and this requires a construction to find the inclination of trace AC to the horizontal. From the construction carried out in Step 3 it is known that the elevation of point c on this trace is 112.5 feet and hence, by setting off a distance of 12.5 feet at right angles to Ac from point c, the inclination of line AC can be established. The horizontal distance between points A and T is determined as shown in Figure A1. The elevation of point T above the horizontal plane is also

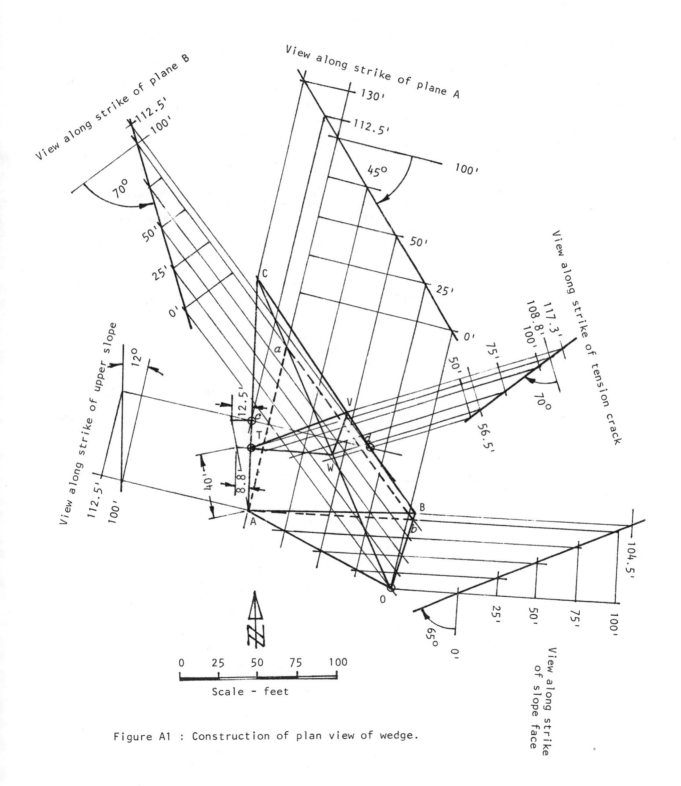

Figure A1 : Construction of plan view of wedge.

found from this construction and a 108.8 foot contour line is drawn through T along the strike of the tension crack.

Step 8.   In the view along the strike of the tension crack, the 50.75 and 100 foot contours are constructed and the intersection of these contours with the equivalent contours on the planes A and B define the traces TW and WV.   The elevations of the points V and W are determined in the view along the strike of the tension crack as 117.3' and 56.5'.

Determination of base areas and volume of wedge

Having completed the plan view of the wedge in Figure A1, the next stage in the analysis is to obtain the base areas of planes A and B, the area of the tension crack surface and the volume of the wedge which will slide.

The simplest solution to this part of the analysis is obtained by determining the true lengths of all the lines of intersection from the lengths of their horizontal projections (given in the plan view) and the difference in elevation between their end points.

Step 1.   On a tracing of the plan view given in Figure A1, the lines of intersection of the different surfaces are numbered as shown in Figure A2.   The sequence of numbering is not important provided that the same sequence is used throughout the analysis.   In order to avoid confusion in Figure A2, the two wedges AOBC and TWVC are dealt with in two separate drawings although, in solving the problem these steps were carried out on the same plan view.

Step 2.   Starting with line 1, the difference in elevation between point A and the toe of the wedge, point 0, has been given as 100 feet.   Setting off a perpendicular line from point A, the length corresponding to 100 feet is scaled off and the line 1' constructed.   The length of the line 1' in Figure A2 gives the true length of the line AC in space.

Step 3.   In the case of line 2, the elevation of point B has been found to be 104.5 feet (see Figure A1).   Setting off a perpendicular line from B with a length corresponding to 104.5 feet gives the line 2'.

A similar process is carried out for all the lines of intersection which define the wedge AOBC.

Step 4.   In the case of the wedge CTWV behind the tension crack, an identical process to that described above is carried out and the true lengths of the lines of intersection 7, 8 and 9 and the lines CT, CW and CV are found as illustrated in the lower drawing of Figure A2.

Step 5.   Having determined the true lengths of all the lines of intersection as shown in Figure A2, the next step is to construct developed views of the two wedges AOBC and CTWV. Starting with the length of one of these lines, say 3', the developed view of triangle AOC is found by setting off the lengths of the adjacent lines 5' and 1' as shown in Figure A3. A similar process is carried out for triangles ABC and COB and for the triangles CVT, CTW, and CWV. The face of the tension crack TVW is also developed in Figure A3 since its area will be required in considering the force due to water in the tension crack.

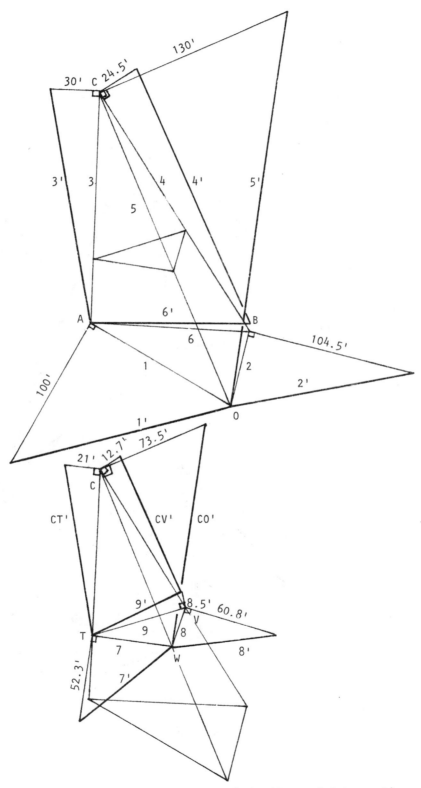

Figure A2: Determination of the lengths of the lines of intersection.

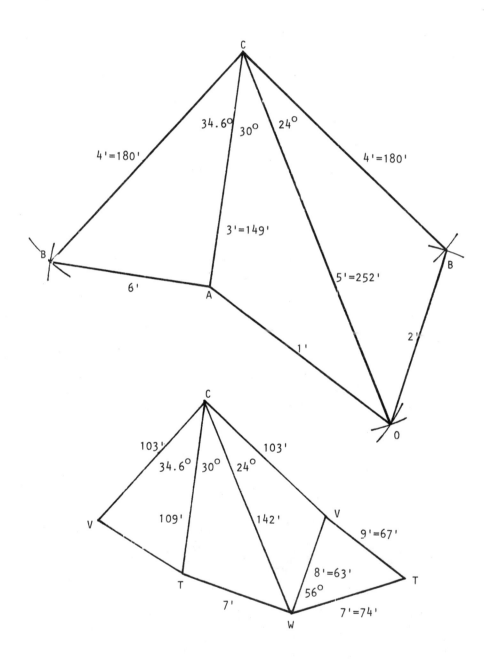

Figure A3: Developed views of wedge surfaces.

Step 6. The areas of the triangles AOC and BOC and of CTW and CWV are found from the relationship : area = one half the product of the lengths of two adjacent sides times the sine of the included angle. Hence, Area AOC = $\frac{1}{2}$ x AC x OC x Sin ACO. The calculated areas are tabulated in Table AII.

Step 7. The volume of the tetrahedrons AOBC and TWVC is obtained from the relationships

$$\text{Vol. AOBC} = \frac{1}{6}.BC.AC.OC.K, \quad \text{Vol.TWVC} = \frac{1}{6}.VC.TC.WC.K$$

where

$$K = (1 - \text{Cos}^2BCA - \text{Cos}^2ACO - \text{Cos}^2OCB + 2.\text{CosBCA.CosACO.CosOCB})^{\frac{1}{2}}$$

Note that, at this stage in the analysis, it is worth constructing a cardboard model of the wedge as illustrated in Figure A4. The developed views of the surfaces given in Figure A3 are traced onto a piece of card or stiff paper and the figures are cut out and folded along the lines of intersection. The authors have found these models to be very useful as aids to understanding the physical significance of steps in the analysis.

Water pressure distribution and uplift forces

In order to analyse the influence of water on the stability of a rock slope containing a wedge such as that defined in Figure A1, the distribution of water pressure on the planes A and B and on the face of the tension crack must be decided upon. Because the type of analysis presented here would apply mainly to hard rock slopes, it is reasonable to assume that the permeability of the rock itself will be very low and hence most of the water in the slope will be transmitted along the discontinuities such as the planes A and B and the faces of the tension crack. The distribution of water pressure along these discontinuities will depend upon the source of the water and its level at various points in the slope. Experience in slope engineering suggests that one of the most critical conditions can arise when the upper surface of the slope is flooded during very heavy rain and when the water is permitted to enter the open tension crack top. This case will be analysed hereunder but it must be emphasised that other water pressure distributions are possible. Each particular slope must be examined in terms of possible water bearing fissures and the water pressures which are likely to arise as a result of the water flow pattern in these fissures. The analysis presented here can be used to obtain a reasonable estimate of the maximum influence of water pressure in the slope.

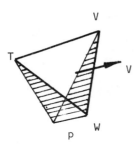

Assuming that the water enters the top of the tension crack along trace TV and that the water pressure increases linearly with depth, the water pressure at the base of the tension crack (point W) will be a product of the average elevation of line TV above point W and the density of water. Hence p = 62.5 x $\frac{1}{2}$((117.3 - 56.5) + (108.8 - 56.5)) = 3534lb/ft^2. The total water force acting at right angles to the face of the tension crack is given by the volume of the pyramid formed by the triangle TVW and the water pressure p acting at point W. Since the area of the tension crack face is 1932 ft^2 (from table AII), the total water force acting on this face is V = $\frac{1}{3}$ x 1932 x 3534 = 2.28 x 10^{6}lb.

Table All -	Areas and volumes	
Plane A:	Area ACO $= \frac{1}{2} \times 149 \times 252 \times 0.5 =$	9387 ft^2
	Area TCW $= \frac{1}{2} \times 109 \times 142 \times 0.5 =$	3870 ft^2
	Area ATWO $= 9387 - 3870 =$	5517 ft^2
Plane B:	Area BCO $= \frac{1}{2} \times 252 \times 180 \times 0.4067 =$	9224 ft^2
	Area VCW $= \frac{1}{2} \times 142 \times 103 \times 0.4067 =$	2974 ft^2
	Area BVWO $= 9224 - 2974 =$	6250 ft^2
Tension crack :	Area TVW $= \frac{1}{2} \times 63 \times 74 \times 0.8290 =$	1932 ft^2
	Volume AOBC $= \frac{1}{6}(180 \times 149 \times 252 \times 0.206) =$	232 046 ft^3
	Volume TWVC $= \frac{1}{6}(103 \times 109 \times 142 \times 0.206) =$	54 735 ft^3
	Volume ATVBOW $= 232\ 046 - 54\ 735 =$	177 311 ft^3

Figure A4 : Cardboard model of wedge constructed from the developed views in Figure A3.

If it is assumed that the water pressure distribution along
the traces TW and VW is the same for the tension crack as
for planes A and B (i.e. there is unrestricted hydraulic
connection between the base of the tension crack and the
planes A and B), then the water pressure distribution on
planes A and B is given by the pressure p acting at point W
with a linear decrease of pressure along the line WO. The
total uplift force U_A acting at right angles to plane A is
given by the volume of the figure formed by the area ATWO
and a distance proportional to the pressure p acting at W.
Hence, $U_A = \frac{1}{3} \times 5517 \times 3534 = 6.50 \times 10^6$ lb.

Summary of forces acting on wedge

Assuming a rock density of 160 lb/ft^3, the total weight of
the wedge AOBC is W = 160 x 232,46 = 37.13 x 10^{6}lb. The
weight W' of the wedge ATVBOW which will slide away from
the tension crack is given by W' = 160 x 177,311 = 28.37 x
10^{6}lb.

A summary of all the forces acting on the slope is given in
Table AIII.

Table AIII - Forces acting on wedge	
W - weight of wedge AOBC	37.12 x 10^{6}lb.
W' - weight of wedge ATVBOW	28.37 x 10^{6}lb.
V - water force on tension crack face	2.28 x 10^{6}lb.
U_A - Uplift force on plane A	6.50 x 10^{6}lb.
U_B - Uplift force on plane B	7.36 x 10^{6}lb.

Resolution of forces acting on wedge

In order to resolve the forces acting on the wedge it is
necessary to construct views perpendicular and parallel to
a vertical plane through OC, as shown in Figure A5. These
views are obtained by projecting the points A, B, C etc. in
a direction perpendicular to OC and parallel to OC and then
establishing the elevation of each of these points along the
projected lines.

Starting with the view perpendicular to the line OC, the
elevations of the points A, B, C, T, V and W are obtained
from Figure AI. In the view along the line OC, the
perpendicular distances between points A and B and the
line OC are determined from the view perpendicular to the
line OC as shown. These distances are 42.5 feet and 70 feet
respectively. This last step is particularly important
because it defines the angle between planes A and B which is
most important in the force resolution. The construction of
the view along the line OC is a frequent source of error in
this analysis and a commom mistake is to assume that the
line AB in this view is horizontal or that the elevations of
points A and B rather than their normal distance from line
OC define their position.

Having constructed the views given in Figure A5, the force
resolution can now be carried out. Because of the large
differences in the magnitudes of the forces acting on the

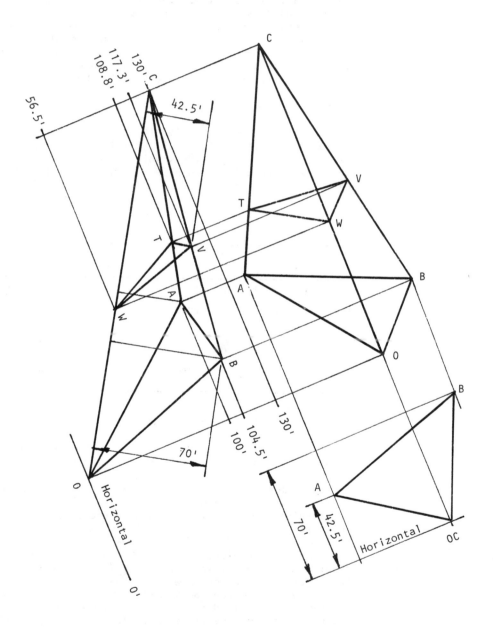

Figure A5: Construction of views perpendicular and parallel
to the line of intersection OC.

wedge, construction of force diagrams using a fixed scale could give rise to large errors. In order to overcome this difficulty, the analysis has been carried out by representing the forces W,V and T as unit lengths (10 inches in the original construction) and expressing all components of these forces as ratios of this unit length. The force diagrams included in Figure A6 have been drawn with a unit length of 1 inch and should not be used for measurement since they are intended to illustrate the method.

Step 1: In a view along the strike of the tension crack, the unit length corresponding to the water force V is set off at right angles to the face of the crack. Note that this is the only view in which the true length of V will be seen.

V is now resolved into its vertical and horizontal components V_v and V_h and these components are transferred onto the plan view. Only the component V_h will be seen in this view but it is important that the vertical component V_v is not forgotten in the next step.

The horizontal component V_h is now resolved into its components V_h' and V_h'' parallel to and normal to the line OC. The horizontal component V_h' and the vertical component V_v are transferred onto the view perpendicular to the line OC while the horizontal component V_h'' is transferred onto the view along the line OC.

In the view perpendicular to the line OC, the components V_v and V_h' are summed to give the total water force V_t in this plane and this total force is then resolved into its two components V_{oc} and V_s which are perpendicular and parallel to the line OC. These components are found to be related to the total water force V as follows:

$$V_{oc} = 0.77V \text{ and}$$
$$V_s = 0.62V$$

Step 2: The external force T is treated in much the same way as the water force V. Starting off with a view perpendicular to the trend of the external force, the unit length T is set off at a dip of 30° as shown. This is resolved through the same stages as outlined in Step 1 giving the final relationship as:

$$T_{oc} = 0.85T$$
$$T_s = 0.44T$$
$$T_A = 0.15T \text{ and}$$
$$T_B = 0.21T$$

Step 3: In order to find the normal and shear components of the weight W of the wedge which slides, the resolution shown in the view perpendicular to the line OC is carried out giving

$$W_{oc} = 0.85W \text{ and}$$
$$S = 0.52W$$

Step 4: The normal reactions N_A and N_B are found by resolving the sum of the forces which act perpendicularly to

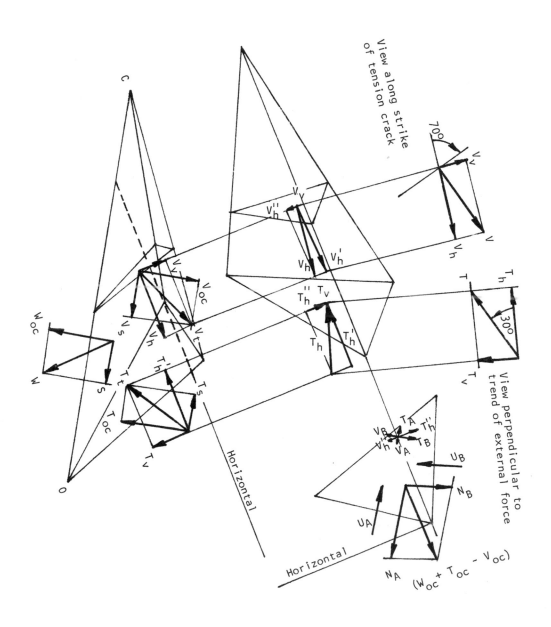

Figure A6: Resolution of forces acting on wedge.

to the line OC(W_{OC}, T_{OC} and V_{OC}) into components normal to planes A and B in the view along the line OC. In order to carry out this step, it is convenient to draw $W_{OC} + T_{OC} - V$ as a unit length and to express N_A and N_B as ratios of this unit. (Note that V_{OC} acts in an opposite direction to W_{OC} and T_{OC} and hence it must be subtracted). This construction gives :

$$N_A = 0.94 \ (W_{OC} + T_{OC} - V_{OC}) \text{ and}$$
$$N_B = 0.56 \ (W_{OC} + T_{OC} - V_{OC}).$$

Finally, the uplift forces U_A and U_B act at right angles to planes A and B in the view along the line OC as shown.

Factor of Safety

All the information for the calculation of the factor of safety of the wedge under various conditions is now available. The factor of safety is given by :

$$F = \frac{c_A \cdot A_A + c_B \cdot A_B + (N_A - U_A + V_A - T_A) \ \text{Tan}\phi_A + (N_B - U_B - V_B + T_B) \ \text{Tan} \ \phi_B}{S + V_S - T_S}$$

where c_A is the cohesive strength of plane A = 500 lb/ft^2

c_B is the cohesive strength of plane B = 1000 lb/ft^2

and A_A and A_B are the areas of planes A and B upon which sliding takes place.

Note that the signs of the various components in the factor of safety equation are most important and must be established by examination of the direction of these forces in Figure A6.

Substituting from the previous section and expressing the areas A_A and A_B in ft^2 and all forces in millions of pounds

$$F = \frac{0.0005A_A + 0.001A_B + 0.566 \ W - 0.364 \ U_A \ 0.577 \ U_B + 0.632 \ T - 0.554 \ V}{0.52 \ W + 0.62 \ V - 0.44 \ T}$$

The factor of safety can now be determined for each of the cases considered in the original problem.

Case a: Slope with no tension crack, no water pressure and no external force.

In this case V = 0, T = 0, $U_A = 0$ and $U_B = 0$

$A_A = 9387 \ ft^2$, $A_B = 9224 \ ft^2$, W = 37.13 x 10^6 lb.

$$F = \frac{0.0005 \times 9387 + 0.001 \times 9224 + 0.566 \times 37.13}{0.52 \times 37.13} = 1.81$$

Case b: Slope with tension crack but with no water and no external force.

In this case, V = 0, T = 0, U_A = 0 and U_B = 0

A_A = 5517 ft^2, A_B = 6250 ft^2, W = W' = 28.37 × 10^{6}lb.

$$F = \frac{0.0005 \times 5517 + 0.001 \times 6250 + 0.566 \times 28.37}{0.52 \times 28.37} = 1.78$$

Case c: Slope with water filled tension crack but no external force.

T = 0, V = 2.28 × 10^{6}lb, U_A = 6.50 × 10^{6}lb,

U_B = 7.36 × 10^{6}lb., A_A = 5517 ft^2, A_B = 6250 ft^2,

W' = 28.37 × 10^{6}lb.

$$F = \frac{0.0005 \times 5517 + 0.001 \times 6250 + 0.566 \times 28.37 - 0.364 \times 6.5 - 0.577 \ 36 - 0.554 \times 2.28}{0.52 \times 28.37 + 0.62 \times 2.28} = 1.06$$

Case d: The force T required to restore the factor of safety to 1.5 in case c is :

$$F = 1.5 = \frac{17.18 + 0.632 \ T}{16.16 - 0.44 \ T} \quad \text{giving} \quad \underline{T = 5.46 \times 10^6 lb.}$$

Note that this external force of 2800 short tons could not be achieved with a single cable but could be applied by 20 × 140 ton cables placed in a pattern on the face of the wedge.

This calculation of the cable tension T assumes that the cables are *pretensioned* and that T reduces the disturbing force tending to induce sliding. There are good theoretical reasons for using alternative methods of calculation to determine T and, in order to avoid confusing the present discussion these methods are discussed fully in Appendix 3.

Part II - Solution Using Spherical Projections

The basic principles of the use of spherical projections in stability analysis have been discussed in Chapter 3. The discussion given in this appendix is concerned with the application of the friction cone concept to stability analysis. Before the solution itself can be discussed, the methods of resolving forces and the basic principle of the friction cone concept must be considered.

Resolution of forces

Consider three force vectors having the following dips, dip directions and magnitudes:

Force	dip	dip direction	magnitude (arbitrary units)
A	54°	240°	3
B	40°	140°	4
C	50°	80°	5

This three-dimensional force system is resolved by splitting it into two two-dimensional stages in which A and B are combined into R_{ab} and then R_{ab} and C are combined into R_{abc}.

Step 1: The plane containing force vectors A and B is represented by the great circle passing through A and B and the angle between the vectors was found to be 64° (see page 46).

The resultant R_{ab} is found by constructing the force diagram as shown. R_{ab} is found to have a magnitude of 6 units and to be inclined at 37° to the force vector A.

Since R_{ab} lies in the same plane as A and B, it will be represented on the stereonet by a point on the great circle passing through A and B. Its position on the great circle is defined by measuring off 37° from A as shown, the tracing having been rotated into the position indicated.

Step 2: The tracing is now rotated until points R_{ab} and C on the tracing lie on the same great circle on the stereonet. The angle between R_{ab} and C is measured as 51° and this information, together with the magnitudes of R_{ab} and C, is used for the construction of the second force diagram.

Step 3: The magnitude of R_{abc} is found, from the force diagram, to be 10 units and its inclination to R_{ab} is 23°. On the stereonet, R_{abc} is represented by a point on the great circle passing through R_{ab} and C and its position on the great circle is found by measuring off 23° as shown.

Step 4: In order to find the dip of R_{abc}, the tracing is rotated until the line joining the point R to the centre of the net lies along the W-E axis. The dip of R_{abc} is measured as 63°

The dip direction of R_{abc} is found to be 128° by rotating the tracing unit until the north mark on the tracing is coincident with the north point of the stereonet.

Summary: The resultant R_{abc} of the three force A, B and C has a magnitude of 10 units, a dip of 63° and a dip direction of 128°.

The friction cone concept

Consider a block of weight W resting on a plane which is inclined at an angle ψ_p to the horizontal. The disturbing force S acting down the plane is given by $S = W.Sin\ \psi_p$ and the normal force N acting across the plane is $N = W.Cos\ \psi_p$.

If the shear strength of the surface between the block and the plane is purely frictional, i.e. the cohesive strength is zero, then the force R_f which resists sliding is given by $R_f = N.Tan\ \phi = W.Cos\ \psi p.Tan\ \phi$, where ϕ is the angle of friction of the surfaces.

Sliding of the block will occur if the disturbing force S is greater than the resisting force R_f or if $W.Sin\ \psi_p > W.Cos\ \psi_p.Tan\ \phi$. This inequality simplifies to $\psi_p > \phi$ as the condition for sliding.

Since the resisting force R_f acts uniformly on the surface between the block and the plane (assuming that the frictional strength of the surfaces is the same in all directions in

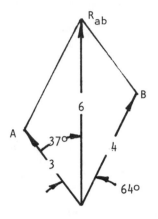

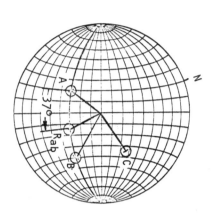

Step 1

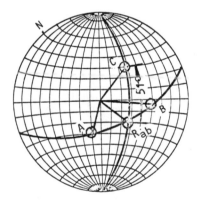

Step 2

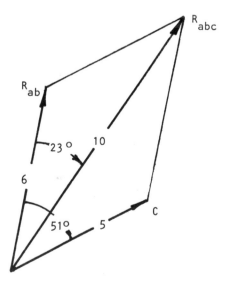

Step 3

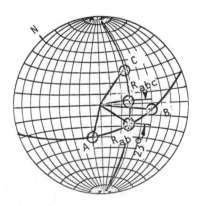

Resolution of forces

the plane), one can imagine a "friction cone" surrounding the normal force N as illustrated in Figure A7. This cone has a base circle radius R_f, a vertical height N and a semi-apical angle. As seen in Figure A7, the condition for sliding ($\psi_p > \phi$) is satisfied if the weight vector W falls outside the friction cone.

The graphical representation of this case is illustrated in Figure A9 in which the projection of the base circle of a cone defined by $\phi = 30^\circ$ is shown surrounding the pole N which defines the normal to the plane. In this example, the plane dips at 40° and it will be seen that the weight vector, W, represented by the centre of the stereonet (dip 90°), falls outside the friction cone and hence the block will slide.

The method of constructing the projection of the friction cone on the stereonet is illustrated in Figure A10.

It is required to construct the projection of the friction cone surrounding a pole to a plane which dips at 70° in a dip direction of 90°. The friction angle $\phi = 30^\circ$.

Referring to Figure A10, the great circle and the pole representing the plane are marked in as shown on diagram a) and angles of 30° are marked off on either side of the pole along the great circle passing through it. Hence a two-dimensional section through the cone has been constructed and is represented by the two points on the great circle.

The tracing is now rotated until the pole lies on another great circle and 30° angles are measured off along this great circle. A second two-dimensional section has now been constructed.

This process is continued with the pole being placed on successive great circles and 30° angles being marked off until a sufficiently large number of points has been marked for the projection to be drawn in with the aid of a template.

Note that in diagram d) the position of the pole is such that a 30° angle measured to the north falls ouside the boundary of the stereonet. This means that one edge of the friction cone has risen above the equatorial plane on which the stereonet is constructed and, in order to retain the information which would otherwise be lost, it is necessary to consider the projection of the cone which surrounds the normal which projects upwards from the plane.

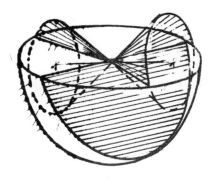

This situation is illustrated pictorially in the margin sketch and it will be seen that the cone surrounding the upper normal will appear on the opposite side of the stereonet from that surrounding the normal which intersects the lower reference hemisphere.

In diagram d) on the next page, the point corresponding to this upper cone is shown arrowed and, in diagram f), the section of the elliptical figure appearing on the opposite side of the stereonet is shown.

The complete projection of the cone is given in diagram f) in which the tracing has been rotated until the north mark is coincident with the north point on the stereonet and the plane and its pole are in their correct positions.

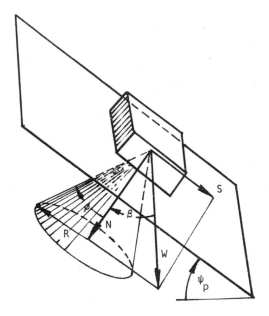

Figure A7 : Block sliding down a plane under its own weight. Sliding of the block occurs when $\psi_p > \phi$ or when the weight vector falls outside the friction cone.

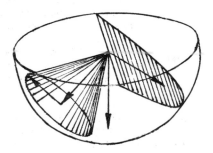

Figure A8 : Pictorial representation of the friction cone concept on the lower reference hemisphere.

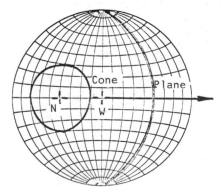

Figure A9 : Representation of the friction cone concept on a stereonet.

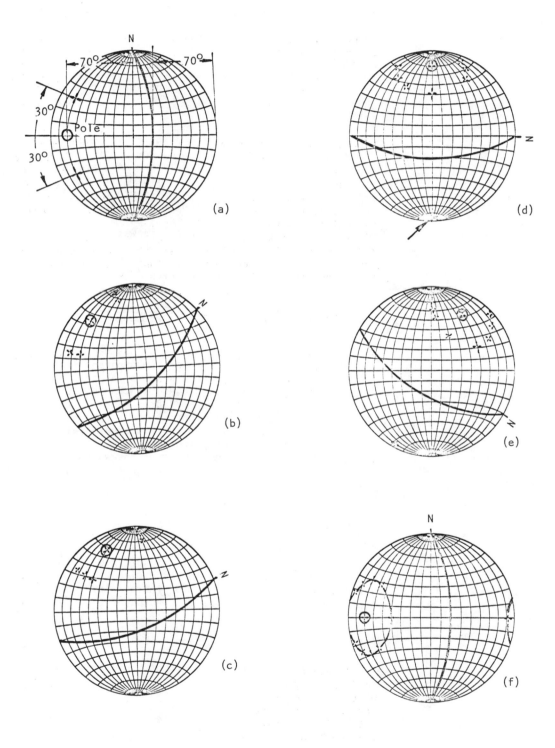

Figure A 10 : Construction of friction cone projection.

Extension of friction cone concept to include cohesion

In most practical slope stability analyses, cohesion is assumed to be zero since it is a very difficult quantity to evaluate and since to ignore it is to err on the side of safety. However, in order to provide a complete solution to the problem studied in this analysis, it is useful to deal with the graphical method for including cohesion in the stability analysis.

The cohesive force resisting sliding of a block or a wedge on a plane is given by $R_c = c.A$, where c is the cohesive strength of the surface and A is the base area of the block or wedge. This force R_c acts uniformly in the plane of sliding (assuming the cohesive strength to be uniform in all directions in the plane), and it can therefore be dealt with by adding it directly to the friction force R_f as shown in Figure A11. This results in a cone with a base circle radius of $R_f + R_c$, a vertical height N and a semi-apical angle of ϕ_a. This apparent friction angle ϕ_a can be determined from the following relationship.

$$\text{Tan } \phi_a = \frac{R_f + R_c}{N} = \text{Tan } \phi + \frac{c.A}{W.\text{Cos}\psi p}$$

Once the apparent friction angle ϕ_a has been determined, the graphical representation of the apparent friction cone is constructed in exactly the same way as the friction cone discussed above. From Figure A11 it can be seen that sliding will occur if S exceeds $R_f + R_c$ or if $\psi_p > \phi_a$. This means that sliding will occur if the weight vector, W, represented by the centre of the stereonet, falls outside the projection of the friction cone. In the case illustrated in Figure A11, it will be seen that the weight vector falls inside the friction cone and hence this particular slope is stable.

Note that the apparent friction angle ϕ_a depends upon the base area A and the weight W of the block and, since both these quantities depend upon the physical dimensions of the block, the solution will only be valid for one particular slope in which the dimensions are specified. This is a significant limitation as compared with the friction only case in which the condition for instability was simply $\psi_p > \phi$ where both ψ_p and ϕ are independent of the physical dimensions of the slope.

Influence of an external force

If, in addition to the self weight W of the block considered in Figures A7 and A11, the block is acted upon by an external force T as shown in Figure A12 the condition for instability discussed above will no longer apply. This external force may be due to water pressure or to a force exerted by a cable anchor or a rockbolt system.

One of the most convenient ways of dealing with this problem is to consider the resultant of the block weight W and the external force T as an effective weight vector W_e as shown in Figure A12. The magnitude, dip and dip direction of this resultant can be determined from a stereonet, with the aid of a supplementary force diagram, as described in the section dealing with the resolution of forces. The dip and dip direction of this effective weight vector W_e, which is no longer vertical, will determine whether it falls outside the friction cone and hence whether the block is unstable. Note that the shear and normal forces S_e and N_e are components of the effective weight vector W_e and that the apparent friction angle ϕ_a is calculated on the basis of W_e and not W as in the previous example.

Consider the example in which a block of weight W rests on a plane dipping at 60° in a dip direction of 120°. This block is acted upon by an external force T which has a magnitude of $\frac{1}{3}W$ and which dips at -20° (i.e. it acts upwards as shown in Figure A13) in a dip direction of 120°. The resultant of W and T is found from the supplementary force diagram in Figure A13 and, as shown, dips at 80° in a dip direction of 120°.

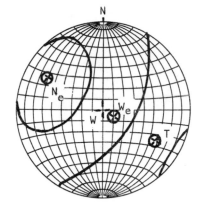

The apparent friction angle ϕ_a, calculated as described above on the basis of W_e, c, A and ϕ, is found to be 40° and the projection of the apparent friction cone is constructed around the normal N_e as shown. The effective weight vector W_e falls well outside the apparent friction cone and hence the block will slide.

The example given above was specifically chosen to illustrate the treatment of an uplift force due to water pressure acting on the under-surface of the block hence, T could have been replaced by U in this example. However, the technique used is not restricted to dealing with uplift and the external force T could have been chosen to act in any other direction in space. A more detailed example will be dealt with later when the problem being analysed in this study is considered and the effective weight vector will be obtained by resolving a number of forces.

Note that the technique above can be extended to deal with the case of the slope being subjected to an acceleration due to a large blast or to an earthquake by calculating the equivalent external force from the relationship $T = W.a$, where a is the applied acceleration.

In some cases it is convenient to treat the cohesive force R_c as an external force rather than to treat it by means of the apparent friction cone concept described earlier. This can be done by reducing the force S in Figure A7 by the amount R_c and determining the position of the resulting weight vector as described above. If this is done, the friction cone is defined by the angle ϕ as shown in Figure A7 and by ϕ_a as in Figure A11.

Figure A13 : Graphical representation and supplementary force diagram for solution of external force problem.

Graphical determination of the Factor of Safety

Consider the example dealt with in Figures A12 and A13. The factor of safety of the block on the inclined place is obtained as follows:

$$F = \frac{R}{S_e} = \frac{W_e.\,Cos\,\eta.\,Tan\,\phi_a}{W_e\,Sin\,\eta} = \frac{Tan\,\phi_a}{Tan\,\eta}$$

Note that, since the angles ϕ_a and η are calculated from relationships which incorporate friction, cohesion, the self weight of the block and the influence of external

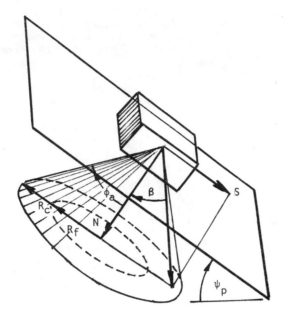

Figure A 11 : Apparent friction angle obtained for resistance due to both friction and cohesion.

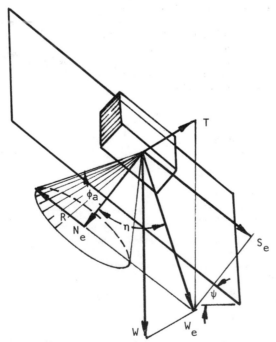

Figure A 12 : Influence of an external force T upon the stability of a block sliding on an inclined plane.

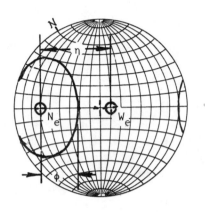

forces including water pressure, this relationship is valid
for the most general case of slope stability.

Determination of the angles ϕ_a and η, in the direction of
potential sliding, is carried out graphically by rotating
the tracing until the points N_e and W_e lie along the W-E
axis of the stereonet as illustrated opposite. The angles
are measured by counting the number of great circles between
N_e and the edge of the friction cone projection and N_e and
the effective weight vector W_e.

In the example illustrated, $\phi_a = 40^\circ$ and $\eta = 70^\circ$ and hence

$$F = \frac{\text{Tan } 40}{\text{Tan } 70} = 0.3$$

*Extension of friction cone concept to the case of two
intersecting planes*

Planes A and B intersect as shown in the pictorial view in
Figure A14. A wedge of rock resting on the two planes must
slide along the line of intersection if it is to remain in
contact with both planes during sliding and hence, in order
to assess its stability, it is necessary to determine the
apparent friction angle ϕ_i which acts in a vertical plane
parallel to the line of intersection.

The resisting force in plane A can be represented by the
force vector Q_a, the resultant of the normal force N_a and
the resisting force R_a which acts parallel to the line of
intersection of the two planes. Similarly, the resisting
force on plane B can be represented by Q_b. If the force
vectors Q_a and Q_b are summed to find their resultant in a
plane parallel to the line of intersection, this resultant
Q_i must lie in the same plane as Q_a and Q_b. The plane Oaib
which contains Q_a, Q_i and Q_b is illustrated in the pictorial
view in Figure A14 and is represented on a stereoplot by a
great circle which passes through points a and b.

To find the point a, the tracing is rotated until the pole
of plane A, marked N_a, lies on the same great circle as the
point I which defines the line of intersection of planes A
and B. The great circle passing through N_a and I, shown
dotted in Figure A14, defines a plane which is parallel to
the line of intersection and the point a is given by the
intersection of this great circle and the friction cone
surrounding the normal N_a. The point b is found by the
same process - rotating the tracing until points N_b and I
lie on the same great circle.

The tracing is now rotated until the points a and b lie on
the same great circle which, in this case, defines the
plane Oaib which contains the three force vectors Q_a, Q_i
and Q_b. The intersection of this great circle with the
line of intersection defines the point i. A great circle
passing through N_a and N_b defines the position of N_i on the
line of intersection. The apparent friction angle ϕ_i,
resulting from the combined resistance of planes A and B is
measured between N_i and i.

The angle η_i between the resultant normal force N_i and the
weight vector (represented by the centre point of the
stereonet) is measured as shown - the tracing having been

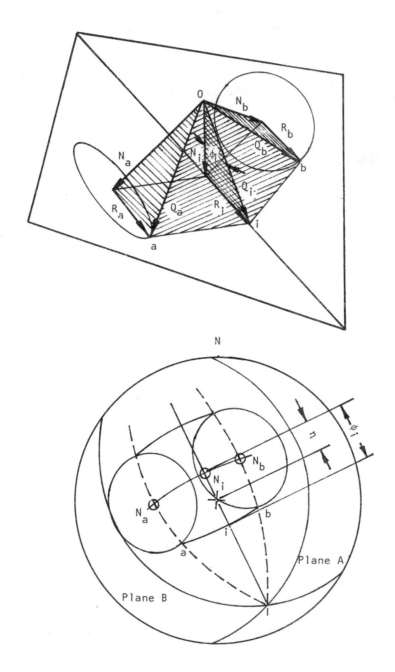

Figure A 14: Pictorial and graphical representation of friction cone concept applied to sliding on two inter-secting planes.

rotated until the line N_iI is coincident with the W-E axis of the stereonet.

The factor of safety for the case illustrated in Figure A14 in which the resistance to sliding is offered by two planes, is given by the same relationship as was established for sliding on a single plane:

$$F = \frac{Tan \; \phi_i}{Tan \; \eta_i}$$

As will be seen from Figure A 14, this particular slope is stable since the weight vector falls inside the friction "cone". If the slope is acted upon by an external force due to water pressure or an anchor load, the position of the effective weight vector is no longer defined by the centre point of the stereonet and the safety of the slope must be assessed in terms of the possible direction of movement.

In Figure A15, the graphical presentation of the example considered in Figure A14 has been redrawn to show the various modes of failure which are defined by the location of the effective weight vector on the stereonet.

Solution of general wedge problem

All the graphical techniques required for the solution of the general wedge problem considered in this study have now been discussed and the analysis is carried out as follows:

Step 1: The various planes which bound the wedge, illustrated pictorially in Figure A16, are plotted as great circles on the stereonet. The resulting tracing is reproduced in Figure A17. It is strongly recommended that the reader plot this information for himself in order to ensure that the technique has been fully understood and that the subsequent steps in the analysis can be followed.

Step 2: The next step in the analysis involves the determination of the areas and volumes of the wedge and this requires the construction of the developed views given in Figure A18. The most convenient procedure to follow in constructing this development is to measure angles off the stereonet as they are required. The detailed steps are as follows:

a) The length of line 1 is established from the fact that the vertical height H_A of point A above the toe of the wedge, point 0, has been specified as 100 feet. The dip of line 1 is measured from Figure 19 as $\psi_1 = 44°$ and hence the length of line 1 is found to be 145 feet.

b) Line 1 is drawn to an appropriate scale and then the angle θ_{13} is measured from Figure A17 (measure along the great circle passing through 1 and 3). This angle is found to be $\theta_{13} = 62°$ and this is set off from point A on line 1 as shown.

c) The angle θ_{15} between lines 1 and 5 is found, from the stereonet, to be $31°$ and this is set off from point 0 on line 1. The intersection of lines 3 and 5 defines the point C on the developed view. The lengths of lines 3 and 5 are measured as 147 feet and 250 feet respectively.

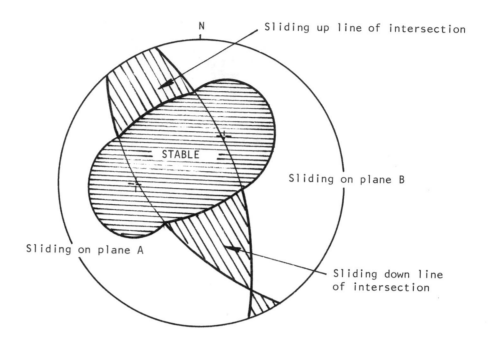

Figure A 15 : Graphical representation of various failure modes depending upon the location of the effective weight vector.

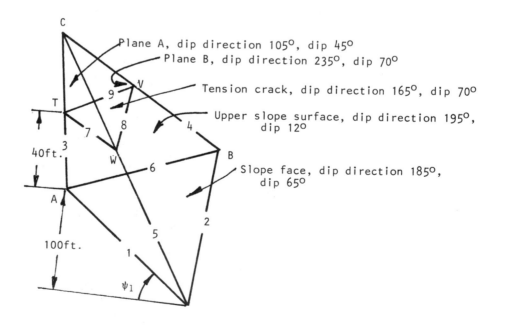

Figure A 16 : Pictorial view of wedge problem to be analysed showing numbering of intersection lines.

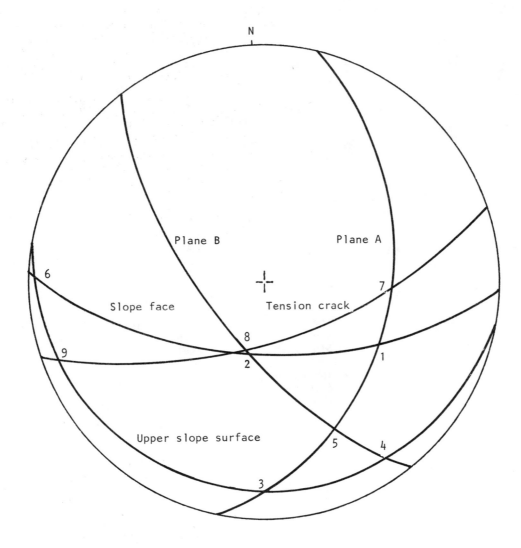

Figure A 17 : Great circle traces representing planes
defined in Figure A 16. Intersection points of great
circles represent lines of intersection of planes.

d) the angles θ_{25} and θ_{45} are measured as 40° and 25° and these are set off from the ends of line 3. Check that the length of line 4 is 176 feet.

e) The angles θ_{36} and θ_{34} are measured as 88° and 35° and these are set off from the ends of line 3. Check that the length of line 4 is 176 feet.

f) The distance AT which locates the tension crack has been specified as 40 feet and hence the length of the portion CT is 147 - 40 = 107 feet. The angles θ_{37} and θ_{39}, 80° and 68° respectively, are set off from point T on line 3 as shown. The lengths of CV and CW are 102 feet and 139 feet respectively.

g) The angle $\theta_{58} = 43°$ and this is set off from point W on line 5. Check that the length CV is the same as that determined in step f. The lengths of lines 7 and 8 are found to be 72.5 and 64 feet respectively.

h) The developed view of the face of the tension crack, the triangle TWV, is constructed from the lengths of lines 7 and 8 and the included angle θ_{78} which is measured from the stereonet is 54°.

i) Calculation of the force due to water pressure in the tension crack necessitates the determination of the average vertical elevation of line 9 above the point W. The dips ψ_7 and ψ_8 of lines 7 and 8 are 44° and 66° respectively and the corresponding vertical elevations H_T and H_V are 50 feet and 58.5 feet repectively. Hence, the vertical elevation of points T and V is 54.3 feet.

The required areas are calculated from the relationship: Area of triangle = one half product of adjacent sides x Sin of the included angle.

The volumes of the tetrahedral wedges AOBC and TWVC are calculated from the relationship: Volume = one sixth of the product of the three adjacent sides x K, where

$$K = (1-\cos^2\theta_{34}-\cos^2\theta_{35}-\cos^2\theta_{45}+2.\cos\theta_{34}.\cos\theta_{35}.\cos\theta_{45}.)^{\frac{1}{2}}$$

Substitution gives K = 0.214

Table AIV - Areas and volumes of wedges	
Plane A: Area ACO=½ x 147 x 250 x 0.5150 =	9463 ft²
Area TCW=½ x 107 x 139 x 0.5150 =	3829 ft²
Area ATWO= 9463 - 3829 =	5634 ft²
Plane B: Area BCO=½ x 250 x 176 x 0.4226 =	9297 ft²
Area VCW=½ x 139 x 102 x 0.4226 =	2996 ft²
Area BVWO= 9297 - 2996 =	6301 ft²
Tension crack area TVW = ½ x 64 x 72.5 x 0.809 =	1877 ft²
Volume AOBC = $\frac{1}{6}$ x 147 x 250 x 176 x 0.214 =	230,692 ft³
Volume TWVC = $\frac{1}{6}$ x 102 x 139 x 107 x 0.214 =	54,108 ft³
Volume ATVBOW = 320,692 - 54,108 =	176,584 ft³

Comparing the values given in Table AIV with the corresponding values in Table AII shows that the results obtained from

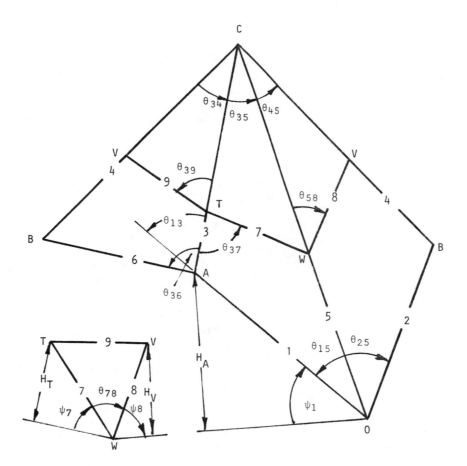

Figure A 18 : Construction of developed views of faces of the wedge in order to determine the lengths of the lines of intersection for the calculation of areas and volumes.

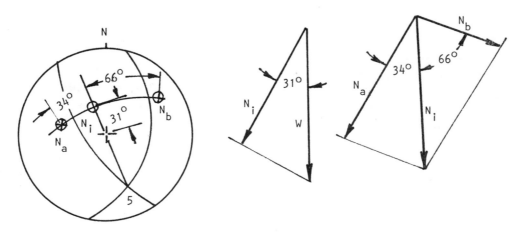

Figure A 19 : Stereonet tracing and supplementary force diagram for the determination of the normal forces N_a and N_b.

the engineering graphics and the stereographic methods are within approximately 3% of each other. No attempt has been made to correct the values in either case and the results are typical of those which should be obtained if reasonable care is taken in the construction. Experience shows that the most serious error in either of the graphical methods arises from inaccurate angle measurement and every effort should be made to work to an accuracy of better that 1°.

Step 3: The next step in the analysis is the calculation of the forces which act on the wedge.

The weight of the wedge AOBC, assuming a rock density of 160 lb/ft^3, is 160 x 230,692 = 36.91 x 10^{6}lb.

The weight of wedge ATVBOW = 160 x 176,584 = 28.25 x 10^{6}lb.

The force V due to water pressure in the tension crack, assuming a linear increase in water pressure over the average vertical elevation of 54.3 feet, is $\frac{1}{3}$ x 62.5 x 54.3 x 1877 = 2.12 x 10^{6}lb.

The uplift force U_A due to water pressure on area ATWO is $U_A = \frac{1}{3}$ x 63.5 x 54.3 x 5634 = 6.37 x 10^{6}lb.

The uplift force U_B due to water pressure on area BVWO is $U_B = \frac{1}{3}$ x 63.5 x 54.3 x 6301 = 7.13 x 10^{6}lb.

Step 4: The forces acting on the wedge, derived in step 3 above, must now be resolved to determine the apparent friction angles and the dip and dip direction of the effective weight vector for the various cases set out in the original question.

The first step is the determination of the normal forces N_a and N_b. Since these forces are normal to the planes, N_a and N_b will be coplanar with N_i (see Figure A14) which is the component, normal to the line of intersection, of the weight vector. The angles between W and N_i and between N_a, N_i and N_b are determined from the stereonet by drawing the great circle passing through N_a, N_i and N_b as shown in Figure A19.

Figure A19 also shows the two supplementary force diagrams which are required to resolve W into N_i and N_i into N_a and N_b. The relationship between the force vectors is as follows:

$$N_i = 0.86W$$
$$N_a = 0.93 \; N_i = 0.80W$$
$$N_b = 0.56 \; N_i = 0.48W$$

Step 5: The apparent friction angles can now be calculated from the following relationships:

$$\text{Tan } \phi_{aA} = \text{Tan } \phi_A + \frac{c_A.A_A}{N_a} = \text{Tan } \phi_A + \frac{c_A.A_A}{0.80W}$$

$$\text{Tan } \phi_{aB} = \text{Tan } \phi_B + \frac{c_B.A_B}{N_b} = \text{Tan } \phi_B + \frac{c_B.A_B}{0.48W}$$

Two cases must be considered:

Case a. The stability of wedge AOBC without the tension crack

Case b. The stability of wedge ATVBOW which lies in front of the tension crack

Substitution of the friction angles and cohesive strength (from Table A) and the areas and volumes (multiplied by the unit weights to give weights) from Table AIV give the apparent friction angles listed in Table AV.

Table AV	-	Apparent friction angles

Case a - Slope without tension crack

$$\text{Tan } \phi_{aA} = \text{Tan } 20^o + \frac{500 \times 9463}{0.8 \times 36.91 \times 10^6} \; ; \; \phi_{aA} = 27.7^o$$

$$\text{Tan } \phi_{aB} = \text{Tan } 30^o + \frac{1000 \times 9297}{0.48 \times 36.91 \times 10^6} \; ; \; \phi_{aB} = 47.8^o$$

Case b - Slope with tension crack

$$\text{Tan } \phi_{aA} = \text{Tan } 20^o + \frac{500 \times 5634}{0.8 \times 28.25 \times 10^6} \; ; \; \phi_{aA} = 26.0^o$$

$$\text{Tan } \phi_{aB} = \text{Tan } 30^o + \frac{1000 \times 6301}{0.48 \times 28.25 \times 10^6} \; ; \; \phi_{aB} = 46.2^o$$

Figure A20 shows the graphical representation of the friction cones surrounding the normals N_a and N_b. Great circles passing through point 5, which defines the lines of intersection of the planes, and points N_a and N_b intersect the friction circles at points a and b (compare Figure A14). Great circles through points a and d define the apparent friction angle along the line of intersection. Note that the upper great circles through a' and b' must be determined from the intersection of the mirror image cone with the friction cone surrounding N_b.

Case a for the slope without the tension crack is represented by the outer dotted friction cones while case b, for the slope with the tension crack, is represented by the inner solid lines.

Step 6: The factors of safety for cases a and b can now be determined from Figure A20. The angle η measured along the lines of intersection, is 31°. The apparent friction angles, also measured along the lines of intersection are 48° and 46° for cases a and b respectively. The factors of safety are:

Case a: $\quad F = \dfrac{\text{Tan } 48^o}{\text{Tan } 31^o} = 1.85$

Case b: $\quad F = \dfrac{\text{Tan } 46^o}{\text{Tan } 31^o} = 1.72$

Step 7; The third case considered in this study is the slope with a water filled tension crack and with the associated uplift forces acting on the base of the wedge. The water in the tension crack results in a force $V = 2.12 \times 10^6$ lb which acts in a dip direction of 165° and has a dip of -20°, i.e. it acts upwards as shown in Figure A22.

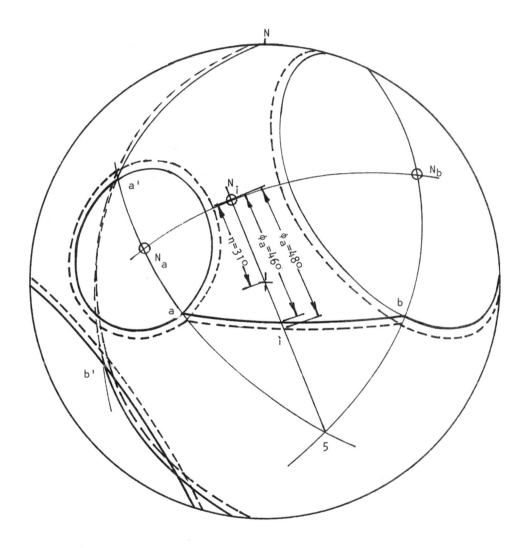

Figure A 20 : Graphical representation of friction cones on planes A and B.

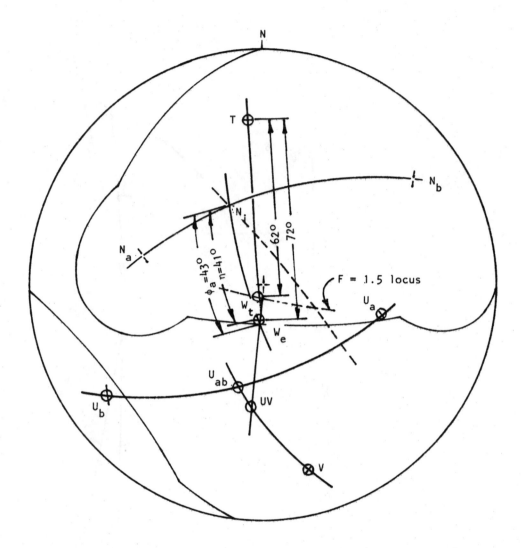

Figure A 21 : Resolution of forces for evaluation of water
pressure influence and for determination of cable tension.

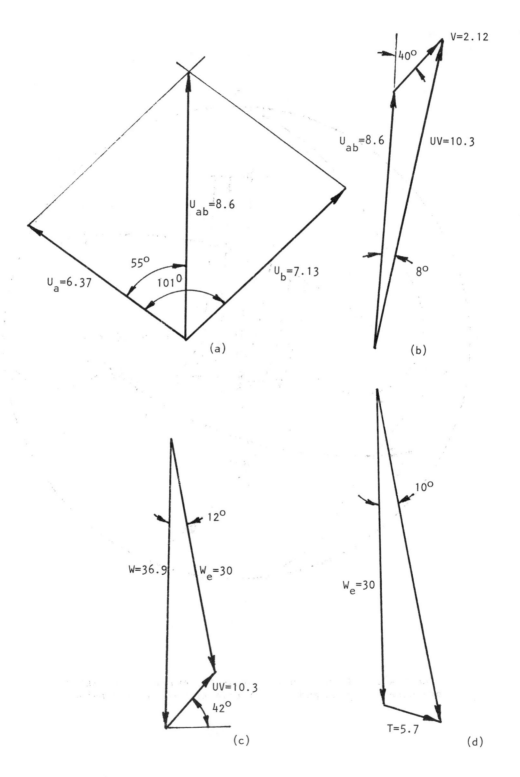

Figure A 22 : Resolution of forces.
Note - all forces in millions of pounds, different scales
have been used in constructing different diagrams.

The point V representing this force is located on the stereonet as shown in the tracing reproduced in Figure A21.

The uplift forces $U_a = 6.37 \times 10^6$ and $U_b = 7.13 \times 10^6$ lb act in the opposite direction to the normals N_a and N_b and are represented by points U_a and U_b on the tracing in Figure A21. The uplift forces U_a and U_b are resolved into their vector sum in the force diagram given in Figure A22a. The angle of 101° between these forces is obtained from Figure A21 by measuring along the great circle passing through points U_a and U_b. The resultant U_{ab} is found to have a magnitude of 8.6×10^6 lb and to be inclined at 55° to U_a. It is located on the great circle passing through U_a and U_b by measuring off 55° from U_a.

A great circle passing through U_{ab} and V gives the angle between these two forces of 40° and the force diagram in Figure A22 gives the magnitude of their resultant Uv as 10.3×10^6 lb. This resultant is inclined at 8° to U_{ab} and this angle, measured along the great circle passing through U_{ab} and V, locates it on the stereonet.

Finally, UV and W are resolved in the force diagram given in Figure A22c, the angle of 41° between them having been measured along the great circle passing through them. The effective weight W_e is found to have a magnitude of 30×10^6 lb and to be inclined at 12° to W.

The angles $\eta = 41^\circ$ and $\phi_a = 43^\circ$, measured along the great circle passing through N_i and W_e, define the factor of safety for this case as:

$$F = \frac{\text{Tan } 43^\circ}{\text{Tan } 41^\circ} = 1.07$$

Step 8: The final case studied in this analysis is that of an external force T dipping at 30° in a dip direction of 355°. It is required to find the magnitude of this force, applied by cable anchors or rockbolts, which will be needed to restore the factor of safety of the slope to 1.5.

The resultant of W_e and T lies along the great circle passing through these two points on the stereonet (Figure A21). The location of the point W_t which represents this resultant is determined by the intersection of the locus of F = 1.5 with the great circle. The construction of the F = 1.5 locus is carried out by rotating the tracing so that successive great circles pass through the point N_i and measuring the angle ϕ_a between N_i and the friction "cone" boundary. One such great circle is shown dotted in Figure A21 and ϕ_a in this case is 56°. Since F = 1.5 = Tan 56°/ Tan η, $\eta = 45^\circ$. Successive determination of the angles η defines the F = 1.5 locus.

The angle between W_e and W_t is found to be 10° (Note that point W_e lies just inside and not on the friction "cone"). The force diagram reproduced in Figure A22 gives the magnitude of T as 5.7×10^6 lb.

Summary of results: Case a: F = 1.85

Case b: F = 1.72

Case c: F = 1.07

Case d: T = 5.7×10^6 lb.

Part III - Analytical solution

Most analytical solutions of the problem of the stability of a tetrahedral wedge are presented in terms of vector methods. While these methods are certainly the most convenient means of dealing with the problem, the applicability of these solutions to practical rock engineering problems is seriously hampered by the fact that many engineers and geologists are not familiar with vector analysis. In order to provide a solution which is more generally understood, the analysis presented on the following pages has been formulated in non-vector terms. The derivation of all equations has been included so that the reader can modify the analysis to suit his own particular needs.

Before discussing the wedge solution, a few basic operations which will be required in the subsequent analysis are considered.

Resolution of a force or a unit length

It is required to resolve a force or a unit length P_a into its normal projection in a different direction in space. Suppose that the force P_a dips at an angle ψ_a in a dip direction α_a and that its projection P_b in a direction defined by a dip ψ_b and a dip direction α_b is required.

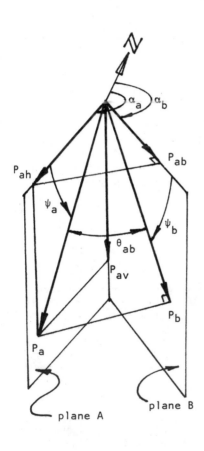

Consider two vertical planes A and B in which the forces P_a and P_b lie. Resolve P_a into its horizontal and vertical components in plane A:

$$P_{ah} = P_a . \cos \psi_a$$
$$P_{av} = P_a . \sin \psi_a \qquad (A1)$$

Projecting the horizontal component P_{ah} onto plane B:

$$P_{ab} = P_{ah} . \cos (\alpha_a - \alpha_b)$$
$$= P_a . \cos \psi_a . \cos (\alpha_a - \alpha_b) \qquad (A2)$$

The required projection P_b, dipping at ψ_b in plane B, is given by the sum of the components of P_{av} and P_{ab} in this direction.

$$P_b = P_{ab} . \cos \psi_b + P_{av} . \sin \psi_b$$

Hence
$$P_b = m_{ab} . P_a$$

$$\boxed{m_{ab} = \cos \psi_a . \cos \psi_b . \cos (\alpha_a - \alpha_b) + \sin \psi_a . \sin \psi_b} \qquad (A3)$$

Note that, if the resolution is carried out in the reverse direction (resolving P_b into P_a), the coefficient m, which is a direction cosine, is the same, i.e.

$$m_{ab} = m_{ba} \qquad (A4)$$

From the diagram it will be seen that $P_b = P_a . \cos \theta_{ab}$ where θ_{ab} is the angle between P_a and P_b. Hence

$$\boxed{\cos \theta_{ab} = \cos \psi_a . \cos \psi_b . \cos (\alpha_a + \alpha_b) + \sin \psi_a . \sin \psi_b}$$
$$(A5)$$

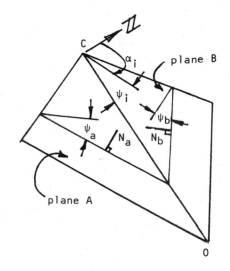

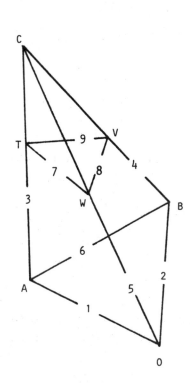

Determination of the line of intersection of two planes

Two planes, A and B, intersect along a line OC and it is required to determine the dip ψ_i and the dip direction α_i of this line.

Consider the normals N_a and N_b to these planes. These normals have dips $\psi_{na} = (\psi_a - 90)$, $\psi_{nb} = (\psi_b - 90)$ repectively while their dip directions are the same as those of the planes, i.e. α_a and α_b.

The coefficient $m_{na.i}$ required to resolve from the direction of the normal N_a to the direction of the line of intersection OC is obtained from equation (A3).

$$m_{na.i} = \cos \psi_a . \cos \psi_i . \cos (\alpha_a - \alpha_i) + \sin \psi_{na} . \sin \psi_i$$

giving

$$m_{na.i} = \sin \psi_a . \cos \psi_i . \cos (\alpha_a - \alpha_i) + \cos \psi_a . \sin \psi_i$$

(A6)

Similarly:

$$m_{nb.i} = \sin \psi_b . \cos \psi_i . \cos (\alpha_b - \alpha_i) + \cos \psi_b . \sin \psi_i$$

(A7)

The normal to a plane is, by definition, perpendicular to every line in that plane and, since the line of intersection OC is common to both planes, the normals N_a and N_b are both perpendicular to it. Since the line has no component at right angles to itself, the coefficients $m_{na.i}$ and $m_{nb.i}$ are both equal to zero. Putting equations (A6) and (A7) equal to zero gives:

$$\boxed{\tan \psi_i = \tan \psi_a . \cos (\alpha_a - \alpha_i) = \tan \psi_b . \cos (\alpha_b - \alpha_i)}$$ (A8)

Expanding equation (A8) and solving for α gives

$$\boxed{\tan \alpha_i = \frac{\tan \psi_a . \cos \alpha_a - \tan \psi_b . \cos \alpha_b}{\tan \psi_b . \sin \alpha_b - \tan \psi_a . \sin \alpha_a}}$$

(A9)

Determination of areas and volumes

In order to determine the forces acting on the wedge, the areas of the surfaces upon which the wedge rests and the volume of the wedge must be calculated. Having established the dips, dip directions and angles between the lines of intersection of the various planes which bound the wedge, from equations (A8), (A9) and (A5), the required areas and volumes can be calculated provided that the length of one of these lines of intersection is known. Any one of the lines can be used as a basis for the calculation but it is obviously advantageous to choose one which can be measured in the field. The traces AC and BC of planes A and B on the upper slope surface are likely to the most accessible in an actual slope problem and hence, the length AC will be used as the basis for the following calculations. The length CT will be used for the calculation of the volume of wedge CTWV.

If the lengths AC and CT have not been measured or specified, their lengths will have to be determined or estimated from whatever information is available as will be demonstrated in the example to be considered later.

The required areas and volumes are calculated as follows:

Plane A:

$$\text{Area ACO} = \tfrac{1}{2}.\text{AC}.\text{CO}.\text{Sin } \theta_{35} \tag{A10}$$

$$\text{Area TCW} = \tfrac{1}{2}.\text{TC},\text{CW}.\text{Sin } \theta_{35} \tag{A11}$$

$$\text{Area ATWO} = \tfrac{1}{2}(\text{AC}.\text{CO} - \text{TC}.\text{CW}) \text{ Sin } \theta_{35} \tag{A12}$$

Plane B:

$$\text{Area BCO} = \tfrac{1}{2}.\text{BC}.\text{CO}.\text{Sin } \theta_{45} \tag{A13}$$

$$\text{Area VCW} = \tfrac{1}{2}.\text{VC}.\text{CW}.\text{Sin } \theta_{45} \tag{A14}$$

$$\text{Area BVWO} = \tfrac{1}{2}(\text{BC}.\text{CO} - \text{VC}.\text{CW})\text{Sin } \theta_{45} \tag{A15}$$

Tension crack:

$$\text{Area TWV} = \tfrac{1}{2}.\text{TV}.\text{TW}.\text{Sin } \theta_{79} \tag{A16}$$

$$\text{Volume ACBO} = \tfrac{1}{6}.\text{AC}.\text{CO}.\text{BC}. \text{ K} \tag{A17}$$

$$\text{Volume TCVW} = \tfrac{1}{6}.\text{TC}.\text{CW}.\text{VC}. \text{ K} \tag{A18}$$

where K is a function of the angles θ_{34}, θ_{35} and θ_{45} which will be defined later.

Application of the sine rule gives:

$$\text{CO} = \text{AC}. \frac{\text{Sin } \theta_{13}}{\text{Sin } \theta_{15}} \tag{A19}$$

$$\text{CW} = \text{TC}. \frac{\text{Sin } \theta_{37}}{\text{Sin } \theta_{57}} \tag{A20}$$

$$\text{BC} = \text{CO}. \frac{\text{Sin } \theta_{25}}{\text{Sin } \theta_{24}} = \text{AC}. \frac{\text{Sin } \theta_{13}.\text{Sin } \theta_{25}}{\text{Sin } \theta_{15}.\text{Sin } \theta_{24}} \tag{A21}$$

$$\text{VC} = \text{CW}. \frac{\text{Sin } \theta_{58}}{\text{Sin } \theta_{48}} = \text{TC}. \frac{\text{Sin } \theta_{37}.\text{Sin } \theta_{58}}{\text{Sin } \theta_{57}.\text{Sin } \theta_{48}} \tag{A22}$$

$$\text{TV} = \text{TC}. \frac{\text{Sin } \theta_{34}}{\text{Sin } \theta_{49}} \tag{A23}$$

$$\text{TW} = \text{TC}. \frac{\text{Sin } \theta_{35}}{\text{Sin } \theta_{57}} \tag{A24}$$

Substitution in equations (A12), (A15), (A16), (A17) and (A18) gives:

Area ATWO on plane A:

$$A_A = \tfrac{1}{2} \left\{ \text{AC}^2. \frac{\text{Sin } \theta_{13}}{\text{Sin } \theta_{15}} - \text{TC}^2. \frac{\text{Sin } \theta_{37}}{\text{Sin } \theta_{57}} \right\} \text{Sin } \theta_{35} \tag{A25}$$

Area BWVO on plane B:

$$A_B = \tfrac{1}{2} \left\{ \text{AC}^2. \frac{\text{Sin}^2\theta_{13}.\text{Sin}\theta_{25}}{\text{Sin}^2\theta_{15}.\text{Sin}\theta_{24}} - \text{TC}^2. \frac{\text{Sin}^2\theta_{37}.\text{Sin}\theta_{58}}{\text{Sin}^2\theta_{57}.\text{Sin}\theta_{48}} \right\} \text{Sin } \theta_{45} \tag{A26}$$

Area TVW of tension crack:

$$A_T = \tfrac{1}{2}. \text{CT}^2. \frac{\text{Sin } \theta_{34}. \text{ Sin } \theta_{35}. \text{ Sin } \theta_{79}}{\text{Sin } \theta_{49}. \text{ Sin } \theta_{57}} \tag{A27}$$

Weight of wedge = $\gamma \cdot$ volume AOBVTW, where γ is unit weight.

$$W = \frac{1}{6} \cdot \gamma \cdot K \left\{ AC^3 \cdot \frac{Sin^2\theta_{13} \cdot Sin\theta_{25}}{Sin^2\theta_{15} \cdot Sin\theta_{24}} - TC^3 \cdot \frac{Sin^2\theta_{37} \cdot Sin\theta_{58}}{Sin^2\theta_{57} \cdot Sin\theta_{48}} \right\} \quad (A28)$$

where

$$K = (1 - Cos^2\theta_{34} - Cos^2\theta_{35} - Cos^2\theta_{45} + 2 \cdot Cos\theta_{34} \cdot Cos\theta_{35} \cdot Cos\theta_{45})^{\frac{1}{2}}$$

(A29)

Forces due to water pressure

In this analysis it has been assumed that water is free to enter the open top of the tension crack (trace TV) and that the water pressure increases linearly with depth to a maximum of p at the point W which defines the intersection of the tension crack with the lines of intersection OC. The magnitude of p is a product of the density of water γ_w and the average vertical elevation of points T and V above W. These elevations are determined from the lengths and dips of lines 7 and 8.

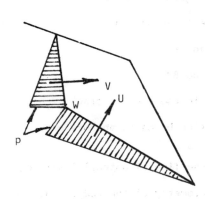

$$TW = TC \cdot \frac{Sin\,\theta_{35}}{Sin\,\theta_{57}} \quad (A30)$$

$$VW = TC \cdot \frac{Sin\,\theta_{35} \cdot Sin\,\theta_{79}}{Sin\,\theta_{57} \cdot Sin\,\theta_{89}} \quad (A31)$$

$$H_T = TC \cdot \frac{Sin\,\theta_{35} \cdot Sin\,\psi_7}{Sin\,\theta_{57}} \quad (A32)$$

$$H_V = TC \cdot \frac{Sin\,\theta_{35} \cdot Sin\,\theta_{79} \cdot Sin\,\psi_8}{Sin\,\theta_{57} \cdot Sin\,\theta_{89}} \quad (A33)$$

Hence, the pressure p is given by:

$$p = \gamma_w \cdot \frac{TC}{2} \cdot \frac{Sin\,\theta_{35}}{Sin\,\theta_{57}} \left\{ Sin\,\psi_7 + \frac{Sin\,\theta_{79}}{Sin\,\theta_{57}} \cdot Sin\,\psi_8 \right\} \quad (A34)$$

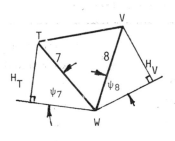

Note that water pressure distributions other than that assumed in deriving equation (A34) are possible and it may be necessary to modify this equation in accordance with the geometry and source of groundwater flow in the particular problem being studied. While no general rules can be given because of the wide range of geometrical combinations possible in actual rock slope problems, it can be stated that the same process as that used in deriving equation (A34) can be applied to most of these problems. The water pressure p given by equation (A34) can be regarded as a realistic maximum value which could occur under conditions of exceptionally heavy rainfall and the factor of safety of the slope calculated on the basis of this value of p can therefore be regarded as the minimum value for that slope.

The force V due to water pressure in the tension crack is given by the volume of the pyramid formed by the pressure p acting at point W (the pressure being zero at T and V) on the area A_T of the tension crack face. Hence

$$V = \frac{1}{3} \cdot p \cdot A_T$$

(A35)

It is assumed that the water pressure is transmitted fully between the tension crack and the faces A and B on which the wedge rests. Hence, the uplift forces U_A and U_B are obtained from the volumes of the figures given by the pressure p acting at point W on the areas A_A and A_B. The pressures along all free faces, i.e. along traces AT, VB, AO and OB, are zero.

$$U_A = \frac{1}{3} \cdot p \cdot A_A \qquad \text{(A36)}$$

$$U_B = \frac{1}{3} \cdot p \cdot A_A \qquad \text{(A37)}$$

Resolution of forces

The forces acting on the wedge are illustrated in the pictorial view in the margin. These forces are:

W   - the weight of the wedge

N_{ae} - the effective normal reaction on plane A

N_{be} - the effective normal reaction on plane B

U_A - the uplift force on plane A

U_B - the uplift force on plane B

V   - the force due to water in the tension crack

T   - the external force applied by cable anchors or rockbolts

S   - the force acting down the line of potential sliding

From a consideration of the geometry of the wedge, the dips and dip directions of these forces are as listed in Table AVI.

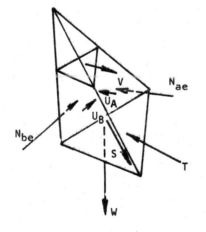

Table AVI - dips and dip directions of forces		
Force	dip	dip direction
W	$90°$	indeterminate
N_{ae}	$(\psi_a - 90°)$	α_a
N_{be}	$(\psi_b - 90°)$	α_b
V	$(\psi_t - 90°)$	α_t
T	ψ_T	α_T
S	ψ_5	α_5

In order to calculate the forces which resist sliding of the wedge, the effective normal reactions N_{ae} and N_{be} are required and these are obtained by resolving in the directions of these normal reactions.

$$(N_{ae} + U_A) + m_{na.nb}(N_{be} + U_B) + m_{W.na}.W + m_{V.na}.V + m_{T.na}.T = 0 \qquad \text{(A38)}$$

$$(N_{be} + U_A) + m_{na.nb}(N_{ae} + U_A) + m_{W.nb}.W + m_{V.nb}.V + m_{T.nb}.T = 0 \qquad \text{(A39)}$$

Solving for N_{ae} and N_{be}:

$$\boxed{N_{ae} = q.W + r.V + s.T - U_A} \qquad \text{(A40)}$$

$$\boxed{N_{be} = x.W + y.V + z.T - U_B} \qquad \text{(41)}$$

where

$$q = (m_{na.nb}.m_{W.nb} - m_{W.na})/(1 - m^2_{na.nb}) \qquad \text{(A42)}$$

$$r = (m_{na.nb}.m_{V.nb} - m_{V.na})/(1 - m^2_{na.nb}) \qquad \text{(A43)}$$

$$s = (m_{na.nb}.m_{T.nb} - m_{T.na})/(1 - m^2_{na.nb}) \qquad \text{(A44)}$$

$$x = (m_{na.nb}.m_{W.na} - m_{W.nb})/(1 - m^2_{na.nb}) \qquad \text{(A45)}$$

$$y = (m_{na.nb}.m_{V.na} - m_{V.nb})/(1 - m^2_{na.nb}) \qquad \text{(A46)}$$

$$z = (m_{na.nb}.m_{T.na} - m_{T.nb})/(1 - m^2_{na.nb}) \qquad \text{(A47)}$$

The disturbing force S acting down the line of intersection OC is given by resolving along OC (numbered 5 in this analysis).

$$\boxed{S = m_{W.5}.W + m_{V.5}.V + m_{T.5}.T} \qquad \text{(A48)}$$

The coefficients required for the solution of equations (A42) to (A48) are listed in Table AVII. These coefficients are obtained by substituting the dips and dip directions listed in Table AVI into equation (A3).

Factor of safety

The factor of safety of the slope is given by the ratio of the resisting force due to friction and cohesion to the disturbing force S. Hence

$$\boxed{F = \frac{c_A A_A + c_B A_B + (q.W + r.V + s.T - U_A)\mathrm{Tan}\phi_A + (x.W + y.V + z.T - U_B)\mathrm{Tan}\phi_B}{m_{W.5}.W + m_{V.5}.V + m_{T.5}.T}} \qquad \text{(A49)}$$

Method of calculation

Because the solution of this problem involves solving a large number of equations, a great deal of time can be wasted and serious errors in calculation can arise if the problem is tackled in a haphazard manner. The authors recommend the use of a calculation table such as that set out on the following pages, modified as required to suit the user's particular needs.

Table AVII - Coefficients for resolution of forces
$m_{na.nb} = Sin\,\psi_a . Sin\,\psi_b . Cos(\alpha_a - \alpha_b) + Cos\,\psi_a . Cos\,\psi_b$
$m_{W.na} = -\,Cos\,\psi_a$
$m_{W.nb} = -\,Cos\,\psi_b$
$m_{V.na} = Sin\,\psi_a . Sin\,\psi_t . Cos(\alpha_a - \alpha_t) + Cos\,\psi_a . Cos\,\psi_t$
$m_{V.nb} = Sin\,\psi_b . Sin\,\psi_t . Cos(\alpha_b - \alpha_t) + Cos\,\psi_b . Cos\,\psi_t$
$m_{T.na} = Cos\,\psi_T . Sin\,\psi_a . Cos(\alpha_T - \alpha_a) - Sin\,\psi_T . Cos\,\psi_a$
$m_{T.nb} = Cos\,\psi_T . Sin\,\psi_b . Cos(\alpha_T - \alpha_b) - Sin\,\psi_T . Cos\,\psi_b$
$m_{W.5} = Sin\,\psi_5$
$m_{V.5} = Cos\,\psi_5 . Sin\,\psi_t . Cos(\alpha_5 - \alpha_t) - Sin\,\psi_5 . Cos\,\psi_t$
$m_{T.5} = Cos\,\psi_5 . Cos\,\psi_T . Cos(\alpha_5 - \alpha_T) + Sin\,\psi_5 . Sin\,\psi_T$

Determination of the minimum cable load for a given factor of safety

Rearranging the factor of safety equation (A49) in terms of the cable load T :

$$T = \frac{c_A A_A + c_B A_B + (qW + rV - U_A)Tan\phi_A + (xW + yV - U_B)Tan\phi_B - F(m_{W.5} + m_{V.5}V)}{F.m_{T.5} - (sTan\phi_A + zTan\phi_B)}$$

(A50)

To find the inclination ψ_T at which the cable should be installed in order to give the minimum cable tension, differentiate for T in terms of ψ_T and equate to zero :

$$Tan\psi_T = \frac{X.Tan\psi_5 - Y.Cot\psi_a - Z.Cot\psi_b}{X.Cos(\alpha_5 - \alpha_T) + Y.Cos(\alpha_T - \alpha_a) + Z.Cos(\alpha_T - \alpha_b)}$$

(A51)

The differential of this expression, with respect to α_T, is equated to zero to give the optimum direction of the cable :

$$Tan\alpha_T = \frac{X.Sin\alpha_5 + Y.Sin\alpha_a + Z.Sin\alpha_b}{X.Cos\alpha_5 + Y.Cos\alpha_a + Z.Cos\alpha_b}$$

(A52)

where

$X = (1 - m_{na.nb}^2)F.Cos\psi_5$

$Y = (Tan\phi_A - m_{na.nb}.Tan\phi_B)Sin\psi_a$

$Z = (Tan\phi_B - m_{na.nb}.Tan\phi_A)Sin\psi_b$

Solving equations A51 and A52 will give several possible values for ψ_T and α_T. The correct values have to be chosen by inspection and are those values which are closest to the following : $\psi_T \simeq (\psi_5 + 180° - \frac{1}{2}(\phi_A + \phi_B))$ and $\alpha_T \simeq \alpha_5 \pm 180°$.

CALCULATION TABLE FOR WEDGE ANALYSIS - SHEET 1	RESULTS
Step 1: Calculation of dip directions of lines of intersection (equation A9)	
Input data:	
ψ_a - dip of plane A	45°
α_a - dip direction of plane A	105°
ψ_b - dip of plane B	70°
α_b - dip direction of plane B	235°
ψ_f - dip of slope face	65°
α_f - dip direction of slope face	185°
ψ_s - dip of upper slope surface	12°
α_s - dip direction of upper slope surface	195°
Calculation:	
$\alpha_1 = \arctan\left\{\dfrac{\mathrm{Tan}\psi_a.\mathrm{Cos}\alpha_a - \mathrm{Tan}\psi_f.\mathrm{Cos}\alpha_f}{\mathrm{Tan}\psi_f.\mathrm{Sin}\alpha_f - \mathrm{Tan}\psi_a.\mathrm{Sin}\alpha_a}\right\}$	121.6°
$\alpha_2 = \arctan\left\{\dfrac{\mathrm{Tan}\psi_b.\mathrm{Cos}\alpha_b - \mathrm{Tan}\psi_f.\mathrm{Cos}\alpha_f}{\mathrm{Tan}\psi_f.\mathrm{Sin}\alpha_f - \mathrm{Tan}\psi_b.\mathrm{Sin}\alpha_b}\right\}$	195.2°
$\alpha_3 = \arctan\left\{\dfrac{\mathrm{Tan}\psi_a.\mathrm{Cos}\alpha_a - \mathrm{Tan}\psi_s.\mathrm{Cos}\alpha_s}{\mathrm{Tan}\psi_s.\mathrm{Sin}\alpha_s - \mathrm{Tan}\psi_a.\mathrm{Sin}\alpha_a}\right\}$	183.0°
$\alpha_4 = \arctan\left\{\dfrac{\mathrm{Tan}\psi_b.\mathrm{Cos}\alpha_b - \mathrm{Tan}\psi_s.\mathrm{Cos}\alpha_s}{\mathrm{Tan}\psi_s.\mathrm{Sin}\alpha_s - \mathrm{Tan}\psi_b.\mathrm{Sin}\alpha_b}\right\}$	148.0°
$\alpha_5 = \arctan\left\{\dfrac{\mathrm{Tan}\psi_a.\mathrm{Cos}\alpha_a - \mathrm{Tan}\psi_b.\mathrm{Cos}\alpha_b}{\mathrm{Tan}\psi_b.\mathrm{Sin}\alpha_b - \mathrm{Tan}\psi_a.\mathrm{Sin}\alpha_a}\right\}$	157.7°
$\alpha_6 = \arctan\left\{\dfrac{\mathrm{Tan}\psi_f.\mathrm{Cos}\alpha_f - \mathrm{Tan}\psi_s.\mathrm{Cos}\alpha_s}{\mathrm{Tan}\psi_s.\mathrm{Sin}\alpha_s - \mathrm{Tan}\psi_f.\mathrm{Sin}\alpha_f}\right\}$	273.9°
$\alpha_7 = \arctan\left\{\dfrac{\mathrm{Tan}\psi_a.\mathrm{Cos}\alpha_a - \mathrm{Tan}\psi_t.\mathrm{Cos}\alpha_t}{\mathrm{Tan}\psi_t.\mathrm{Sin}\alpha_t - \mathrm{Tan}\psi_a.\mathrm{Sin}\alpha_a}\right\}$	96.1°
$\alpha_8 = \arctan\left\{\dfrac{\mathrm{Tan}\psi_b.\mathrm{Cos}\alpha_b - \mathrm{Tan}\psi_t.\mathrm{Cos}\alpha_t}{\mathrm{Tan}\psi_t.\mathrm{Sin}\alpha_t - \mathrm{Tan}\psi_b.\mathrm{Sin}\alpha_b}\right\}$	200.0°
$\alpha_9 = \arctan\left\{\dfrac{\mathrm{Tan}\psi_s.\mathrm{Cos}\alpha_s - \mathrm{Tan}\psi_t.\mathrm{Cos}\alpha_t}{\mathrm{Tan}\psi_t.\mathrm{Sin}\alpha_t - \mathrm{Tan}\psi_s.\mathrm{Sin}\alpha_s}\right\}$	252.6°
Step 2: Calculation of dips of lines of intersection equation (A8). Input data from step 1.	
Calculation:	
$\psi_1 = \arctan(\mathrm{Tan}\,\psi_a.\mathrm{Cos}\,(\alpha_a - \alpha_1))$	43.8°
$\psi_2 = \arctan(\mathrm{Tan}\,\psi_b.\mathrm{Cos}\,(\alpha_b - \alpha_2))$	64.7°
$\psi_3 = \arctan(\mathrm{Tan}\,\psi_a.\mathrm{Cos}\,(\alpha_a - \alpha_3))$	11.5°
$\psi_4 = \arctan(\mathrm{Tan}\,\psi_b.\mathrm{Cos}\,(\alpha_b - \alpha_4))$	8.2°
$\psi_5 = \arctan(\mathrm{Tan}\,\psi_a.\mathrm{Cos}\,(\alpha_a - \alpha_5))$	31.2°

CALCULATION TABLE FOR WEDGE ANALYSIS - SHEET 2	RESULTS
Step 2 continued:	
$\psi_6 = \arctan\ (\text{Tan}\ \psi_f.\text{Cos}\ (\alpha_f - \alpha_6))$	2.3°
$\psi_7 = \arctan\ (\text{Tan}\ \psi_a.\text{Cos}\ (\alpha_a - \alpha_7))$	44.6°
$\psi_8 = \arctan\ (\text{Tan}\ \psi_b.\text{Cos}\ (\alpha_b - \alpha_8))$	66.0°
$\psi_9 = \arctan\ (\text{Tan}\ \psi_s.\text{Cos}\ (\alpha_s - \alpha_9))$	6.5°

Step 3: Calculation of angles between lines of intersection. (Equation A5)

Input data: from steps 1 and 2.

Calculalation :

	RESULTS
$\theta_{13} = \arccos\ (\text{Cos}\psi_1.\text{Cos}\psi_3.\text{Cos}\ (\alpha_1 - \alpha_3) + \text{Sin}\psi_1.\text{Sin}\psi_3$	61.6°
$\theta_{15} = \arccos\ (\text{Cos}\psi_1.\text{Cos}\psi_5.\text{Cos}\ (\alpha_1 - \alpha_5) + \text{Sin}\psi_1.\text{Sin}\psi_5$	31.0°
$\theta_{24} = \arccos\ (\text{Cos}\psi_2.\text{Cos}\psi_4.\text{Cos}\ (\alpha_2 - \alpha_4) + \text{Sin}\psi_2.\text{Sin}\psi_4$	65.4°
$\theta_{25} = \arccos\ (\text{Cos}\psi_2.\text{Cos}\psi_5.\text{Cos}\ (\alpha_2 - \alpha_5) + \text{Sin}\psi_2.\text{Sin}\psi_5$	40.7°
$\theta_{34} = \arccos\ (\text{Cos}\psi_3.\text{Cos}\psi_4.\text{Cos}\ (\alpha_3 - \alpha_4) + \text{Sin}\psi_3.\text{Sin}\psi_4$	34.6°
$\theta_{35} = \arccos\ (\text{Cos}\psi_3.\text{Cos}\psi_5.\text{Cos}\ (\alpha_3 - \alpha_5) + \text{Sin}\psi_3.\text{Sin}\psi_5$	30.6°
$\theta_{37} = \arccos\ (\text{Cos}\psi_3.\text{Cos}\psi_7.\text{Cos}\ (\alpha_3 - \alpha_7) + \text{Sin}\psi_3.\text{Sin}\psi_7$	79.8°
$\theta_{45} = \arccos\ (\text{Cos}\psi_4.\text{Cos}\psi_5.\text{Cos}\ (\alpha_4 - \alpha_5) + \text{Sin}\psi_4.\text{Sin}\psi_5$	24.7°
$\theta_{48} = \arccos\ (\text{Cos}\psi_4.\text{Cos}\psi_8.\text{Cos}\ (\alpha_4 - \alpha_8) + \text{Sin}\psi_4.\text{Sin}\psi_8$	68.2°
$\theta_{49} = \arccos\ (\text{Cos}\psi_4.\text{Cos}\psi_9.\text{Cos}\ (\alpha_4 - \alpha_9) + \text{Sin}\psi_4.\text{Sin}\psi_9$	103.4°
$\theta_{57} = \arccos\ (\text{Cos}\psi_5.\text{Cos}\psi_7.\text{Cos}\ (\alpha_5 - \alpha_7) + \text{Sin}\psi_5.\text{Sin}\psi_7$	49.1°
$\theta_{58} = \arccos\ (\text{Cos}\psi_5.\text{Cos}\psi_8.\text{Cos}\ (\alpha_5 - \alpha_8) + \text{Sin}\psi_5.\text{Sin}\psi_8$	43.1°
$\theta_{79} = \arccos\ (\text{Cos}\psi_7.\text{Cos}\psi_9.\text{Cos}\ (\alpha_7 - \alpha_9) + \text{Sin}\psi_7.\text{Sin}\psi_9$	124.7°
$\theta_{89} = \arccos\ (\text{Cos}\psi_8.\text{Cos}\psi_9.\text{Cos}\ (\alpha_8 - \alpha_9) + \text{Sin}\psi_8.\text{Sin}\psi_9$	69.6°

Step 4: Calculation of areas of planes (equations A25, A26 and A27)

Input data: from step 3 and from measured lengths AC and TC.

Supplementary calculation of lengths AC and TC

Vertical elevation of A above O = 100 ft. and $\psi_1 = 43.8°$, hence

$AO = 100/\text{Sin}\ 43.8° = 144.5$ ft.

$AC = AO.\dfrac{\text{Sin}\ \theta_{15}}{\text{Sin}\ \theta_{35}} =$ 146.2 ft.

$TC = AC - 40 =$ 106.2 ft.

CALCULATION TABLE FOR WEDGE ANALYSIS - SHEET 3	RESULTS
Calculation: $A_A = \frac{1}{2}\left\{AC^2 \cdot \frac{Sin\theta_{13}}{Sin\theta_{15}} - TC^2 \cdot \frac{Sin\theta_{37}}{Sin\theta_{57}}\right\} Sin\,\theta_{35}$ for $TC = 0$	$5672\ ft^2$ $9291\ ft^2$
$A_B = \frac{1}{2}\left\{AC^2 \cdot \frac{Sin^2\theta_{13} \cdot Sin\theta_{25}}{Sin^2\theta_{15} \cdot Sin\theta_{24}} - TC^2 \cdot \frac{Sin^2\theta_{37} \cdot Sin\theta_{58}}{Sin^2\theta_{57} \cdot Sin\theta_{48}}\right\} Sin\theta_{45}$ for $TC = 0$	$6497\ ft^2$ $9344\ ft^2$
$A_T = \frac{1}{2} \cdot TC^2 \cdot \frac{Sin\theta_{34} \cdot Sin\theta_{35} \cdot Sin\theta_{79}}{Sin\theta_{49} \cdot Sin\theta_{57}}$	$1822\ ft^2$
Step 5: Calculation of weight of wedge (equations A29 and A28) Input data: Angles from step 3 plus lengths AC and TC. Density of rock $\gamma = 160\ lb/ft^3$ Calculation: $K = (1-Cos^2\theta_{34} - Cos^2\theta_{35} - Cos^2\theta_{45} + 2Cos\theta_{34}\,Cos\theta_{35}Cos\theta_{45})^{\frac{1}{2}}$ $W = \frac{1}{6}\gamma K\left\{AC^3 \cdot \frac{Sin^2\theta_{13} \cdot Sin\theta_{25}}{Sin^2\theta_{15} \cdot Sin\theta_{24}} - TC^3 \cdot \frac{Sin^2\theta_{37} \cdot Sin\theta_{58}}{Sin^2\theta_{57} \cdot Sin\theta_{48}}\right\}$ for $TC = 0$	0.208 $28.30 \times 10^6 lb$ $36.18 \times 10^6 lb$
Step 6: Calculation of forces due to water pressure. Input data: From steps 2, 3 and 4. Density of water $\gamma_W = 62.5\ lb/ft^2$ Calculation: $p = \gamma_W \cdot \frac{TC \cdot Sin\theta_{35}}{2 \cdot Sin\theta_{57}}\left\{Sin\psi_7 + \frac{Sin\theta_{79}}{Sin\theta_{89}} \cdot Sin\psi_8\right\}$ $V = \frac{1}{3} \cdot p \cdot A_T$ $U_A = \frac{1}{3} \cdot p \cdot A_A$ $U_B = \frac{1}{3} \cdot p \cdot A_B$	$3360\ lb/ft^2$ $2.04 \times 10^6 lb.$ $6.35 \times 10^6 lb.$ $7.28 \times 10^6 lb.$

CALCULATION TABLE FOR WEDGE ANALYSIS - SHEET 4	RESULTS

Step 7: Calculation of coefficients for resolution of forces (Table AVII)

Input data: original data on planes and step 2.

Calculation:

$m_{na.nb}$	$= \mathrm{Sin}\psi_a.\mathrm{Sin}\psi_b.\mathrm{Cos}(\alpha_a - \alpha_b) + \mathrm{Cos}\psi_a.\mathrm{Cos}\psi_b$	-0.185
$m_{W.na}$	$= -\mathrm{Cos}\psi_a$	-0.707
$m_{W.nb}$	$= -\mathrm{Cos}\psi_b$	-0.342
$m_{V.na}$	$= \mathrm{Sin}\psi_a.\mathrm{Sin}\psi_t.\mathrm{Cos}(\alpha_a - \alpha_t) + \mathrm{Cos}\psi_a.\mathrm{Cos}\psi_t$	0.574
$m_{V.nb}$	$= \mathrm{Sin}\psi_b.\mathrm{Sin}\psi_t.\mathrm{Cos}(\alpha_b - \alpha_t) + \mathrm{Cos}\psi_b.\mathrm{Cos}\psi_t$	0.419
$m_{T.na}$	$= \mathrm{Cos}\psi_T.\mathrm{Sin}\psi_a.\mathrm{Cos}(\alpha_T - \alpha_a) - \mathrm{Sin}\psi_T.\mathrm{Cos}\psi_a$	-0.563
$m_{T.nb}$	$= \mathrm{Cos}\psi_T.\mathrm{Sin}\psi_b.\mathrm{Cos}(\alpha_T - \alpha_b) - \mathrm{Sin}\psi_T.\mathrm{Cos}\psi_b$	-0.578
$m_{W.5}$	$= \mathrm{Sin}\psi_5$	0.518
$m_{V.5}$	$= \mathrm{Cos}\psi_5.\mathrm{Sin}\psi_t.\mathrm{Cos}(\alpha_5 - \alpha_t) - \mathrm{Sin}\psi_5.\mathrm{Cos}\psi_t$	0.620
$m_{T.5}$	$= \mathrm{Cos}\psi_5.\mathrm{Cos}\psi_T.\mathrm{Cos}(\alpha_5 - \alpha_T) + \mathrm{Sin}\psi_5.\mathrm{Sin}\psi_T$	-0.448

Step 8: Calculation of coefficients for determination of effective normal reactions N_{ae} and N_{be}. (Equations A42 to A47)

Input data: from step 7

Calculation:

$q = (m_{na.nb}.m_{W.nb} - m_{W.na}) / (1 - m^2_{na.nb})$	0.80
$r = (m_{na.nb}.m_{V.nb} - m_{V.na}) / (1 - m^2_{na.nb})$	-0.67
$s = (m_{na.nb}.m_{T.nb} - m_{T.na}) / (1 - m^2_{na.nb})$	0.69
$x = (m_{na.nb}.m_{W.na} - m_{W.nb}) / (1 - m^2_{na.nb})$	0.49
$y = (m_{na.nb}.m_{V.na} - m_{V.nb}) / (1 - m^2_{na.nb})$	-0.54
$z = (m_{na.nb}.m_{T.na} - m_{T.nb}) / (1 - m^2_{na.nb})$	0.71

Step 9: Calculation of Factor of Safety (Equation A49)

$$F = \frac{c_A A_A + c_B A_B + (qW + rV + sT - U_A)\mathrm{Tan}\,\phi_A + (xW + yV + zT - U_B)\mathrm{Tan}\phi_B}{m_{W.5}W + m_{V.5}V + m_{T.5}T}$$

$$F = \frac{0.0005\,A_A + 0.001\,A_B + 0.57\,W - 0.56\,V + 0.66\,T - 0.36\,U_A - 0.58\,U_B}{0.52\,W + 0.62\,V - 0.45\,T}$$

CALCULATION TABLE FOR WEDGE ANALYSIS – SHEET 5	RESULTS

Calculation of Factor of Safety:

Input data: $c_A = 500 \ lb/ft^2$ $c_B = 1000 \ lb/ft^2$

$\quad\quad\quad\quad Tan\,\phi_A = 0.364$ $Tan\,\phi_B = 0.577$

<u>Case a</u> : Dry slope with no tension crack.

$$V = T = U_A = U_B = 0$$

$$W = 36.18 \times 10^6 \ lb, \quad A_A = 9291 \ ft^2, \quad A_B = 9344 \ ft^2$$

$$F = \frac{0.0005 \times 9291 + 0.001 \times 9344 + 0.57 \times 36.18}{0.52 \times 36.18}$$

1.84

<u>Case b</u>: Dry slope with tension crack

$$V = T = U_A = U_B = 0$$

$$W = 28.30 \times 10^6 \ lb., \quad A_A = 5672 \ ft^2, \quad A_B = 6497 \ ft^2$$

$$F = \frac{0.0005 \times 5672 + 0.001 \times 6497 + 0.57 \times 28.3}{0.52 \times 28.3}$$

1.73

<u>Case c</u>: Slope with tension crack and water pressure.

$T = 0$, $W = 28.30 \times 10^6 \ lb$, $A_A = 5672 \ ft^2$,

$A_B = 6497 \ ft^2$, $V = 2.03 \times 10^6 \ lb.$, $U_A = 6.34 \times 10^6 \ lb$

$U_B = 7.27 \times 10^6 \ lb.$

$$F = \frac{2.84 + 6.50 + 16.13 - 0.56 \times 2.03 - 0.36 \times 6.34 - 0.58 \times 7.27}{14.72 + 0.62 \times 2.03}$$

1.10

<u>Case d</u>:   External force T required to restore
factor of safety for case c to 1.5.

$$F = 1.5 = \frac{17.82 + 0.66T}{16.0 - 0.45T}$$

$$T = (24 - 17.82)/(0.66 + 0.675) =$$

4.64×10^6
lb.

Appendix 2 : Wedge solution for rapid computation

Introduction

The solution of the wedge problem presented in Appendix 1 was designed for teaching purposes rather than for convenience of calculation. In this appendix, two solutions designed for maximum speed and efficiency of calculation* are given. These solutions are :

1. A short solution for a wedge with a horizontal slope crest and with no tension crack. Each plane may have a different friction angle and cohesive strength and the influence of water pressure on each plane is included in the solution. The influence of an external force is not included in this solution.

2. A comprehensive solution which included the effects of a superimposed load, a tension crack and an external force such as that applied by a tensioned cable.

The short solution is suitable for programming on a pocket calculator such as a Hewlett-Packard 67 or a Texas Instruments SR52. It can also be used with a non-programmable calculator such as a Hewlett-Packard 21 and a typical problem would require about 30 minutes of calculation on such a machine.

The comprehensive solution is 4 to 5 times longer than the short solution and would normally be programmed on a desk top calculator or in a computer.

SHORT SOLUTION

Scope of solution

The solution presented is for the computation of the factor of safety for translational slip of a tetrahedral wedge formed in a rock slope by two intersecting discontinuities, the slope face and the upper ground surface. It does not take account of rotational slip or toppling, nor does it include a consideration of those cases in which more than two intersecting discontinuities isolate tetrahedral or tapered wedges of rock. In other words, the influence of a tension crack is not considered in this solution.

The solution allows for different strength parameters and water pressures on the two planes of weakness. It is assumed that the slope crest is horizontal, ie the upper ground surface is either horizontal or dips in the same direction as the slope face or at 180° to this direction.

When a pair of discontinuities are selected at random from a set of field data, it is not known whether :

a) the planes could form a wedge (the line of intersection may plunge too steeply to daylight in the slope face or it may be too flat to intersect the upper ground surface).

---

* These solutions are based upon the work of one of the authors (J.W.B) with assistance from Dr E.T.Brown at Imperial College, London.

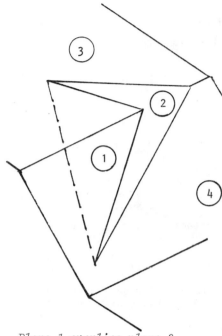

Plane 1 overlies plane 2

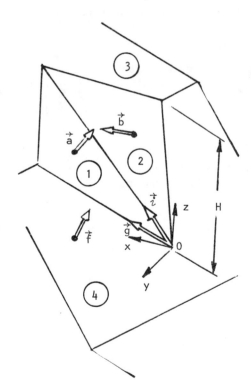

b) one of the planes overlies the other (this affects the calculation of the normal reactions on the planes)

c) one of the planes lies to the right or the left of the other plane when viewed from the bottom of the slope.

In order to resolve these uncertainties, the solution has been derived in such a way that either of the planes may be labelled 1 (or 2) and allowance has been made for one plane overlying the other. In addition, a check on whether the two planes do form a wedge is included in the solution at an early stage. Depending upon the geometry of the wedge and the magnitude of the water pressure acting on each plane, contact may be lost on either plane and this contingency is provided for in the solution.

Notation

The geometry of the problem is illustrated in the margin sketch. The discontinuities are denoted by 1 and 2, the upper ground surface by 3 and the slope face by 4. The data required for the solution of the problem are the unit weight of the rock γ, the height H of the crest of the slope above the intersection 0, the dip ψ and dip direction α of each plane , the cohesion c and the friction angle ϕ for planes 1 and 2 and the average water pressure u on each of the planes 1 and 2*. If the slope face overhangs the toe of the slope, the index η is assigned the value of -1; if the slope does not overhang, η = +1.

Other terms used in the solution are :

F = factor of safety against wedge sliding calculated as the ratio of the resisting to the actuating shear forces

A = area of a face of the wedge

W = weight of the wedge

N = effective normal reaction on a plane

S = actuating shear force on a plane

x,y,z = co-ordinate axes with origin at 0. The z axis is directed vertically upwards, the y axis is in the dip direction of plane 2

$\vec{a}$ = unit vector in the direction of the normal to plane 1 with components (a_x, a_y, a_z)

$\vec{b}$ = unit vector in the direction of the normal to plane 2 with components (b_x, b_y, b_z)

$\vec{f}$ = unit vector in the direction of the normal to plane 4 with components (f_x, f_y, f_z)

$\vec{g}$ = vector in the direction of the line of intersection of planes 1 and 4 with components (g_x, g_y, g_z)

$\vec{i}$ = vector in the direction of the line of intersection of planes 1 and 2 with components (i_x, i_y, i_z)

---

*If it is assumed that the discontinuities are completely filled with water and that the water pressure varies from zero at the free faces at a maximum at some point on the line of intersection, then $u_1 = u_2 = \gamma_w H_w/6$ where H_w is the overall height of the wedge.

$i = -i_z$

q = component of $\vec{g}$ in the direction of $\vec{b}$

r = component of $\vec{a}$ in the direction of $\vec{b}$

$k = |\vec{i}|^2 = i_x^2 + i_y^2 + i_z^2$

$l = W/A_2$

$p = A_1/A_2$

$n_1 = N_1/A_2$

$n_2 = N_2/A_2$ $\left.\begin{array}{l}\\ \\ \\\end{array}\right\}$ Assuming contact on both planes

$|li|/\sqrt{k} = SA_2$

$m_1 = N_1/A_2$

denominator of $F = S_1/A_2$ $\left.\begin{array}{l}\\ \\\end{array}\right\}$ contact on plane 1 only

$m_2 = N_2/A_2$

denominator of $F = S_2/A_2$ $\left.\begin{array}{l}\\ \\\end{array}\right\}$ contact on plane 2 only

Sequence of calculations

The factor of safety of a tetrahedral wedge against sliding along a line of intersection may be calculated as follows :

1. $(a_x, a_y, a_z) = \{\mathrm{Sin}\psi_1.\mathrm{Sin}(\alpha_1 - \alpha_2),\ \mathrm{Sin}\psi_1.\mathrm{Cos}(\alpha_1 - \alpha_2),\ \mathrm{Cos}\psi_1\}$

2. $(f_x, f_y, f_z) = \{\mathrm{Sin}\psi_4.\mathrm{Sin}(\alpha_4 - \alpha_2),\ \mathrm{Sin}\psi_4.\mathrm{Cos}(\alpha_4 - \alpha_2),\ \mathrm{Cos}\psi_4\}$

3. $b_y = \mathrm{Sin}\psi_2$

4. $b_z = \mathrm{Cos}\psi_2$

5. $i = a_x b_y$

6. $g_z = f_x a_y - f_y a_x$

7. $q = b_y(f_z a_x - f_x a_z) + b_z g_z$

8. If $\eta q/i > 0$, or if $\mathrm{Tan}\psi_3 > \sqrt{1 - f_z^2}/(f_z - \eta q/i)$, and $\alpha_3 = \eta\alpha_4$, no wedge is formed and the calculation should be terminated.

9. $r = a_y b_y + a_z b_z$

10. $k = 1 - r^2$

11. $l = (\gamma H q)/(3 g_z)$

12. $p = -b_y f_x/g_z$

13. $n_1 = \{(l/k)(a_z - r b_z) - p u_1\}.p/|p|$

14. $n_2 = \{(l/k)(b_z - r a_z) - u_2\}$

15. $m_1 = (l a_z - r u_2 - p u_1).p/|p|$

16. $m_2 = (l b_z - r p u_1 - u_2)$

17. a) If $n_1 > 0$ and $n_2 > 0$, there is contact on both planes and

$$F = (n_1.\mathrm{Tan}\phi_1 + n_2.\mathrm{Tan}\phi_2 + |p|c_1 + c_2)\sqrt{k}/|li|$$

b) If $n_2 < 0$ and $m_1 > 0$, there is contact on plane 1

only and

$$F = \frac{m_1 . \text{Tan } \phi_1 + |p| c_1}{\{l^2(1 - a_z^2) + ku_2^2 + 2(ra_z - b_z)lu_2\}^{\frac{1}{2}}}$$

c) If $n_1 < 0$ and $m_2 > 0$, there is contact on plane 2 only and

$$F = \frac{m_2 . \text{Tan} \phi_2 + c_2}{\{l^2 b_y^2 + kp^2 u_1^2 + 2(rb_z - a_z)plu_1\}^{\frac{1}{2}}}$$

d) If $m_1 < 0$ and $m_2 < 0$, contact is lost on both planes and the wedge floats as a result of water pressure acting on planes 1 and 2. In this case, the factor of safety falls to zero.

Example

Calculate the factor of safety against wedge failure of a slope for which the following data applies :

Plane	1	2	3	4
ψ°	47	70	10	65
α°	052	018	045	045

$\gamma = 25$ kN/m^3 , H = 20m , $c_1 = 25$ kN/m^2 , $c_2 = 0$,
$\phi_1 = 30^\circ$, $\phi_2 = 35^\circ$, $u_1 = u_2 = 30$ kN/m^2, $\eta = +1$

1. $(a_x, a_y, a_z) = (0.40897, 0.60632, 0.68200)$

2. $(f_x, f_y, f_z) = (0.41146, 0.80753, 0.42262)$

3. $b_y = 0.93969$

4. $b_z = 0.34202$

5. $i = 0.38431$

6. $g_z = -0.08078$

7. $q = -0.12981$

8. $\eta q/i < 0$, $\text{Tan}\psi_3 = 0.17633$; $\sqrt{1 - f_z^2}/(f_z - \eta q/i) = 1.19558$,
   A wedge is formed, continue calculation sequence.

9. $r = 0.80301$

10. $k = 0.35517$

11. $l = 265.969$

12. $p = 4.78639$

13. $n_1 = 161.456$

14. $n_2 = -183.988$

15. $m_1 = 13.7089$

16. $m_2 = -54.3389$

17. $n_2 < 0$ and $m_1 > 0$, hence there is contact on plane 1 only ;

    c) gives $F = 0.626$ and hence the slope is unstable.

Note that water pressure acting on the planes has a significant influence upon the solution to this problem. If $u_1 = u_2 = 0$, $F = 1.154$.

COMPREHENSIVE SOLUTION

Scope of solution

As in the previous solution, this solution is for computation of the factor of safety for translational slip of a tetrahedral wedge formed in a rock slope by two intersecting discontinuities, the slope face and the upper ground surface. In this case, the influence of a tension crack is included in the solution. The solution does not take account of rotational slip or toppling.

The solution allows for different strength parameters and water pressures on the two planes of weakness and for water pressure in the tension crack. There is no restriction on the inclination of the crest of the slope. The influence of an external load E and a cable tension T are included in the analysis and supplementary sections are provided for the examination of the minimum factor of safety for a given external load (eg a blast acceleration acting in a known direction) and for minimising the cable force required for a given factor of safety.

Part of the input data is concerned with the average values of the water pressure on the failure planes (u_1 and u_2) and on the tension crack (u_5). These may be estimated from field data or by using some form of analysis. In the absence of precise information on the water bearing fissures, one cannot hope to make accurate predictions but two simple methods for obtaining approximate estimates were suggested in Appendix 1 of this book. In both methods it is assumed that extreme conditions of very heavy rainfall occur, and that in consequence the fissures are completely full of water. Again , in both methods, it is assumed that the pressure varies from zero at the free faces to a maximum value at some point on the line of intersection of the two failure planes. The first method treats the case where no tension crack exists and gives the result $u_1 = u_2 = \gamma_w H_w/6$, where H_w is the total height of the wedge. The second method allows for the presence of a tension crack and gives $u_1 = u_2 = u_5 = \gamma_w H_{5w}/3$, where H_{5w} is the depth of the bottom vertex of the tension crack below the upper ground surface.

As in the short solution, allowance is made for the following :

a)   interchange of planes 1 and 2

b)   the possibility of one of the planes overlying the other

c)   the situation where the crest overhangs the base of the slope (in which case $\eta = -1$)

d)   the possibility of contact being lost on either plane.

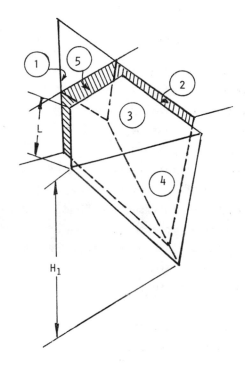

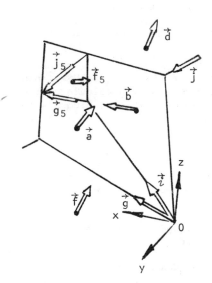

In addition to detecting whether or not a wedge can form, the solution also examines how the tension crack intersects the other planes and only accepts those cases where the tension crack truncates the wedge in the manner shown in the margin sketch.

Notation

The geometry of the problem is illustrated in the margin sketch. The failure surfaces are denoted by 1 and 2, the upper ground surface by 3, the slope face by 4 and the tension crack by 5. The following input data are required for the solution :

ψ, α = dip and dip direction of plane or plunge and trend of force

H_1 = slope height referred to plane 1

L = distance of tension crack from crest, measured along the trace of plane 1

u = average water pressure on face of wedge

c = cohesive strength of each failure plane

ϕ = angle of friction of each failure plane

γ = unit weight of rock

γ_w = unit weight of water

T = cable or bolt tension

E = external load

η = -1 if slope is overhanging and +1 if slope does not overhang

Other terms used in the solution are

F = factor of safety against sliding along the line of intersection or on plane 1 or plane 2

A = area of face of wedge

W = weight of wedge

V = water thrust on tension crack face

N_a = total normal force on plane 1

S_a = shear force on plane 1

Q_a = shear resistance on plane 1

F_1 = factor of safety

} when contact is maintained on plane 1 only

N_b = total normal force on plane 2

S_b = shear force on plane 2

Q_b = shear resistance on plane 2

F_2 = factor of safety

} when contact is maintained on plane 2 only

N_1, N_2 = effective normal reactions

S = total shear force on 1,2

Q = total shear resistance on 1,2

F_3 = factor of safety

} when contact is maintained on both planes 1 and 2

N_1', N_2' = values of N_1, N_2, S etc when T = 0
S', etc

N_1'', N_2'' = values of N_1, N_2, S etc when E = 0
S'', etc

$\vec{a}$ = unit normal vector for plane 1

$\vec{b}$ = unit normal vector for plane 2

$\vec{d}$ = unit normal vector for plane 3

$\vec{f}$ = unit normal vector for plane 4

$\vec{f}_5$ = unit normal vector for plane 5

$\vec{g}$ = vector in direction of intersection line of 1,4

$\vec{g}_5$ = vector in direction of intersection line of 1,5

$\vec{i}$ = vector in direction of intersection line of 1,2

$\vec{j}$ = vector in direction of intersection line of 3,4

$\vec{j}_5$ = vector in direction of intersection line of 3,5

$\vec{k}$ = vector in plane 2 normal to $\vec{i}$

$\vec{l}$ = vector in plane 1 normal to $\vec{i}$

R   = magnitude of vector $\vec{i}$

G   = square of magnitude of vector $\vec{g}$

G_5 = square of magnitude of vector $\vec{g}_5$

Note 1 : The computed value of V is negative when the
tension crack dips away from the toe of the
slope but this does not indicate a tensile force.

   2 : The expressions for α_i, α_{e1}, α_{e2}, α_{e3}, α_{t1}, α_{t2} & α_{t3}
which occur later in the solution are normally
evaluated by ATAN2 of Fortran, or by "Rectangular
to Polar" operation on desk top calculators. For
this reason, $-\nu$ should not be cancelled out in
equation 47 for α_i.

Sequence of calculations

1. *Calculation of factor of safety when the forces T and E
are either zero or completely specified in magnitude
and direction.*

a. Components of unit vectors in directions of normals
to planes 1 to 5, and of forces T and E.

(a_x, a_y, a_z) = $(Sin\psi_1 Sin\alpha_1,\ Sin\psi_1 Cos\alpha_1,\ Cos\psi_1)$ 1

(b_x, b_y, b_z) = $(Sin\psi_2 Sin\alpha_2,\ Sin\psi_2 Cos\alpha_2,\ Cos\psi_2)$ 2

(d_x, d_y, d_z) = $(Sin\psi_3 Sin\alpha_3,\ Sin\psi_3 Cos\alpha_3,\ Cos\psi_3)$ 3

(f_x, f_y, f_z) = $(Sin\psi_4 Sin\alpha_4,\ Sin\psi_4 Cos\alpha_4,\ Cos\psi_4)$ 4

(f_{5x}, f_{5y}, f_{5z}) = $(Sin\psi_5 Sin\alpha_5,\ Sin\psi_5 Cos\alpha_5,\ Cos\psi_5)$ 5

(t_x, t_y, t_z) = $(Cos\psi_t Sin\alpha_t,\ Cos\psi_t Cos\alpha_t,\ -Sin\psi_t)$ 6

(e_x, e_y, e_z) = $(Cos\psi_e Sin\alpha_e,\ Cos\psi_e Cos\alpha_e,\ -Sin\psi_e)$ 7

b. Components of vectors in the direction of the lines of intersection of various planes .

$$(g_x, g_y, g_z) = (f_y a_z - f_z a_y), (f_z a_x - f_x a_z), (f_x a_y - f_y a_x) \qquad 8$$

$$(g_{5x}, g_{5y}, g_{5z}) = (f_{5y} a_z - f_{5z} a_y), (f_{5z} a_x - f_{5x} a_z), (f_{5x} a_y - f_{5y} a_x) \qquad 9$$

$$(i_x, i_y, i_z) = (b_y a_z - b_z a_y), (b_z a_x - b_x a_z), (b_x a_y - b_y a_x) \qquad 10$$

$$(j_x, j_y, j_z) = (f_y d_z - f_z d_y), (f_z d_x - f_x d_z), (f_x d_y - f_y d_x) \qquad 11$$

$$(j_{5x}, j_{5y}, j_{5z}) = (f_{5y} d_z - f_{5z} d_y), (f_{5z} d_x - f_{5x} d_z), (f_{5x} d_y - f_{5y} d_x) \qquad 12$$

$$(k_x, k_y, k_z) = (i_y b_z - i_z b_y), (i_z b_x - i_x b_z), (i_x b_y - i_y b_x) \qquad 13$$

$$(l_x, l_y, l_z) = (a_y i_z - a_z i_y), (a_z i_x - a_x i_z), (a_x i_y - a_y i_x) \qquad 14$$

c. Numbers proportional to cosines of various angles

$$m = g_x d_x + g_y d_y + g_z d_z \qquad 15$$

$$m_5 = g_{5x} d_x + g_{5y} d_y + g_{5z} d_z \qquad 16$$

$$n = b_x j_x + b_y j_y + b_z j_z \qquad 17$$

$$n_5 = b_x j_{5x} + b_y j_{5y} + b_z j_{5z} \qquad 18$$

$$p = i_x d_x + i_y d_y + i_z d_z \qquad 19$$

$$q = b_x g_x + b_y g_y + b_z g_z \qquad 20$$

$$q_5 = b_x g_{5x} + b_y g_{5y} + b_z g_{5z} \qquad 21$$

$$r = a_x b_x + a_y b_y + a_z b_z \qquad 22$$

$$s = a_x t_x + a_y t_y + a_z t_z \qquad 23$$

$$v = b_x t_x + b_y t_y + b_z t_z \qquad 24$$

$$w = i_x t_x + i_y t_y + i_z t_z \qquad 25$$

$$s_e = a_x e_x + a_y e_y + a_z e_z \qquad 26$$

$$v_e = b_x e_x + b_y e_y + b_z e_z \qquad 27$$

$$w_e = i_x e_x + i_y e_y + i_z e_z \qquad 28$$

$$s_5 = a_x f_{5x} + a_y f_{5y} + a_z f_{5z} \qquad 29$$

$$v_5 = b_x f_{5x} + b_y f_{5y} + b_z f_{5z} \qquad 30$$

$$w_5 = i_x f_{5x} + i_y f_{5y} + i_z f_{5z} \qquad 31$$

$$\lambda = i_x g_x + i_y g_y + i_z g_z \qquad 32$$

$$\lambda_5 = i_x g_{5x} + i_y g_{5y} + i_z g_{5z} \qquad 33$$

$$\varepsilon = f_x f_{5x} + f_y f_{5y} + f_z f_{5z} \qquad 34$$

d. Miscellaneous factors

$$R = \sqrt{1 - r^2} \qquad 35$$

$$\rho = \frac{1}{R^2} \cdot \frac{nq}{|nq|} \qquad 36$$

$$\mu = \frac{1}{R^2} \cdot \frac{mq}{|mq|} \qquad 37$$

$$\nu = \frac{1}{R} \cdot \frac{p}{|p|} \qquad 38$$

$$G = g_x^2 + g_y^2 + g_z^2 \qquad 39$$

$$G_5 = g_{5x}^2 + g_{5y}^2 + g_{5z}^2 \qquad 40$$

$$M = (Gp^2 - 2|mp|\lambda + m^2 R^2)^{\frac{1}{2}} \qquad 41$$

$$M_5 = (G_5 p^2 - 2|m_5 p|\lambda_5 + m_5{}^2 R^2)^{\frac{1}{2}} \qquad 42$$

$$h = H_1/|g_z| \qquad 43$$

$$h_5 = (Mh - |p|L)/M_5 \qquad 44$$

$$B = (\text{Tan}^2\phi_1 + \text{Tan}^2\phi_2 - 2(\mu r/\rho)\text{Tan}\phi_1\text{Tan}\phi_2)/R^2 \qquad 45$$

e. Plunge and trend of line of intersection of planes 1 & 2

$$\psi_{\hat{\imath}} = \text{Arcsin}(\nu i_z) \qquad 46$$

$$\alpha_{\hat{\imath}} = \text{Arctan}(-\nu i_x/-\nu i_y) \qquad 47$$

f. Check on wedge geometry

No wedge is formed, terminate computation.
$$\begin{cases} \text{If } p i_z < 0 \text{ or} & 48 \\ \text{if } nq i_z < 0 & 49 \end{cases}$$

Tension crack invalid, terminate computation.
$$\begin{cases} \text{if } \varepsilon n q_5 i_z < 0 \text{, or} & 50 \\ \text{if } h_5 < 0 \text{, or} & 51 \\ \text{if } \left|\dfrac{m_5 h_5}{mh}\right| > 1 \text{, or} & 52 \\ \text{if } \left|\dfrac{n q_5 m_5 h_5}{n_5 q m h}\right| > 1 & 53 \end{cases}$$

g. Areas of faces and weight of wedge

$$A_1 = (|mq|h^2 - |m_5 q_5|h_5{}^2)/2|p| \qquad 54$$

$$A_2 = (|q/n|m^2 h^2 - |q_5/n_5|m_5{}^2 h_5{}^2)/2|p| \qquad 55$$

$$A_5 = |m_5 q_5|h_5{}^2/2|n_5| \qquad 56$$

$$W = \gamma(q^2 m^2 h^3/|n| - q_5{}^2 m_5{}^2 h_5{}^3/|n_5|)/6|p| \qquad 57$$

h. Water pressures
   i) With no tension crack

$$u_1 = u_2 = \gamma_w h|m i_z|/6|p| \qquad 58$$

   ii) With tension crack

$$u_1 = u_2 = u_5 = \gamma_w h_5|m_5|/3d_z \qquad 59$$

$$V = u_5 A_5 n \varepsilon/|\varepsilon| \qquad 60$$

i. Effective normal reactions on planes 1 and 2 assuming contact on both planes

$$N_1 = \rho\{Wk_z + T(rv - s) + E(rv_e - s_e) + V(rv_5 - s_5)\} - u_1 A_1 \qquad 61$$

$$N_2 = \mu\{Wl_z + T(rs - v) + E(rs_e - v_e) + V(rs_5 - v_5)\} - u_2 A_2 \qquad 62$$

j. Factor of safety when $N_1 < 0$ and $N_2 < 0$ (contact is lost on both planes)

$$F = 0 \qquad 63$$

k. If $N_1 > 0$ and $N_2 < 0$, contact is maintained on plane 1 only and the factor of safety is calculated as follows:

$$N_a = Wa_z - Ts - Es_e - Vs_5 - u_2A_2r \qquad\qquad 64$$

$$S_x = -(Tt_x + Ee_x + N_aa_x + Vf_{5x} + u_2A_2b_x) \qquad\qquad 65$$

$$S_y = -(Tt_y + Ee_y + N_aa_y + Vf_{5y} + u_2A_2b_y) \qquad\qquad 66$$

$$S_z = -(Tt_z + Ee_z + N_aa_z + Vf_{5z} + u_2A_2b_z) + W \qquad\qquad 67$$

$$S_a = (S_x^2 + S_y^2 + S_z^2)^{\frac{1}{2}} \qquad\qquad 68$$

$$Q_a = (N_a - u_1A_1)Tan\phi_1 + c_1A_1 \qquad\qquad 69$$

$$F_1 = Q_a/S_a \qquad\qquad 70$$

l. If $N_1 < 0$ and $N_2 > 0$, contact is maintained on plane 2 only and the factor of safety is calculated as follows:

$$N_b = Wb_z - Tv - Ev_e - Vv_5 - u_1A_1r \qquad\qquad 71$$

$$S_x = -(Tt_x + Ee_x + N_bb_x + Vf_{5x} + u_1A_1a_x) \qquad\qquad 72$$

$$S_y = -(Tt_y + Ee_y + N_bb_y + Vf_{5y} + u_1A_1a_y) \qquad\qquad 73$$

$$S_z = -(Tt_z + Ee_z + N_bb_z + Vf_{5z} + u_1A_1a_z) + W \qquad\qquad 74$$

$$S_b = (S_x^2 + S_y^2 + S_z^2) \qquad\qquad 75$$

$$Q_b = (N_b - u_2A_2)Tan\phi_2 + c_2A_2 \qquad\qquad 76$$

$$F_2 = Q_b/S_b \qquad\qquad 77$$

m. If $N_1 > 0$ and $N_2 > 0$, contact is maintained on both planes and the factor of safety is calculated as follows:

$$S = v(Wi_z - Tw - Ew_e - Vw_5) \qquad\qquad 78$$

$$Q = N_1Tan\phi_1 + N_2Tan\phi_2 + c_1A_1 + c_2A_2 \qquad\qquad 79$$

$$F_3 = Q/S \qquad\qquad 80$$

2. *Minimum factor of safety produced when load E of given magnitude is applied in the worst direction.*

a) Evaluate N_1'', N_2'', S'', Q'', F_3'' by use of equations 61, 62, 78, 79 and 80 with $E = 0$.

b) If $N_1'' < 0$ and $N_2'' < 0$, even before E is applied, then $F = 0$, terminate computation.

c) $D = \{ (N_1'')^2 + (N_2'')^2 + 2\dfrac{mn}{|mn|} N_1''N_2''r\}^{\frac{1}{2}} \qquad\qquad 81$

$$\psi_e = Arcsin\left(-\frac{1}{G}(\left|\frac{m}{m}\right| \cdot N_1''a_z + \left|\frac{n}{n}\right| \cdot N_2''b_z)\right) \qquad\qquad 82$$

$$\alpha_e = Arctan\left\{ \frac{\left|\frac{m}{m}\right| \cdot N_1''a_x + \left|\frac{n}{n}\right| \cdot N_2''b_x}{\left|\frac{m}{m}\right| \cdot N_1''a_y + \left|\frac{n}{n}\right| \cdot N_2''b_y} \right\} \qquad\qquad 83$$

If $E > D$, and E is applied in the direction ψ_e, α_e, or within a certain range encompassing this direction, then contact is lost on both planes and $F = 0$. Terminate computation.

d) If $N_1'' > 0$ and $N_2'' < 0$, assume contact on plane 1 only after application of E.

Determine S_x'', S_y'', S_z'', S_a'', Q_a'', F_1'' from equations 65 to 70 with $E = 0$.

If $F_1'' < 1$, terminate computation.

If $F_1'' > 1$:

$$F_1 = \frac{S_a'' Q_a'' - E\{(Q_a'')^2 + ((S_a'')^2 - E^2)Tan^2\phi_1\}^{\frac{1}{2}}}{(S_a'')^2 - E^2} \qquad 84$$

$$\psi_{e1} = Arcsin(S_z''/S_a'') - Arctan(Tan\phi_1/F_1) \qquad 85$$

$$\alpha_{e1} = Arctan(S_x''/S_y'') + 180^o \qquad 86$$

e) If $N_1'' < 0$ and $N_2'' > 0$, assume contact on plane 2 only after application of E.

Determine S_x'', S_y'', S_z'', S_b'', Q_b'' and F_2'' from equations 72 to 77 with $E = 0$.

If $F_2'' < 1$, terminate computation.

If $F_2'' > 1$:

$$F_2 = \frac{S_b'' Q_b'' - E\{(Q_b'')^2 + ((S_b'')^2 - E^2)Tan^2\phi_2\}^{\frac{1}{2}}}{(S_b'')^2 - E^2} \qquad 87$$

$$\psi_{e2} = Arcsin(S_z''/S_b'') - Arctan(Tan\phi_2/F_2) \qquad 88$$

$$\alpha_{e2} = Arctan(S_x''/S_y'') + 180^o \qquad 89$$

f) If $N_1'' > 0$ and $N_2'' > 0$, assume contact on both planes after application of E.

If $F_3'' < 1$, terminate computation.

If $F_3'' > 1$:

$$F_3 = \frac{S'' Q'' - E\{(Q'')^2 + B((S'')^2 - E^2)\}^{\frac{1}{2}}}{(S'')^2 - E^2} \qquad 90$$

$$\chi = \sqrt{B + F_3^2} \qquad 91$$

$$e_x = -(F_3 \nu i_x - \rho k_x Tan\phi_1 - \mu l_x Tan\phi_2)/\chi \qquad 92$$

$$e_y = -(F_3 \nu i_y - \rho k_y Tan\phi_1 - \mu l_y Tan\phi_2)/\chi \qquad 93$$

$$e_z = -(F_3 \nu i_z - \rho k_z Tan\phi_1 - \mu l_z Tan\phi_2)/\chi \qquad 94$$

$$\psi_{e3} = Arcsin(-e_z) \qquad 95$$

$$\alpha_{e3} = Arctan(e_x/e_y) \qquad 96$$

Compute s_e and v_e using equations 26 and 27

$$N_1 = N_1'' + E\rho(rv_e - s_e) \qquad 97$$

$$N_2 = N_2'' + E\mu(rs_e - v_e) \qquad 98$$

Check that $N_1 \geqq 0$ and $N_2 \geqq 0$

3. *Minimum cable or bolt tension* T_{min} *required to raise the factor of safety to some specified value* F.

a). Evaluate N_1', N_2', S', Q' by means of equations 61, 62, 78, 79 with $T = 0$.

b) If $N_2' < 0$, contact is lost on plane 2 when $T = 0$. Assume contact on plane 1 only, after application of T. Evaluate S_x', S_y', S_z', S_a' and Q_a' using equations 65 to 69 with $T = 0$.

$$T_1 = (FS_a' - Q_a')/\sqrt{F^2 + Tan^2\phi_1} \qquad\qquad 99$$

$$\psi_{t1} = Arctan(Tan\phi_1/F) - Arcsin(S_z'/S_a') \qquad 100$$

$$\alpha_{t1} = Arctan(S_x'/S_y') \qquad\qquad 101$$

c) If $N_1' < 0$, contact is lost on plane 1 when $T = 0$. Assume contact on plane 2 only, after application of T. Evaluate S_x', S_y', S_z', S_b' and Q_b' using equations 72 to 76 with $T = 0$.

$$T_2 = (FS_b' - Q_b')/\sqrt{F^2 + Tan^2\phi_2} \qquad\qquad 102$$

$$\psi_{t2} = Arctan(Tan\phi_1/F) - Arcsin(S_z'/S_b') \qquad 103$$

$$\alpha_{t2} = Arctan(S_x'/S_y') \qquad\qquad 104$$

d) All cases. No restriction on values of N_1' and N_2'. Assume contact on both planes after application of T.

$$\chi = (F^2 + B)^{\frac{1}{2}} \qquad\qquad 105$$

$$T_3 = (FS' - Q')/\chi \qquad\qquad 106$$

$$t_x = (Fvi_x - \rho k_x Tan\phi_1 - \mu l_x Tan\phi_2)/\chi \qquad 107$$

$$t_y = (Fvi_y - \rho k_y Tan\phi_1 - \mu l_y Tan\phi_2)/\chi \qquad 108$$

$$t_z = (Fvi_z - \rho k_z Tan\phi_1 - \mu l_z Tan\phi_2)/\chi \qquad 109$$

$$\psi_{t3} = Arcsin(-t_z) \qquad\qquad 110$$

$$\alpha_{t3} = Arctan(t_x/t_y) \qquad\qquad 111$$

Compute s and v using equations 23 and 24

$$N_1 = N_1' + T_3\rho(rv - s) \qquad\qquad 112$$

$$N_2 = N_2' + T_3\mu(rs - v) \qquad\qquad 113$$

If $N_1 < 0$ or $N_2 < 0$, ignore the results of this section

If $N_1' > 0$ and $N_2' > 0$, $T_{min} = T_3$

If $N_1' > 0$ and $N_2' < 0$, T_{min} = smallest of T_1, T_3

If $N_1' < 0$ and $N_2' > 0$, T_{min} = smallest of T_2, T_3

If $N_1' < 0$ and $N_2' < 0$, T_{min} = smallest of T_1, T_2, T_3

Example

Calculate the factor of safety for the following wedge :

Plane	1	2	3	4	5	
ψ	45	70	12	65	70	$\eta = +1$
α	105	235	195	185	165	

$H_1 = 100'$, $L = 40'$, $c_1 = 500$ lb/ft^2, $c_2 = 1000$ lb/ft^2

$\phi_1 = 20°$, $\phi_2 = 30°$, $\gamma = 160$ lb/ft^3.

1a) T = 0, E = 0, $u_1 = u_2 = u_5$ from equation 59.

(a_x, a_y, a_z) = (0.68301, - 0.18301, 0.70711)

(b_x, b_y, b_z) = (-0.76975, -0.53899, 0.34202)

(d_x, d_y, d_z) = (-0.05381, -0.20083, 0.97815)

(f_x, f_y, f_z) = (-0.07899, -0.90286, 0.42262)

(f_{5x}, f_{5y}, f_{5z}) = (0.24321, -0.90767, 0.34202)

(g_x, g_y, g_z) = (-0.56107, 0.34451, 0.63112)

(g_{5x}, g_{5y}, g_{5z}) = (-0.57923, 0.061627, 0.57544)

(i_x, i_y, i_z) = (-0.31853, 0.77790, 0.50901)

(j_x, j_y, j_z) = (-0.79826, 0.05452, -0.03272)

(j_{5x}, j_{5y}, j_{5z}) = (-0.81915, -0.25630, -0.09769)

(k_x, k_y, k_z) = (0.54041, -0.28287, 0.77047)

(l_x, l_y, l_z) = (-0.64321, -0.57289, 0.47302)

m = 0.57833

m_5 = 0.58166

n = 0.57388

n_5 = 0.73527

p = 0.35880

q = 0.46206

q_5 = 0.60945

r = -0.18526

s_5 = 0.57407

v_5 = 0.41899

w_5 = -0.60945

λ = 0.76796

λ_5 = 0.52535

ε = 0.65574

R = 0.98269

ρ = 1.03554

μ = 1.03554

ν = 1.01762

G = 0.83180

G_5 = 0.67044

M = 0.33371

M_5 = 0.44017

h = 158.45

h_5 = 87.521

B = 0.56299

ψ_i = 31.20°

α_i = 157.73°

$pi_z > 0$ }
$nqi_z > 0$ } Wedge is formed

$\varepsilon nq_5 i_z > 0$
$h_5 > 0$
$|m_5 h_5| / |mh| = 0.55554 < 1$
$|nq_5 m_5 h_5| / |n_5 qmh| = 0.57191 < 1$ } Tension crack valid

$A_1 = 5565.0$ ft^2
$A_2 = 6428.1$ ft^2
$A_5 = 1846.6$ ft^2
$W = 2.8272 \times 10^7$ lb
$u_1 = u_2 = u_5 = 1084.3$ lb/ft^2 ; $V = 2.0023 \times 10^6$ lb
$N_1 = 1.5171 \times 10^7$ lb } Both positive therefore contact
$N_2 = 5.7892 \times 10^6$ lb } on planes 1 and 2.
$S = 1.5886 \times 10^7$ lb
$Q = 1.8075 \times 10^7$ lb
$F = 1.1378$ - *Factor of safety.*

This value can be compared with the value of 1.10 obtained in case c of Appendix 1 - note that small differences in calculated areas and water pressures account for the difference in the factors of safety.

1b) $T = 0$, $E = 0$, dry slope, $u_1 = u_2 = u_5 = 0$.

As in 1a) except as follows:

$V = 0$
$N_1 = 2.2565 \times 10^7$ lb } Both positive therefore contact
$N_2 = 1.3853 \times 10^7$ lb } on both planes 1 and 2
$S = 1.4644 \times 10^7$ lb
$Q = 2.5422 \times 10^7$ lb
$F_3 = 1.7360$ - *Factor of safety*

Compare this value with the value of 1.73 obtained in case b of Appendix 1.

2) As in 1b) except $E = 8 \times 10^6$ lb. Find the value of F_{min}.

Values of N_1'', N_2'', S'', Q'', F_3'' as given in 1b).
$N_1'' > 0$, $N_2'' > 0$, $F_3'' > 1$, continue calculation.
$B = 0.56299$
$F_3 = 1.04$ - F_{min} *(Minimum factor of safety)*
$\chi = 1.2798$
$e_x = 0.12128$
$e_y = -0.99226$
$e_z = 0.028243$

ψ_{e3} = -1.62°   – *Plunge of force (upwards)*

α_{e3} = 173.03°   – *Trend of force*

N_1 = 1.9517 x 10^7 lb   Both positive therefore contact

N_2 = 9.6793 x 10^6 lb   maintained on both planes.

3). As in 1a) except that the minimum cable tension T_{min} required to increase the factor of safety to 1.5 is to be determined.

N_1', N_2', S' and Q' – as given in 1a)

χ = 1.6772

T_3 = 3.4307 x 10^6 lb – T_{min} *(Minimum cable tension)*

t_x = -0.18205

t_y = 0.97574

t_z = 0.12148

ψ_{t3} = -6.98°   – *Plunge of cable (upwards)*

α_{t3} = 349.43°   – *Trend of cable*

Note that the optimum plunge and trend of the cable are approximately :

$$\psi_{t3} \simeq \psi_i + 180° - \tfrac{1}{2}(\phi_1 + \phi_2) = 31.2 + 180 - 25$$
$$= -6.2° \text{ (upwards)}$$

and $\alpha_{t3} \simeq \alpha_i \pm 180° = 157.73 + 180 = 337.73°$

In other words, a practical rule of thumb for the best direction in which to install the cables to reinforce a wedge is :

The cable should be aligned with the line of intersection of the two planes, viewed from the bottom of the slope, and it should be inclined at the average friction angle to the line of intersection.

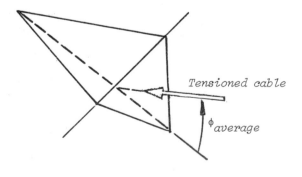

Optimum cable direction for reinforcement of a wedge.

Appendix 3: Factors of safety for reinforced rock slopes

Throughout this book, the factor of safety of a reinforced rock slope (for both plane and wedge failure) has been defined as :

$$F = \frac{\text{Resisting force}}{\text{Disturbing force} - T.\text{Sin}\,\theta} \qquad 1$$

where T is the force applied to the rock by the reinforcing member.

In other words, the force T is assumed to act in such a manner as to *decrease* the *disturbing force*. Pierre Londe*, in a personal communication to the authors, suggested that a second definition is equally applicable :

$$F = \frac{\text{Resisting force} + T}{\text{Disturbing force}} \qquad 2$$

In this definition, the force T *increases* the *resisting force*.

Which definition should be used ? Londe suggests that there is some justification for using equation 1 when T is an *active* force, ie the cable is tensioned *before* any movement of the rock block or wedge has taken place. On the other hand, if T is a *passive* force, applied by un-tensioned bars or cables, the resisting force can only be developed *after* some movement has taken place. In this case, Londe suggests that equation 2 is more appropriate.

In fact, since one never knows the exact sequence of loading and movement in a rock slope, the choice becomes arbitrary. However, in considering this problem, a second and more significant problem arises and this relates to the degree of confidence attached to the values of the shear strengths and water pressures used in the stability analysis. The method of solution described below, based upon a suggestion by Londe, is designed to overcome both of the problems raised here.

If, for the moment, it is assumed that the frictional and cohesive strengths of a rock surface are known with a high degree of precision and the water pressures have been measured by means of piezometers, one may be led to believe that a high degree of confidence can be attached to the calculated driving and resisting forces and hence the factor of safety of the slope. While this confidence would be justified in the case of an unreinforced slope, that same could not be said of a reinforced slope. This is because the response of the various elements to displacement in the slope is not the same. The development of the full frictional strength, due to ϕ, and the cohesive strength c require a finite displacement on the sliding surface and this dis-placement may be incompatible with that imposed by the application of the cable tension T. Similarly, water pressures in the fissures are sensitive to displacement and may increase or decrease, depending upon the manner in which the cables are installed. Consequently, it cannot be assumed that the cable tension T, the frictional strength due to ϕ, the cohesive strength c and the various water pressures are all fully mobilised at the same time.

---

*Technical director, Coyne & Bellier, Paris, France.

Londe suggests that, instead of using a single factor of safety to define the stability of the slope, different factors of safety should be used, depending upon the degree of confidence which the designer has in the particular parameter being considered. High factors of safety can be applied to ill-defined parameters (such as water pressures and cohesive strengths) while low factors of safety can be used for those quantities (such as the weight of a wedge) which are known with a greater degree of precision. For a typical problem, Londe suggests :

$f_c = 1.5$ for cohesive strengths (c)

$f_\phi = 1.2$ for frictional strengths (ϕ)

$f_u = 2.0$ for water pressures

$f_w = 1.0$ for weights and forces.

Using these values, the conditions of limiting equilibrium expressed in equation 12 on page 77 , for a block sliding down a plane, can be expressed as :

$$W.\text{Sin}\psi + 2V - T.\text{Cos}\beta = \frac{cA}{1.5} + (W.\text{Cos}\psi - 2U + T.\text{Sin}\beta)\frac{\text{Tan}\phi}{1.2} \quad 3$$

Note that the factors of safety (given in italics) for the parameters corresponding to the resisting forces (c and ϕ) are decreasing factors while they are increasing factors for the driving forces (U and V).

Solving equation 3 for T gives :

$$T = \frac{W(\text{Sin}\psi - 0.83.\text{Cos}\psi.\text{Tan}\phi) + 2V + 1.67U\text{Tan}\phi - 0.67cA}{0.83\text{Sin}\beta.\text{Tan}\phi + \text{Cos}\beta} \quad 4$$

This is the cable tension required to satisfy the factors of safety assigned to each of the components of the driving and resisting force terms.

Consider the practical example discussed in Case d) of Appendix 1 of this book. It was required to determine the cable tension T needed to increase the factor of safety of a rock wedge to 1.5. In this case, the factor of safety was applied uniformly to c_A, c_B, $\text{Tan}\phi_A$ and $\text{Tan}\phi_B$ while a factor of safety of 1 was applied to U_A, U_B, V and W. The required value of T was found to be 4.64×10^6 lb.

Using the alternative method proposed by Londe and substituting the factors of safety suggested by him, equation A49 of Appendix 1 can be rewritten as :

$$\frac{1}{1.5}(c_A A_A + c_B A_B) + (qW + 2rV + sT - 2U_A)\frac{\text{Tan}\phi_A}{1.2}$$

$$+(xW + 2yV + zT - 2U_B)\frac{\text{Tan}\phi_B}{1.2} = m_{w5}W - 2m_{v5}V - m_{T5}T \quad 5$$

Solving for T, using the same values as in case d) of Appendix 1, gives $T = 9.8 \times 10^6$ lb.

This value is approximately twice that obtained by using a single value of F = 1.5 and Londe considers it to be more realistic in view of the uncertainties associated with the water pressures and the simultaneous mobilisation of T, c and ϕ. The reader should not be alarmed unduly by the discrepancy between the two calculated values of T since

these depend upon the values of the factors of safety which
have been chosen for this illustrative example. On the other
hand, the authors are in full agreement with Londe that the
assignment of different factors of safety for different
parameters, depending upon the degree of confidence in the
value of the parameter in question, is a more realsitic
way of treating the problem of slope reinforcement. It is
recommended that, when a critical slope is to be reinforced,
the method described in this appendix should be used to
calculate the required bolt or cable tension. The reader
is also referred to the detailed discussions on this problem
which have been published by Londe and his co-workers[190-192].

Appendix 4 : Trigonometric table

Angle°	Sin	Cos	Tan	Cot	Angle°	Sin	Cos	Tan	Cot
0	0.0000	1.0000	0.0000		60	.8660	.5000	1.7321	.5774
1	.0175	.9998	.0175	57.2900	61	.8746	.4848	1.8040	.5543
2	.0349	.9994	.0349	28.6363	62	.8829	.4695	1.8807	.5317
3	.0523	.9986	.0524	19.0811	63	.8910	.4540	1.9626	.5095
4	.0698	.9976	.0699	14.3007	64	.8988	.4384	2.0503	.4877
5	.0872	.9962	.0875	11.4301	65	.9063	.4226	2.1445	.4663
6	.1045	.9945	.1051	9.5144	66	.9135	.4067	2.2460	.4452
7	.1219	.9925	.1228	8.1443	67	.9205	.3907	2.3559	.4245
8	.1392	.9903	.1405	7.1154	68	.9272	.3746	2.4751	.4040
9	.1564	.9877	.1584	6.3138	69	.9336	.3584	2.6051	.3839
10	.1736	.9848	.1763	5.6713	70	.9397	.3420	2.7475	.3640
11	.1908	.9816	.1944	5.1446	71	.9455	.3256	2.9042	.3443
12	.2079	.9781	.2126	4.7046	72	.9511	.3090	3.0777	.3249
13	.2250	.9744	.2309	4.3315	73	.9563	.2924	3.2709	.3057
14	.2419	.9703	.2493	4.0108	74	.9613	.2756	3.4874	.2867
15	.2588	.9659	.2679	3.7321	75	.9659	.2588	3.7321	.2679
16	.2756	.9613	.2867	3.4874	76	.9703	.2419	4.0108	.2493
17	.2924	.9563	.3057	3.2709	77	.9744	.2250	4.3315	.2309
18	.3090	.9511	.3249	3.0777	78	.9781	.2079	4.7046	.2126
19	.3256	.9455	.3443	2.9042	79	.9816	.1908	5.1446	.1944
20	.3420	.9397	.3640	2.7475	80	.9848	.1736	5.6713	.1763
21	.3584	.9336	.3839	2.6051	81	.9877	.1564	6.3138	.1584
22	.3746	.9272	.4040	2.4751	82	.9903	.1392	7.1154	.1405
23	.3907	.9205	.4245	2.3559	83	.9925	.1219	8.1443	.1228
24	.4067	.9135	.4452	2.2460	84	.9945	.1045	9.5144	.1051
25	.4226	.9063	.4663	2.1445	85	.9962	.0872	11.4301	.0875
26	.4384	.8988	.4877	2.0503	86	.9976	.0698	14.3007	.0699
27	.4540	.8910	.5095	1.9626	87	.9986	.0523	19.0811	.0524
28	.4695	.8829	.5317	1.8807	88	.9994	.0349	28.6363	.0349
29	.4848	.8746	.5543	1.8040	89	.9998	.0175	57.2900	.0175
30	.5000	.8660	.5774	1.7321	90	1.0000	0.0000		.0000
31	.5150	.8572	.6009	1.6643	91	.9998	-.0175	-57.2900	-.0175
32	.5299	.8480	.6249	1.6003	92	.9994	-.0349	-28.6363	-.0349
33	.5446	.8387	.6494	1.5399	93	.9986	-.0523	-19.0811	-.0524
34	.5592	.8290	.6745	1.4826	94	.9976	-.0698	-14.3007	-.0699
35	.5736	.8192	.7002	1.4281	95	.9962	-.0872	-11.4301	-.0875
36	.5878	.8090	.7265	1.3764	96	.9945	-.1045	-9.5144	-.1051
37	.6018	.7986	.7536	1.3270	97	.9925	-.1219	-8.1443	-.1228
38	.6157	.7880	.7813	1.2799	98	.9903	-.1392	-7.1154	-.1405
39	.6293	.7771	.8098	1.2349	99	.9877	-.1564	-6.3138	-.1584
40	.6428	.7660	.8391	1.1918	100	.9348	-.1736	-5.6713	-.1763
41	.6561	.7547	.8693	1.1504	101	.9816	-.1908	-5.1446	-.1944
42	.6691	.7431	.9004	1.1106	102	.9781	-.2079	-4.7046	-.2126
43	.6820	.7314	.9325	1.0724	103	.9744	-.2250	-4.3315	-.2309
44	.6947	.7193	.9657	1.0355	104	.9703	-.2419	-4.0108	-.2493
45	.7071	.7071	1.0000	1.0000	105	.9659	-.2588	-3.7321	-.2679
46	.7193	.6947	1.0355	.9657	106	.9613	-.2756	-3.4874	-.2867
47	.7314	.6820	1.0724	.9325	107	.9563	-.2924	-3.2709	-.3057
48	.7431	.6691	1.1106	.9004	108	.9511	-.3090	-3.0777	-.3249
49	.7547	.6561	1.1504	.8693	109	.9455	-.3256	-2.9042	-.3443
50	.7660	.6428	1.1918	.8391	110	.9397	-.3420	-2.7475	-.3640
51	.7771	.6293	1.2349	.8098	111	.9336	-.3584	-2.6051	-.3839
52	.7880	.6157	1.2799	.7813	112	.9272	-.3746	-2.4751	-.4040
53	.7986	.6018	1.3270	.7536	113	.9205	-.3907	-2.3559	-.4245
54	.8090	.5878	1.3764	.7265	114	.9135	-.4067	-2.2460	-.4452
55	.8192	.5736	1.4281	.7002	115	.9063	-.4226	-2.1445	-.4663
56	.8290	.5592	1.4826	.6745	116	.8938	-.4384	-2.0503	-.4877
57	.8387	.5446	1.5399	.6494	117	.8910	-.4540	-1.9626	-.5095
58	.8480	.5299	1.6003	.6249	118	.8829	-.4695	-1.8807	-.5317
59	.8572	.5150	1.6643	.6009	119	.8746	-.4848	-1.8040	-.5543

Angle°	Sin	Cos	Tan	Cot	Angle°	Sin	Cos	Tan	Cot
120	.8660	-.5000	-1.7321	-.5774	180	0.0000	-1.0000	0.0000	
121	.8572	-.5150	-1.6643	-.6009	181	-.0175	-.9998	.0175	57.2900
122	.8480	-.5299	-1.6003	-.6249	182	-.0349	-.9994	.0349	28.6363
123	.8387	-.5446	-1.5399	-.6494	183	-.0523	-.9986	.0524	19.0811
124	.8290	-.5592	-1.4826	-.6745	184	-.0698	-.9976	.0699	14.3007
125	.8192	-.5736	-1.4281	-.7002	185	-.0872	-.9962	.0875	11.4301
126	.8090	-.5878	-1.3764	-.7265	186	-.1045	-.9945	.1051	9.5144
127	.7986	-.6018	-1.3270	-.7536	187	-.1219	-.9925	.1228	8.1443
128	.7880	-.6157	-1.2799	-.7813	188	-.1392	-.9903	.1405	7.1154
129	.7771	-.6293	-1.2349	-.8098	189	-.1564	-.9877	.1584	6.3138
130	.7660	-.6428	-1.1918	-.8391	190	-.1736	-.9848	.1763	5.6713
131	.7547	-.6561	-1.1504	-.8693	191	-.1908	-.9816	.1944	5.1446
132	.7431	-.6691	-1.1106	-.9004	192	-.2079	-.9781	.2126	4.7046
133	.7314	-.6820	-1.0724	-.9325	193	-.2250	-.9744	.2309	4.3315
134	.7193	-.6947	-1.0355	-.9657	194	-.2419	-.9703	.2493	4.0108
135	.7071	-.7071	-1.0000	-1.0000	195	-.2588	-.9659	.2679	3.7321
136	.6947	-.7193	-.9657	-1.0355	196	-.2756	-.9613	.2867	3.4874
137	.6820	-.7314	-.9325	-1.0724	197	-.2924	-.9563	.3057	3.2709
138	.6691	-.7431	-.9004	-1.1106	198	-.3090	-.9511	.3249	3.0777
139	.6561	-.7547	-.8693	-1.1504	199	-.3256	-.9455	.3443	2.9042
140	.6428	-.7660	-.8391	-1.1918	200	-.3420	-.9397	.3640	2.7475
141	.6293	-.7771	-.8098	-1.2349	201	-.3584	-.9336	.3839	2.6051
142	.6157	-.7880	-.7813	-1.2799	202	-.3746	-.9272	.4040	2.4751
143	.6018	-.7986	-.7536	-1.3270	203	-.3907	-.9205	.4245	2.3559
144	.5878	-.8090	-.7265	-1.3764	204	-.4067	-.9135	.4452	2.2460
145	.5736	-.8192	-.7002	-1.4281	205	-.4226	-.9063	.4663	2.1445
146	.5592	-.8290	-.6745	-1.4826	206	-.4384	-.8988	.4877	2.0503
147	.5446	-.8387	-.6494	-1.5399	207	-.4540	-.8910	.5095	1.9626
148	.5299	-.8480	-.6249	-1.6003	208	-.4695	-.8829	.5317	1.8807
149	.5150	-.8572	-.6009	-1.6643	209	-.4848	-.8746	.5543	1.8040
150	.5000	-.8660	-.5774	-1.7321	210	-.5000	-.8660	.5774	1.7321
151	.4848	-.8746	-.5543	-1.8040	211	-.5150	-.8572	.6009	1.6643
152	.4695	-.8829	-.5317	-1.8807	212	-.5299	-.8480	.6249	1.6003
153	.4540	-.8910	-.5095	-1.9626	213	-.5446	-.8387	.6494	1.5399
154	.4384	-.8988	-.4877	-2.0503	214	-.5592	-.8290	.6745	1.4826
155	.4226	-.9063	-.4663	-2.1445	215	-.5736	-.8192	.7002	1.4281
156	.4067	-.9135	-.4452	-2.2460	216	-.5878	-.8090	.7265	1.3764
157	.3907	-.9205	-.4245	-2.3559	217	-.6018	-.7986	.7536	1.3270
158	.3746	-.9272	-.4040	-2.4751	218	-.6157	-.7880	.7813	1.2799
159	.3584	-.9336	-.3839	-2.6051	219	-.6293	-.7771	.8098	1.2349
160	.3420	-.9397	-.3640	-2.7475	220	-.6428	-.7660	.8391	1.1918
161	.3256	-.9455	-.3443	-2.9042	221	-.6561	-.7547	.8693	1.1504
162	.3090	-.9511	-.3249	-3.0777	222	-.6691	-.7431	.9004	1.1106
163	.2924	-.9563	-.3057	-3.2709	223	-.6820	-.7314	.9325	1.0724
164	.2756	-.9613	-.2867	-3.4874	224	-.6947	-.7193	.9657	1.0355
165	.2588	-.9659	-.2679	-3.7321	225	-.7071	-.7071	1.0000	1.0000
166	.2419	-.9703	-.2493	-4.0108	226	-.7193	-.6947	1.0355	.9657
167	.2250	-.9744	-.2309	-4.3315	227	-.7314	-.6820	1.0724	.9325
168	.2079	-.9781	-.2126	-4.7046	228	-.7431	-.6691	1.1106	.9004
169	.1908	-.9816	-.1944	-5.1446	229	-.7547	-.6561	1.1504	.8693
170	.1736	-.9848	-.1763	-5.6713	230	-.7660	-.6428	1.1918	.8391
171	.1564	-.9877	-.1584	-6.3138	231	-.7771	-.6293	1.2349	.8098
172	.1392	-.9903	-.1405	-7.1154	232	-.7880	-.6157	1.2799	.7813
173	.1219	-.9925	-.1228	-8.1443	233	-.7986	-.6018	1.3270	.7536
174	.1045	-.9945	-.1051	-9.5144	234	-.8090	-.5878	1.3764	.7265
175	.0872	-.9962	-.0875	-11.4301	235	-.8192	-.5736	1.4281	.7002
176	.0698	-.9976	-.0699	-14.3007	236	-.8290	-.5592	1.4826	.6745
177	.0523	-.9986	-.0524	-19.0811	237	-.8387	-.5446	1.5399	.6494
178	.0349	-.9994	-.0349	-28.6363	238	-.8480	-.5299	1.6003	.6249
179	.0175	-.9998	-.0175	-57.2900	239	-.8572	-.5150	1.6643	.6009

Angle°	Sin	Cos	Tan	Cot	Angle°	Sin	Cos	Tan	Cot
240	-.8660	-.5000	1.7321	.5774	300	-.8660	.5000	-1.7321	-.5774
241	-.8746	-.4848	1.8040	.5543	301	-.8572	.5150	-1.6643	-.6009
242	-.8829	-.4695	1.8807	.5317	302	-.8480	.5299	-1.6003	-.6249
243	-.8910	-.4540	1.9626	.5095	303	-.8387	.5446	-1.5399	-.6494
244	-.8988	-.4384	2.0503	.4877	304	-.8290	.5592	-1.4826	-.6745
245	-.9063	-.4226	2.1445	.4663	305	-.8192	.5736	-1.4281	-.7002
246	-.9135	-.4067	2.2460	.4452	306	-.8090	.5878	-1.3764	-.7265
247	-.9205	-.3907	2.3559	.4245	307	-.7986	.6018	-1.3270	-.7536
248	-.9272	-.3746	2.4751	.4040	308	-.7880	.6157	-1.2799	-.7813
249	-.9336	-.3584	2.6051	.3839	309	-.7771	.6293	-1.2349	-.8098
250	-.9397	-.3420	2.7475	.3640	310	-.7660	.6428	-1.1918	-.8391
251	-.9455	-.3256	2.9042	.3443	311	-.7547	.6561	-1.1504	-.8693
252	-.9511	-.3090	3.0777	.3249	312	-.7431	.6691	-1.1106	-.9004
253	-.9563	-.2924	3.2709	.3057	313	-.7314	.6820	-1.0724	-.9325
254	-.9613	-.2756	3.4874	.2867	314	-.7193	.6947	-1.0355	-.9657
255	-.9659	-.2588	3.7321	.2679	315	-.7071	.7071	-1.0000	-1.0000
256	-.9703	-.2419	4.0108	.2493	316	-.6947	.7193	-.9657	-1.0355
257	-.9744	-.2250	4.3315	.2309	317	-.6820	.7314	-.9325	-1.0724
258	-.9781	-.2079	4.7046	.2126	318	-.6691	.7431	-.9004	-1.1106
259	-.9816	-.1908	5.1446	.1944	319	-.6561	.7547	-.8693	-1.1504
260	-.9848	-.1736	5.6713	.1763	320	-.6428	.7660	-.8391	-1.1918
261	-.9877	-.1564	6.3138	.1584	321	-.6293	.7771	-.8098	-1.2349
262	-.9903	-.1392	7.1154	.1405	322	-.6157	.7880	-.7813	-1.2799
263	-.9925	-.1219	8.1443	.1228	323	-.6018	.7986	-.7536	-1.3270
264	-.9945	-.1045	9.5144	.1051	324	-.5878	.8090	-.7265	-1.3764
265	-.9962	-.0872	11.4301	.0875	325	-.5736	.8192	-.7002	-1.4281
266	-.9976	-.0698	14.3007	.0699	326	-.5592	.8290	-.6745	-1.4826
267	-.9986	-.0523	19.0811	.0524	327	-.5446	.8387	-.6494	-1.5399
268	-.9994	-.0349	28.6363	.0349	328	-.5299	.8480	-.6249	-1.6003
269	-.9998	-.0175	57.2900	.0175	329	-.5150	.8572	-.6009	-1.6643
270	-1.0000	0.0000		0.0000	330	-.5000	.8660	-.5774	-1.7321
271	-.9998	.0175	-57.2900	-.0175	331	-.4848	.8746	-.5543	-1.8040
272	-.9994	.0349	-28.6363	-.0349	332	-.4695	.8829	-.5317	-1.8807
273	-.9986	.0523	-19.0811	-.0524	333	-.4540	.8910	-.5095	-1.9626
274	-.9976	.0698	-14.3007	-.0699	334	-.4384	.8988	-.4877	-2.0503
275	-.9962	.0872	-11.4301	-.0875	335	-.4226	.9063	-.4663	-2.1445
276	-.9945	.1045	-9.5144	-.1051	336	-.4067	.9135	-.4452	-2.2460
277	-.9925	.1219	-8.1443	-.1228	337	-.3907	.9205	-.4245	-2.3559
278	-.9903	.1392	-7.1154	-.1405	338	-.3746	.9272	-.4040	-2.4751
279	-.9877	.1564	-6.3138	-.1584	339	-.3584	.9336	-.3839	-2.6051
280	-.9848	.1736	-5.6713	-.1763	340	-.3420	.9397	-.3640	-2.7475
281	-.9816	.1908	-5.1446	-.1944	341	-.3256	.9455	-.3443	-2.9042
282	-.9781	.2079	-4.7046	-.2126	342	-.3090	.9511	-.3249	-3.0777
283	-.9744	.2250	-4.3315	-.2309	343	-.2924	.9563	-.3057	-3.2709
284	-.9703	.2419	-4.0108	-.2493	344	-.2756	.9613	-.2867	-3.4874
285	-.9659	.2588	-3.7321	-.2679	345	-.2588	.9659	-.2679	-3.7321
286	-.9613	.2756	-3.4874	-.2867	346	-.2419	.9703	-.2493	-4.0108
287	-.9563	.2924	-3.2709	-.3057	347	-.2250	.9744	-.2309	-4.3315
288	-.9511	.3090	-3.0777	-.3249	348	-.2079	.9781	-.2126	-4.7046
289	-.9455	.3256	-2.9042	-.3443	349	-.1908	.9816	-.1944	-5.1446
290	-.9397	.3420	-2.7475	-.3640	350	-.1736	.9848	-.1763	-5.6713
291	-.9336	.3584	-2.6051	-.3839	351	-.1564	.9877	-.1584	-6.3138
292	-.9272	.3746	-2.4751	-.4040	352	-.1392	.9903	-.1405	-7.1154
293	-.9205	.3907	-2.3559	-.4245	353	-.1219	.9925	-.1228	-8.1443
294	-.9135	.4067	-2.2460	-.4452	354	-.1045	.9945	-.1051	-9.5144
295	-.9063	.4226	-2.1445	-.4663	355	-.0872	.9962	-.0875	-11.4301
296	-.8988	.4384	-2.0503	-.4877	356	-.0698	.9976	-.0699	-14.3007
297	-.8910	.4540	-1.9626	-.5095	357	-.0523	.9986	-.0524	-19.0811
298	-.8829	.4695	-1.8807	-.5317	358	-.0349	.9994	-.0349	-28.6363
299	-.8746	.4848	-1.8040	-.5543	359	-.0175	.9998	-.0175	-57.2900

Appendix 5: Conversion factors

		Imperial	Metric	SI
Length	1 mile	1.609 km	1.609 km
	1 ft	0.3048 m	0.3048 m
	1 in	2.54 cm	25.40 mm
Area	1 mile^2	2.590 km^2	2.590 km^2
	1 acre	0.4047 hectare	4046.9 m^2
	1 ft^2	0.0929 m^2	0.0929 m^2
	1 in^2	6.452 cm^2	6.452 cm^2
Volume	1 yd^3	0.7646 m^3	0.7646 m^3
	1 ft^3	0.0283 m^3	0.0283 m^3
	1 ft^3	28.32 litres	0.0283 m^3
	1 UK gal.	4.546 litres	4546 cm^3
	1 US gal.	3.785 litres	3785 cm^3
	1 in^3	16.387 cm^3	16.387 cm^3
Mass	1 ton	1.016 tonne	1.016 Mg
	1 lb	0.4536 kg	0.4536 kg
	1 oz	28.352 g	28.352 g
Density	1 lb/ft^3	16.019 kg/m^3	16.019 kg/m^3
Unit weight	1 lbf/ft^3	16.019 kgf/m^3	0.1571 kN
Force	1 ton f	1.016 tonne f	9.964 kN
	1 lb f	0.4536 kg f	4.448 N
Pressure or stress	1 ton f/in^2	157.47 kg f/cm^2	15.44 MPa
	1 ton f/ft^2	10.936 tonne f/m^2	107.3 kPa
	1 lb f/in^2	0.0703 kg f/cm^2	6.895 kPa
	1 lb f/ft^2	4.882 kg f/m^2	0.04788 kPa
	1 standard atmosphere	1.033 kg f/m^2	101.325 kPa
	14.495 lb f/in^2	1.019 kg f/cm^2	1 bar
	1 ft water	0.0305 kg f/cm^2	2.989 kPa
	1 in mercury	0.0345 kg f/cm^2	3.386 kPa
Permeability	1 ft/year	0.9659×10^{-6} cm/s	0.9659×10^{-8} m/s
Rate of flow	1 ft^3/s	0.02832 m^3/s	0.02832 m^3/s
Moment	1 lbf ft	0.1383 kgf m	1.3558 Nm
Energy	1 ft lbf	1.3558 J	1.3558 J
Frequency	1 c/s	1 c/s	1 Hz

SI unit prefixes

Prefix	tera	giga	mega	kilo	milli	micro	nano	pico
Symbol	T	G	M	k	m	μ	n	p
Multiplier	10^{12}	10^{9}	10^{6}	10^{3}	10^{-3}	10^{-6}	10^{-9}	10^{-12}

SI symbols and definitions

N = Newton = kg m/s^2
Pa = Pascal = N/m^2
J = Joule = m.N